Signals and Communication Technology

The series "Signals and Communications Technology" is devoted to fundamentals and applications of modern methods of signal processing and cutting-edge communication technologies. The main topics are information and signal theory, acoustical signal processing, image processing and multimedia systems, mobile and wireless communications, and computer and communication networks. Volumes in the series address researchers in academia and industrial R&D departments. The series is application-oriented. The level of presentation of each individual volume, however, depends on the subject and can range from practical to scientific.

More information about this series at http://www.springer.com/series/4748

Alan J. Sangster

Compact Slot Array Antennas for Wireless Communications

Springer

Alan J. Sangster
Edinburgh, UK

ISSN 1860-4862 ISSN 1860-4870 (electronic)
Signals and Communication Technology
ISBN 978-3-030-01752-1 ISBN 978-3-030-01753-8 (eBook)
https://doi.org/10.1007/978-3-030-01753-8

Library of Congress Control Number: 2018959269

This Springer imprint is published by the registered company Springer Nature Switzerland AG
The registered company address is: Gewerbestrasse 11, 6330 Cham, Switzerland

To Hamish

Whatever you choose to be, be the very best you can. That way lies fulfilment.

Preface

At about the time when this book was being compiled, it was hardly controversial to suggest that the predominant evolutionary course of electronic engineering had been characterised by an inexorable drive toward the miniaturisation of components, devices, and systems, facilitated by the unprecedented advances in integrated circuit (IC) technology. The obvious example is the computer, the hardware for which could fill a modest laboratory in the 1960's, yet by 2015 a device of comparable computational power was small enough to be held in the palm of the hand—Moore's law gives this trend a quantitive basis. However, for practitioners in the discipline of radio frequency communications the 'fly in the oitment', to this generally desirable miniaturisation drift, has been the interface with 'free-space', namely the antenna. This has been particularly true in the burgeoning area of handheld mobile devices.

Unlike virtually all other components in an electronic circuit the antenna is dimensionally constrained by the unyielding laws of physics. For reasons of radiation efficiency, at least one major dimension (L) of the antenna has to be, in size, close to, or greater than, a half of the free space wavelength (λ_o) at the frequency of operation ($L \geq \lambda_0/2$). Needless to say, in the realms of portable and handheld mobile communications devices, or in the domain of implantable devices for monitoring medical conditions, circumventing this restraint has been a major focus of recent compact antenna research. The problem can be simply stated. To create an efficient compact antenna, at a prescribed frequency, any electrical dimension which governs radiation strength must be preserved yet remain within the bounds of a physical package that is required to be everywhere smaller. These contradictory requirements are usually satisfied, with varying degrees of success, by securing electrical length in a dimensionally restricted package through the adoption of 'tricks' such as meandering, corkscrewing and twisting the key dimensional element of the shrunk antenna. However, not surprisingly, these distortions can have deleterious effects on other important antenna properties and compensating techniques are generally necessary to achieve viability.

The purpose of this text lies in describing and providing design guidelines for antennas which pursue compactness by recruiting the slot radiator as the fundamental building block. Initial research on the so called slot radiator can be traced back in time to before the Second World War. It was strenuously propelled by military pressures to advance the detection capabilities of radar. As a consequence of this early 'baptism' the literature is comprehensive and extensive, requiring patient time-consuming study by antenna designers, and by research engineers, with a need to claim mastery of the topic. Recent progress has been largely triggered by the burgeoning of modern communications techniques demanding increasingly compact, lighter, more versatile antennas, offering large bandwidth to the user. For the design engineer due diligence requires a thorough sifting of this literature, which is inevitably time consuming and expensive when it is so extensive.

While the primary threads in the historical evolution of compact array antennas are identified, the main focus is to provide an electrical engineering text which summarises the extensive compact antenna related literature. More particularly, the aim is to illuminate, in relation to the now ubiquitous slot radiator, how despite significant antenna size reductions, seemingly defying physics, desirable radiation pattern properties can be preserved. This leads to an examination of recent advances in frequency selective surfaces or metamaterials which can, with caution, be used to facilitate physics defying dimensional restrictions.

Slot radiators, excited from waveguide, have for some time provided basic elements for creating array antennas most commonly for applications in radar and in satellite communications. On the other hand, in stripline, microstrip or coplanar waveguide formats slot radiators, either directly excited or through a patch transition, additionally form fundamental components in compact antennas and conformal antennas, particularly in the realm of mobile communications. The range of possible formats is briefly illustrated in Chap. 1.

In constructing the eleven chapters, focused both on already established and on continuing developments in the realm of compact antennas for wireless communications, the intention has been to restrict the content to topics deemed salient to antennas of the compact variety and particularly is this so in addressing theory, modelling and design. This stance has been achieved by relegating some general, but useful, background material to appendices where a copious bibliography is provided directing the reader to standard texts which provide further elaboration.

The early chapters of the book (Chaps. 1–5) focus on the radiation mechanisms and characteristics of the slot radiator when treated as an aperture antenna. The progressive incorporation of the slot radiator into antenna technology developments and in particular those directed toward compactness are both introduced and presented schematically in Chap. 1. This chapter also provides a guide to more detailed explorations and expositions of compact antenna evolution which appear in later chapters. In Chap. 2 the electromagnetic fundamentals of the radiation mechanism are introduced using analogies from electrostatics and magnetostatics to construct the key equations without undue mathematical toil. With these indispensable equations established, Chap. 3 uses them to construct the now conventional mathematical representations of the linear aperture which can be viewed as the building block for electromagnetic modelling of more complex antenna geometries. It also introduces the reader to basic antenna pattern forms, definitions and trade-offs.

The reader is introduced to the method of moments in Chaps. 4 and 5 as arguably the predominant route to efficient electromagnetic modelling of antenna forms containing slot radiators. The fundamental Helmholtz inhomogeneous differential equations for radiation generated by electric and magnetic current sources are tabulated in Chap. 4, alongside the equivalent Green's equations and the relevant Green's function solutions. Armed with these tools the moment method is constructed. It is consolidated in Chap. 5 by applying it to the solution of two typical slot radiation problems, namely the conventional shunt slot in rectangular waveguide and the less common polarisation agile T-slot radiator in bifurcated rectangular waveguide.

The second half of the book (Chaps. 6–11) delves into the topic of array antennas. The evolving compact forms of such arrays, particularly the partially reflective periodic surface varieties, have been attracting increasing interest from the wireless communications community in recent years. To set the scene, the fundamental nature of arrays is first examined in Chap. 6, by analysing traditional linear phased arrays comprising independent isotropic radiators. This is accomplished by following a well established analysis route based on a polynomial representation of such an array. The procedure permits the illumination of basic relationships between array geometry, in particular element excitation levels and element spacing, and performance specifications such as directivity, beamwidth, side-lobe levels, grating lobes levels, mainbeam scanning and bandwidth limitations.

Slot array manifestations which are potentially much more compact and cost effective than phased array alternatives are broached in Chaps. 7 and 8, with travelling-wave array examples in the former, and scanned and leaky-wave array versions in the latter. Design principles for conventional waveguide fed travelling-wave arrays are presented in depth in Chap. 7 with the results used to establish the basic trade-offs, firstly between slot orientation and positioning, and secondly between achievable radiation patterns and the undesirable distortions imposed by frequency deviation from the design value. While frequency deviation is represented as a problem for conventional travelling-wave array antennas, this need not be the case as is demonstrated in Chap. 8. There, it is shown that intentional frequency scanning of the primary radiating beam of a travelling-wave array can be significantly amplified by resorting to phase delay enhancement techniques. The technology is shown to have major implications for radar. The reader is thereby introduced to the periodic structure and its complex modal solutions and hence to the importance of slow-waves and surface waves. It is demonstrated that the key to the formulation of field solutions which satisfy periodic boundary conditions are the Floquet harmonics. With this armoury, it is shown that the application of electromagnetic field boundary conditions at a periodic interface leads to an important determinantal equation for computing the phase coefficient for the fundamental mode. A quasi-Green's function for periodic waveguide is also developed. This permits a slot radiator in a periodic waveguide feed to be modelled using the moment method, in a manner not too unlike the slot discontinuity in conventional waveguide. The chapter concludes with an examination of an antenna which is physically similar to the travelling-wave array but is really quite different electromagnetically—namely the leaky-wave array. The differences between this array and the conventional travelling-wave array are emphasised.

The state-of-the-art in the evolution for communication systems of compact planar antennas, particularly those which utilize periodic array structures but without any phased array connotations, is broached in Chaps. 9, 10 and 11. Arguably, the least radical departure from established antenna practice is the planar reflect-array which is addressed in Chap. 9. In essence, the reflect-array aims to replace the parabolic reflector in a conventional high gain antenna with a planar array geometry. To do this the surface is made to replicate the optics of the replaced parabola by providing equivalent element-to-element phase profiling over the flat digitized array surface.

A planar array antenna which arguably represents a more significant departure from the preceding technology is provided in Chap. 10 by the introduction of periodic surfaces into the design process. The substrates thus formed are variously termed, frequency selective surfaces, partially reflective surfaces, textured surfaces or occasionally metamaterials. In planar arrangements the periodic layer usually forms a sandwich with a ground plane thus creating a parallel plate structure which acts as either a Fabry-Perot resonator or as a leaky-wave array. The chapter outlines the operational modes which lead to a high gain fixed beam antenna option or to a frequency scanned leak-wave antenna alternative. Reconfigurability of the periodic surface to achieve leaky-wave scanning at fixed frequency is also studied.

Finally in Chap. 11 an emerging technology is examined which could exercise considerable influence on compact array antenna developments intended for satellite communication systems and for devices incorporated into mobile platforms. An old technique, initially developed for enhancing the radar visibility of, in particular, small sea or ocean going vessels, has been re-engineered for incorporation into array antennas. Modern applications for this technology have grown out of the need to create antennas which can automatically track a signal source when relative movement exists between the transmitter and the receiver. A relevant example is secure communications with a low earth orbit satellite.

Naturally all views, assertions, claims, calculations and items of factual information contained in this book have been selected or generated by myself, and any errors therein are my responsibility. However, the book would not have seen the light of day without numerous personal interactions (too many to identify), with family, with friends, and with colleagues at the Heriot-Watt University, on the topic of array antenna developments to meet the burgeoning needs of wireless communications. So if I have talked to you on this topic, I thank you for your contribution, and the stimulus it may have provided for the creation of this book. A special thank you is extended to Dr. Hanyang Wang for figures in Chap. 3. I would, also, particularly like to thank the members of staff, at the Heriot-Watt University library, who have been very helpful in ensuring that I was able to access a wide range of written material, the contents of some of which have been germane to the realisation of this project.

Edinburgh, UK

Professor Alan J. Sangster
C.Eng., FIET. B.Sc. (Eng), M.Sc., Ph.D.

Contents

Abbreviations

A	Amps
AC	Alternating current
AMC	Artificial magnetic conductor
BGAN	Broadband global area network
BPF	Bandpass filter
BW	Bandwidth
BWFN	Beamwidth for nulls
C	Capacitance (Farads)
CBCPW	Conductor backed co-planar waveguide
CEM	Classical electromagnetism
CRA	Cavity resonator antenna
CW	Continuous wave
DC	Direct current
DUT	Device under test
EM	Electromagnetism
EMW	Electromagnetic wave
FEM	Finite element method
FET	Field effect transistor
FSS	Frequency selective surface
HD	High definition
HDMI	High definition multimedia interface
HFSS	High frequency structure simulator
HIS	High impedance surface
HPBW	Half power beamwidth
IF	Intermediate frequency
IQ	In-phase/quadrature-phase
L	Inductance (Henries)
LEO	Low earth orbit
LWA	Leaky-wave antenna
MF	Medium frequency

MKS	Metre/kilogram/second
MMIC	Monolithic microwave integrated circuit
MoM	Method of Moments
PC	Printed circuit
PCC	Phase conjugating circuit
PMC	Perfect magnetic conductor
PPL	Phase locked loop
PRS	Partially reflective surface
Q	Quality factor (of resonance)
RF	Radio frequency
RFID	Radio frequency identification
RLP	Ring loaded patch
SIW	Substrate integrated waveguide
SLL	Side lobe level
SP	Square patch
SR	Square ring
TE	Transverse electric
TEM	Transverse electromagnetic
TEN	Transmission line equivalent network
TM	Transverse magnetic
UWB	Ultra-wideband
V	Volt
VCO	Voltage controlled oscillator
VHF	Very high frequency
VLF	Very low frequency
WG	Waveguide
WLAN	Wireless local area network
WPAN	Wireless personal area network

List of Figures

Chapter 1
Evolution of Compact Slot Antennas

1.1 Free-Space Waves and Transmission Lines

The aim of this chapter is to introduce to readers the idea that even quite complex geometrical arrangements of slots or patches in compact array antennas can be evaluated and understood, at least initially, by empirical or heuristic methods, based on a good understanding of properties of the most basic electromagnetic wave solution governed by the Maxwell equations. The form of this solution is generally referred to as the plane wave or the transverse electromagnetic (TEM) wave. In fact this understanding need not extend beyond the underlying physics of plane waves in order to permit satisfactory comprehension of radiation sources and the influence of boundaries and antenna geometries on radiation patterns. This empirical exercise actually represents a very useful preliminary stage in the antenna design process, particularly since today novel designs are generally procured through the implementation of simulations on full-wave software analysis tools such as HFSS, COMSOL, or CONCERTO for example. Of course the efficient application of these tools, is greatly enhanced if the user has acquired a thorough appreciation of the mathematical procedures which are incorporated into such simulations. This recommendation is addressed in much more detail in Chaps. 4 and 5.

Readers of this text are assumed to have some background in electromagnetic theory, at least to the level of familiarity with the formulation of electromagnetic wave solutions from the Maxwell's equations for passive environments such as free-space, simple transmission lines and regular waveguides. For those requiring a memory boost and who have limited access to alternative texts (see Bibliography), Appendices A and B contain summaries of the requisite material.

A. J. Sangster, *Compact Slot Array Antennas for Wireless Communications*, Signals and Communication Technology, https://doi.org/10.1007/978-3-030-01753-8_1

1.2 Plane Electromagnetic Wave in Free-Space

The plane electromagnetic wave is, in words, a radio-wave, or even a wave at light frequencies, in open space where reflecting and refracting boundaries are absent. The essence of the wave's field structure is depicted in Fig. 1.1.

In Fig. 1.1, **E** and **H** represent the electric and magnetic field intensities in the region of interest. We can also define **D** = ε**E** where **D** is the electric flux density and **B** = μ**H** where **B** is the magnetic flux density. Note that in vacuum, or air, permittivity $\varepsilon = \varepsilon_0$ and permeability $\mu = \mu_0$, while the phase coefficient $k = k_0 = \omega\sqrt{\mu_0\varepsilon_0} = \omega/c$. Fuller parameter definitions are provided in Appendix A.

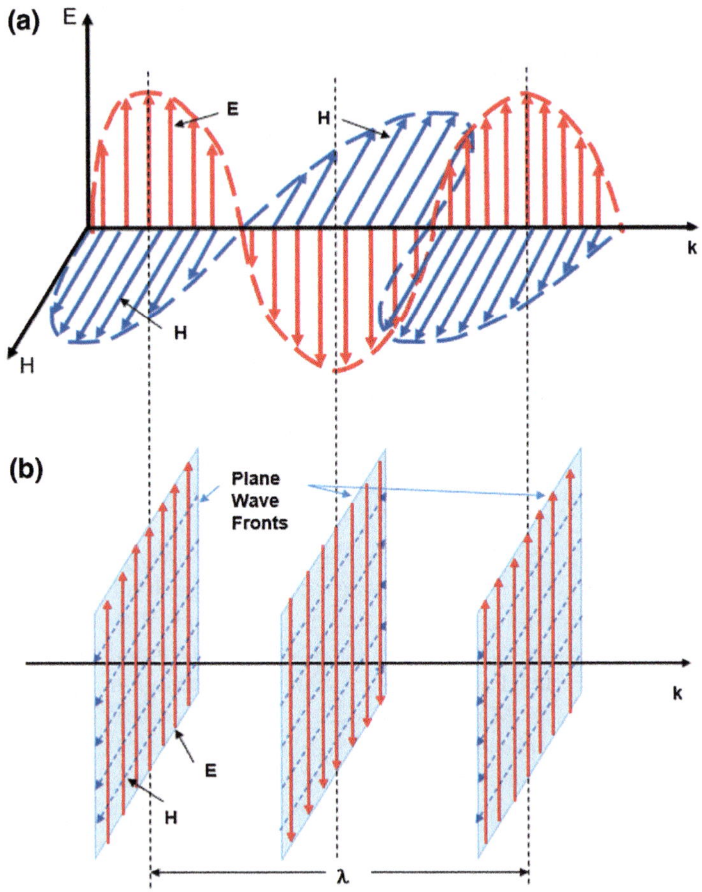

Fig. 1.1 Representation of a simple sinusoidal plane electromagnetic wave showing the fundamental orthogonal relationships between the vector electric (**E**) and magnetic (**H**) fields and the direction of propagation (**k**)

The mathematical expression for the general free-space wave (see Appendix A) is:

$$E_x = Af(z - ct) + Bf(z + ct) \tag{1.1}$$

where f is an unspecified function for a wave travelling in the z-direction which is aligned with **k** (Fig. 1.1). For a sinusoidal wave, Eq. (1.1) is replaced by:

$$E_x = A\,\sin(z - ct) + B\,\sin(z + ct) \tag{1.2}$$

The first term in Eq. (1.1) represents a wave travelling in the $+z$ direction while the second allows for a reflected wave, if such exists. In this case a standing wave is formed on the mismatched line. Given that velocity is the rate of change of z with respect to time, it is evident that c represents velocity (actually phase velocity) of the electromagnetic wave in 'free space'. For vacuum it is equal to 3×10^8 m/sec. The application of Maxwell's equations also gives $H_z = 0$ and:

$$H_y = \frac{A}{\eta_0} f(z - ct) + \frac{B}{\eta_0} f(z + ct) \tag{1.3}$$

Also

$$\frac{E_x}{H_y} = \pm\sqrt{\frac{\mu_0}{\varepsilon_0}} = \pm\eta_0 \tag{1.4}$$

η_0 is termed the free space wave impedance, which for air or vacuum has the value $120\pi\Omega$. The resultant solution is a plane electromagnetic wave, also termed a transverse electromagnetic (TEM) wave, for which E and H are transverse to the direction of propagation and orthogonal to each other. E and H are also in time phase, as Eqs. (1.1) and (1.3) attest (see Fig. 1.1). If time phase between E and H prevails, which it does in lossless free-space, the power flow density along the direction **k** is given by:

$$p = \frac{|E||H|}{2}\ \text{Watts/m}^2 \tag{1.5}$$

The plane wave fronts in Fig. 1.1b lie transverse to **k** and, in theory, cover the entire, theoretically infinite, transverse plane. On any given plane, **E** and **H** are of constant magnitude, represented by equally spaced vectors. Further, the magnitudes of E and H vary in accordance with Eqs. (1.2) and (1.3) respectively, as we move from one plane to the next.

While this plane wave solution to the Maxwell equations evidently provides a useful guide to the nature of electromagnetic waves in free-space, the importance of this wave structure, from an antenna perspective, is that it also describes electromagnetic wave propagation in a variety of transmission systems which are important antenna feed systems. The obvious reason for this beneficial position lies in the fact that guided electromagnetic waves can be procured by simply trapping the TEM wave [see Bibliography] between a pair of flat conducting surfaces that lie normal to the electric field vectors. Surface charge on the conductors instantaneously redistributes itself, thus ensuring that the boundary conditions on both normal D and tangential H are satisfied. That is the charge adjustment ensures that (using unit vector notation):

$$\hat{\mathbf{n}} \cdot \mathbf{D} = \rho \quad \text{and} \quad \hat{\mathbf{n}} \times \mathbf{H} = \mathbf{J} \tag{1.6}$$

where ρ and $\mathbf{J}$ are respectively the surface charge density (Coulomb/m^2) and the surface current density (A/m) on the metal surface. Following convention, the unit vector is directed normal to the boundary surface.

1.3 TEM Transmission Lines

The most common TEM transmission line alternatives are shown in Fig. 1.2. Parallel plate waveguide (Fig. 1.2a) provides only limited guidance in directions other than normal to the plates, but clearly shows how the insertion of smooth conducting planes normal to the E-field has negligible effect on the propagation conditions for the TEM mode. The resultant surface currents, dictated by the boundary conditions, are represented by blue dashed lines. Full trapping is provided by coaxial line (Fig. 1.2b) but at the expense of phase velocity reduction associated with the unavoidable dielectric 'separator' between the inner and outer conductors. The potential for power loss in this dielectric medium is a drawback particularly in flexible coaxial cables. The axial current flow on the conductor surfaces are again represented by blue dashed lines. Stripline (Fig. 1.2c) is essentially 'flattened' coaxial line and has the advantage of ease of fabrication of intricate transmission networks using printed circuit board (pcb) techniques. In coaxial line and in stripline the dielectric separator usually displays a relative permittivity (ε_r) of between 2 and 3. By increasing this to between 6 and 10 in microstrip (Fig. 1.2d), it bccomes possible to dispense with the upper ground plane and create an open structure into which microwave components can relatively easily be inserted. The primary disadvantage, as the schematic suggests, is high dielectric losses and radiation losses.

1.3.1 TEM Transmission Line Fed Radiating Elements

The location of surface currents in the TEM transmission lines of Fig. 1.2 has been highlighted for antenna purposes. In order to create or enhance radiation from such lines it is generally necessary to disrupt the current flow directions. This is illustrated perfectly in Fig. 1.3 for microstrip line. This is the version from which radiation is most easily extracted.

The slot is located in the ground plane symmetrically and directly above (or below) the microstrip line and orthogonal to it, although this need not be the case.

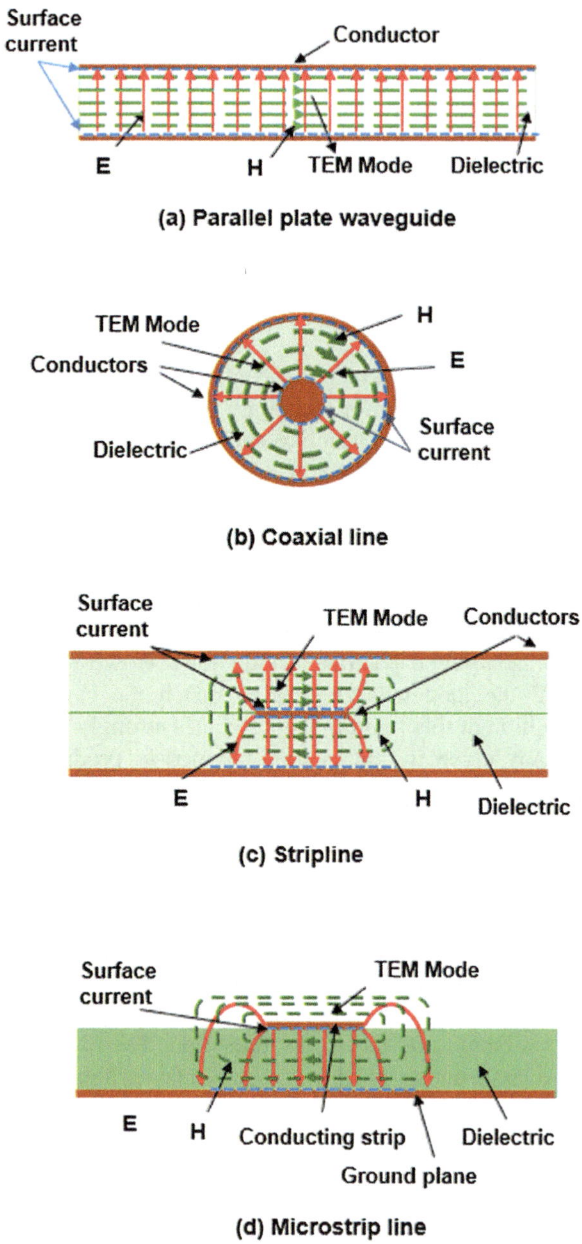

Fig. 1.2 Trapped TEM mode transmission lines systems

For illustrative purposes, the axial TEM mode currents are assumed to be approximately constant over a length of line which can be presumed to be less than $\lambda/4$ long, where λ is the TEM wavelength. The currents on the ground plane are strongest directly above the strip and consequently the slot exercises maximum

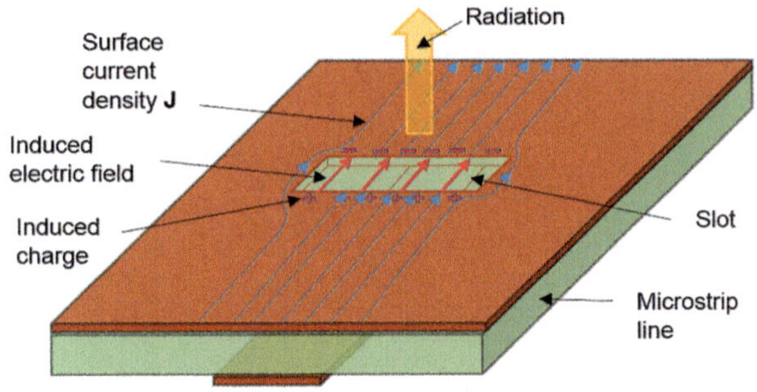

Fig. 1.3 Slot radiator located in the ground plane of microstripline

disruptive effect. Near the centre of the slot the current is totally impeded and induces a build-up of charge on the slot rims, shown by the positive and negative symbols. This is not unlike current flow into a capacitor, and as with a capacitor, electric field is formed in the gap. On the other hand, near the slot extremities, current paths can continue to flow by ‘bending’ around the slot ends. This increase in current path length, as in a coil, introduces inductance into the equation. Hence the slot can be modelled electrically as a two wire transmission line shunted by a parallel combination of a capacitor and an inductor. If the slot length is much less than half the free-space wavelength ($<\lambda_o/2$) the inductance (L) is negligible and the slot is largely capacitive (C) in nature. However, as the slot length increases toward $\lambda_o/2$ the inductance begins to match the capacitance and resonance occurs. In addition at this point the slot radiates strongly with the E and H vectors in the slot, at any given instant in time, aligned to produce an outwardly directed Poynting vector (Appendix A—Eq. A.19). In modelling terms, a resistive component (R_{rad}) may be included along with the L and the C for long slots.

An alternative scheme for inducing radiation from a microstrip line, which is, perhaps, more in sympathy with its basic electrical characteristics, is the microstrip patch antenna. In its simplest and original form the usually 50Ω characteristic impedance of the feed line is significantly reduced by abruptly widening the line in a step transition (see Fig. 1.4a). If the wide portion is then terminated in an open-circuit, a mechanism for resonance by reflection is created between the open-circuit and the abrupt transition. The transition, viewed from within the very low impedance patch, also presents a high impedance reflective discontinuity. Consequently, if the distance between the open-circuit termination and the step junction is set to $\lambda/2$, where λ is the TEM wavelength within the microstrip, the first standing wave resonance is trapped in the patch as shown in Fig. 1.4b, c. The patch is usually chosen to be approximately $\lambda_o/2$ in width for optimum radiation from the patch edges. The two ‘edges’ can be viewed as ‘patch slots’. In directions normal to the patch face the slot pair radiate in phase developing maximum power density as shown in Fig. 1.4b. Note that the feed line can be matched to the resonant patch by means of rectangular inserts as shown in Fig. 1.4a. The technique is not difficult for a radiating patch. In this case, in the absence of a reflection, the line current or voltage amplitude is unvarying in the feed line as shown. Finally relative to the

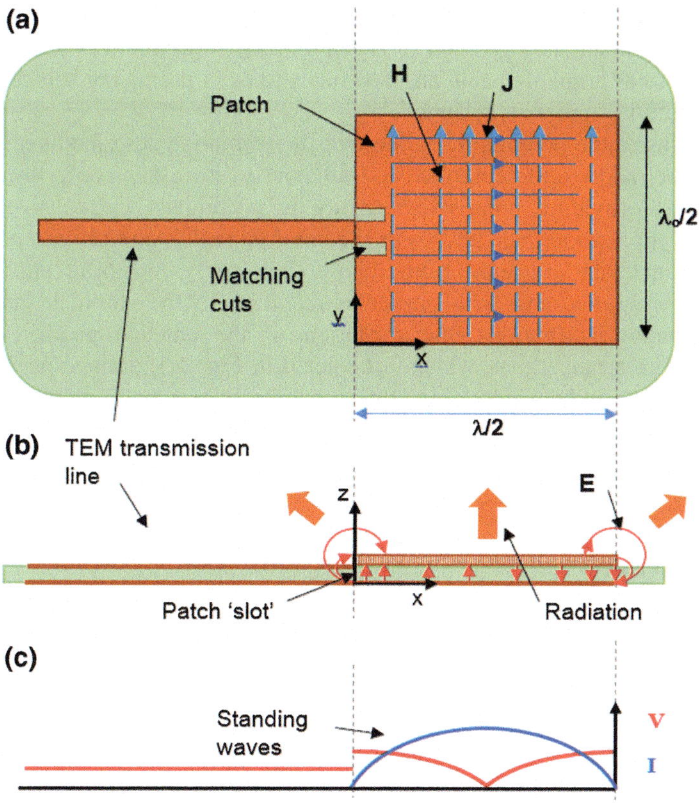

Fig. 1.4 Fundamental mode of operation for patch antenna, **a** top view of patch antenna, **b** side view showing trapped E-field—note edge radiation can be viewed as radiation from a two element slot array, **c** voltage and current variation in patch and line

propagation direct (*x* in this case) the trapped mode under the patch displays a one half-cycle variation in *x*, zero in *y* and zero in *z*, so the mode is termed TE_{100}. Note that if the *x*-direction were chosen to be normal to the patch surface then the same mode would be termed TM_{100}.

1.3.2 Radiating Element with Polarisation Agility

Hopefully, it has become clear from the above discussion, that it is possible to evaluate the radiation mechanism associated with TEM transmission line fed compact antennas, by performing a heuristic examination of the disruption caused to current flows and to E and H field directions in the TEM line. Examples of applying this kind of cogitative process are becoming quite common in the literature assisted by visual field presentations constructed from electromagnetic simulations [1–3]. In reference [4] a wide, but by no means exclusive, range of possible compact antenna geometries based on microstrip line technology has been developed. It has arguably been, and continues to be, a promising source of stimuli for novel compact antenna designs.

To emphasise the potential usefulness of the technique described here we apply it to a geometrically quite complex slot radiator fed from microstrip line, which is envisaged as providing polarisation agility in a compact inexpensive radiating element for application in an array antenna environment. It comprises a polarisation agile, T-shaped slot powered from microstripline. By insightful application of boundary conditions, and with particular attention to the slot field and surface current density distributions, the evolution of the antenna geometry is fully explained. The arrangement, which is depicted in Fig. 1.5, aims to achieve polarisation agility by independently exciting a pair of orthogonally aligned slots from microstrip. Thus the slot pair can radiate axial or transverse linear polarisations if excited separately (see Fig. 1.6), or if excited simultaneously and in phase

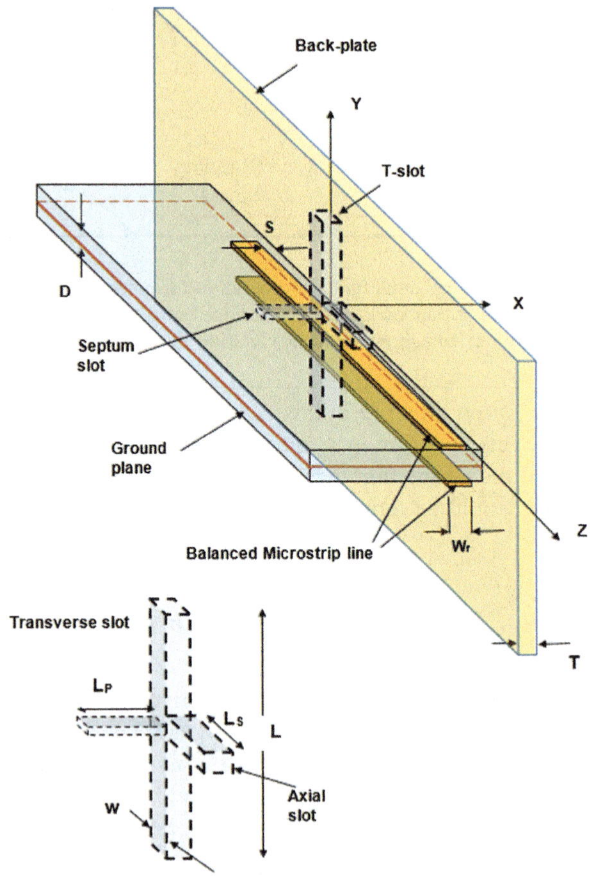

Fig. 1.5 T-slot radiator fed from balanced microstrip line

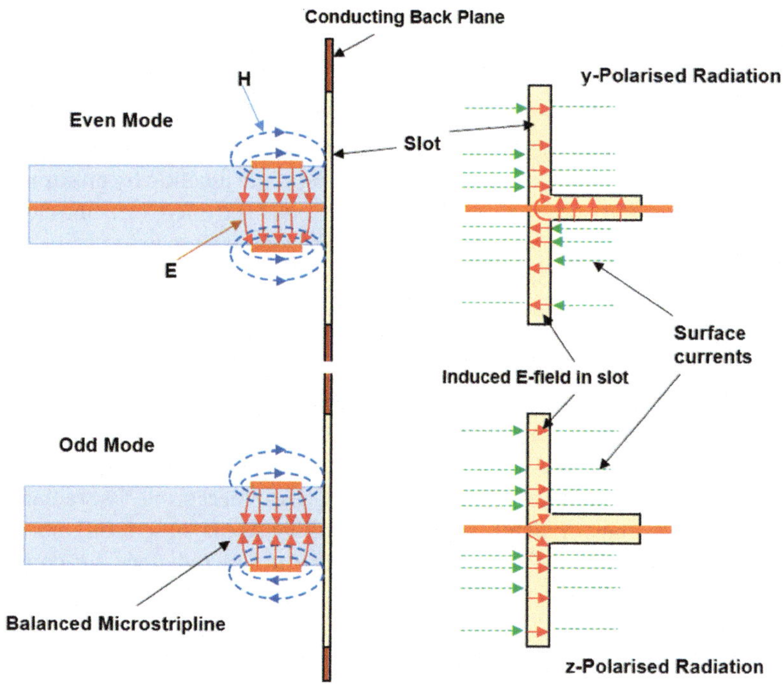

Fig. 1.6 Mechanism for delivering polarisation agility based on microstrip

quadrature circularly polarised radiation is produced [5]. The T-shaped slot is machined into the metal back-plate depicted in the Fig. 1.5. This plate is located in close proximity to back-to-back 'balanced' microstrip lines so that the slots are immersed in the fringing edge magnetic fields of these lines.

The fringing fields from the balanced microstrip lines induce z-directed current flows in the back-plate and hence an electric field in the transverse leg of the T-slot which in turn initiates the mechanism for radiation, as illustrated in Fig. 1.6. The balanced microstrip line can be operated in two modes, namely an even mode with the TEM mode E-fields in phase across the ground plane and an odd mode when they are in antiphase. The resultant fields in the slot are depicted in Fig. 1.6 which shows that the odd mode sets up transversely (y-directed) polarised radiation while the even mode generates axial (z-directed) polarisation. It should be note that the incident TEM mode, whether odd or even, directly excites only the 'vertical' y-directed, or transverse, component of the T-slot essentially through the fringing magnetic field components and the resultant z-directed current which is interrupted only by this component of the T-slot. The axial portion of the T-slot cannot be excited directly be a TEM mode.

In the case of even-mode excitation, while the upper half of the transverse slot component is, at the instant shown, immersed in a downward (y-directed) magnetic field, the opposite is the case for the lower half of the slot. The E-fields in the upper and lower portions of the transverse slot are in antiphase so that the slot as a whole is predicted to radiate very weakly. On the other hand, these antiphase slot fields in the transverse slot can combine at the junction with the axial slot, producing a strong parasitic excitation of this slot, if it is long enough. Radiation which displays y-directed polarisation is the result. The septum slot, or clearance slot, in the microstrip ground plane is introduced to minimise E-field suppression at the junction by proximity of the ground plane edge. For odd-mode excitation of the balanced microstrip line the E-fields in the upper and lower portions of the transverse slot are predicted to be in phase (Fig. 1.6) with negligible parasitic coupling into the axial slot. Radiation is therefore almost entirely from the transverse slot generating z-directed polarisation. In principle, the above radiator is capable of generating circularly polarised radiations, in addition to the linear possibilities already indicated. To achieve this the exciting TEM mode components will require to be in quadrature phase.

Based on a moment method study [5] of the proposed T-slot geometry the potential of this novel and compact slot to generate effective polarisation agility has been established quantitatively, by modelling and measuring its radiation and polarisation properties. Electromagnetic modelling tools based on the moment method and focused on the scattering properties of slot radiators are comprehensively developed in Chaps. 4 and 5.

1.4 Waveguide Feed System

It is perhaps not too unreasonable to suggest from a cursory scanning of the literature that compact antennas based on slots radiators are much more commonly fed from TE mode waveguide than from TEM transmission lines. So, following the same empirical route pursued in our examination of TEM line excited radiators, we shall now develop, from a plane wave perspective, the modal field and current patterns for rectangular waveguides, which are essentially hollow metallic pipes. These patterns are subsequently studied with the aim of conducting empirical explorations of a range of typical waveguide excited slot radiators.

Snell's laws can be used to explore the processes behind electromagnetic wave trapping, or guidance, concepts which are needed in later chapters. While at low frequencies, as we have seen, TEM waves can be guided by a pair of conductors, such as in power lines, in parallel wire telephone lines, or in coaxial lines, high frequency wave trapping in hollow conducting pipes is not so easy to comprehend. Such waveguides are increasingly being used in many of the compact antenna configurations employed in mobile communications. This method of guidance is very efficient, and is especially applicable to high power transmission at microwave frequencies. It relies on the nature of plane wave interference patterns and can, perhaps, best be explained by consideration of Figs. 1.7 and 1.8.

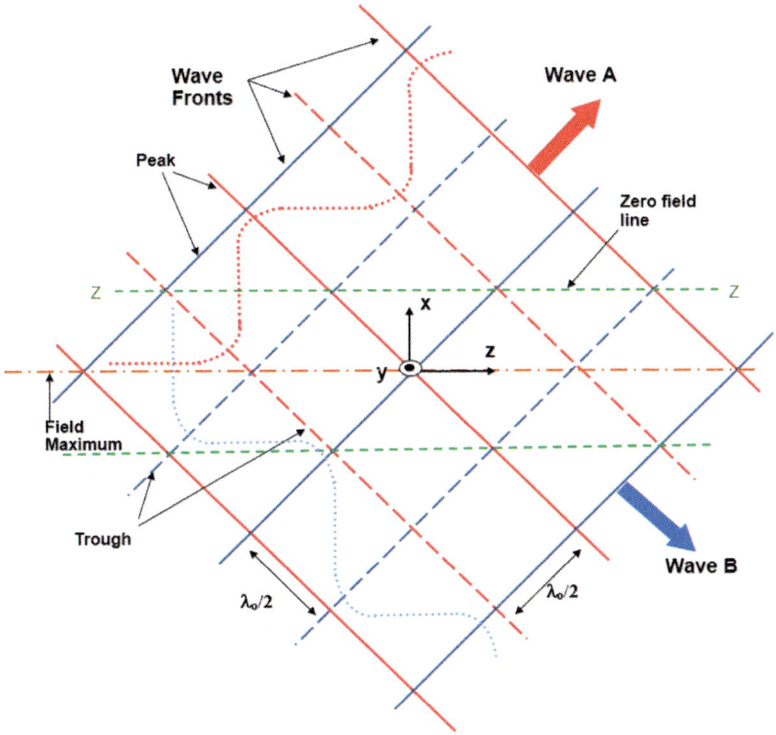

Fig. 1.7 Plane wave interference pattern

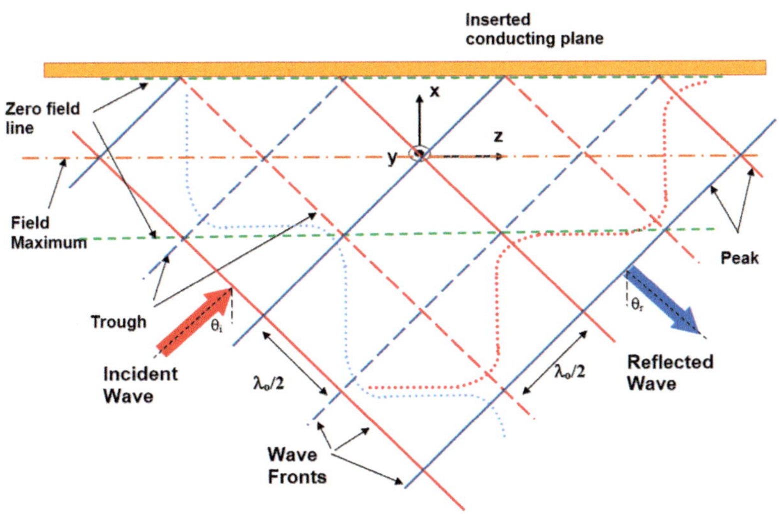

Fig. 1.8 Interference patterns of plane electromagnetic waves reflecting off a flat perfectly conducting surface

1.4.1 Trapped TE/TM Waveguide Modes

Figure 1.7 depicts (in two dimensions for simplicity) a pair of plane electromagnetic waves (TEM waves) of equal magnitude travelling in different directions A and B. The waves are represented by their wavefronts, with the wave peaks in each case denoted by solid transverse lines (planes in 3-D) and wave troughs by dashed lines. The distance between a wave peak and wave trough is, of course, half of the free-space wavelength ($\lambda_o/2$). The waves are travelling at the velocity of light [c] in the directions of the large arrows. On examination of this wave pattern it is not too difficult to observe that along the horizontal chain-dotted line (or in three dimensions—the y-z plane) peaks of wave A coincide with those of wave B, and troughs coincide with troughs—and this is independent of the movement of the waves. This line (or plane) represents a stationary (in the x-direction) field maximum 'independent of time' while the waves continue to exist.

In contrast, along the green dashed line, peaks of wave A coincide with troughs of wave B, and vice versa, resulting in a stationary field null at these positions. Consequently, if a perfectly conducting sheet of infinite extent, orientated normal to the x-axis, is located at the stationary null position (Z-Z) the field pattern remains unchanged (see Fig. 1.8). For a sheet at the upper dashed line, the red direction arrow then represents an incident wave and the blue arrow a reflected wave, which according to Snell's laws at a perfect mirror, reflects with a magnitude equal to the incident wave and at an angle such that $\theta_r = \theta_i$, as is required to retain the pattern. For perpendicularly polarised plane waves with the E-field confined to the y-direction, the E-field pattern forms a cosine distribution between the null planes. This pattern can be trapped or guided by introducing a second conducting sheet at the lower null locus in Fig. 1.8 (green dashed line). The trapped pattern travels in the z-direction with a phase velocity:

$$v_p = c/\sin\theta \tag{1.7}$$

and a wavelength:

$$\lambda_p = \lambda_o/\sin\theta \tag{1.8}$$

The phase coefficient for this trapped wave $\beta_p = 2\pi/\lambda_p$ and hence

$$\beta_p = 2\pi\sin\theta/\lambda_o = k_o\sin\theta \tag{1.9}$$

c is the speed of light, λ_o is the TEM wavelength in free-space, k_o is the phase coefficient for the constituent plane wave and $\theta = \theta_i = \theta_r$. The magnetic field distribution can easily be deduced by applying trigonometrical rules and the total E-H pattern is termed a TE (Transverse Electric) guided wave. Alternative TM (Transverse Magnetic) guided wave can be formed by commencing with parallel polarised TEM components (For mathematical details see Appendix B).

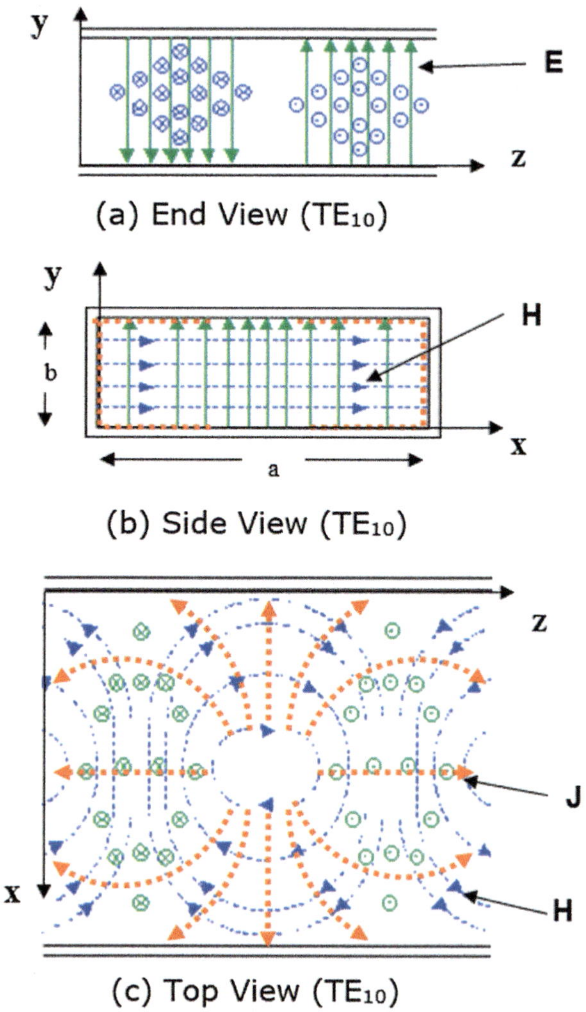

Fig. 1.9 TE_{10} mode field patterns for rectangular waveguide: width a, height b, **a** E and H-fields in the y-z plane, **b** E-field in the x-y plane, **c** E and H-fields in the x-z plane (E-field = green solid lines: H-field = blue dashed lines: current density J = orange chain-dotted lines)

A TE wave between parallel conducting planes separated by a distance *a* is illustrated in Fig. 1.9. The sinusoidal field variations in x are clearly shown in cases (b) and (c). The relationship between plane separation *a* and wavelength λ_o can again be deduced from trigonometry and yields:

$$\cos\theta = \frac{m\lambda_o}{2a} \tag{1.10}$$

m is the number of half-sinusoids of field pattern between the null planes. In Fig. 1.8 it is not necessary to choose the nearest null planes to create a trapped pattern. Equation (1.10) only has meaning for $m\lambda_o < 2a$, so that for m = 1 the case depicted in Fig. 1.9, the free-space wavelength must be less than $2a$ for propagation to occur. The corollary is that the frequency of the wave f (=$1/\lambda_o$) must be greater than a certain critical value or cut-off value corresponding to the cut-off wavelength $\lambda_c = 2a$. Furthermore if $a < \lambda_o < 2a$ the m = 2, 3, 4, solutions all yield the impossible requirement that $\cos\theta > 1$. This means that in the prescribed frequency range only the m = 1 solution is possible. The solution is termed the dominant mode for the parallel plane waveguide of separation a, and is defined as the TE_{10} mode—with one E-field variation in x and zero variation in y. This mode is shown in Fig. 1.9a–c. Perfectly conducting 'lids' can be introduced at y = 0 and y = b to form a rectangular waveguide, without altering the pattern, because the E-field is normal to these walls, and the H-field induces surface current density flows (J) which ensure magnetic field boundary conditions are satisfied. These current flows are shown in Fig. 1.9b, c. The b-dimension is usually chosen to be approximately half the a-dimension to maximise bandwidth.

Equations (1.7), (1.8) and (1.9) embody the essential features of the frequency characteristics of the dominant TE_{10} mode. Equation (1.9) implies that at cut-off ($\theta = 0$) for the TE_{10} mode $\beta_p = \beta_{10} = 0$, while at very high frequencies well above cut-off where $\theta = 90°$ then $\beta_{10} \to k_o$. At the same time from Eq. (1.7) we observe that $v_{p10} \to c$. This behaviour is summarised in the Brillouin diagram presented in Fig. 1.10, which for a waveguide of width a shows the TE_{10} mode cutting-on at $k_o a = \pi$. Thereafter it grows in an parabolic manner (see Appendix A.2) becoming increasingly asymptotic to the $v_p = c$ line. Close to cut-off where β_{10} approaches

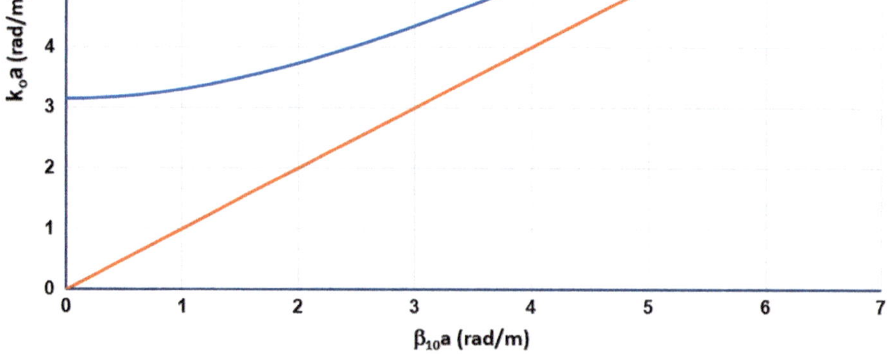

Fig. 1.10 Brillouin diagram showing graphically the frequency characteristics of the TE_{10} mode in rectangular waveguide (orange line denotes $\beta_{10} = k_o$ and thus $v_p = c$ for all frequencies: blue solid line = β_{10} for a rectangular waveguide of width a)

zero the guide wavelength $\lambda_{g10} \to \infty$ making wavelength-dependent component design impossible. Consequently this region is avoided in engineering practice. The practical frequency range is generally assumed to run from $k_o a = 1.25\pi$ ($f = 1.25 f_{c10}$) to $k_o a = 2\pi$ where the TE_{20} mode starts to propagate. Note that in this region the TE_{10} mode plot is relatively linear thus minimising the frequency dispersion which can be an awkward feature, in communication terms, of hollow metal waveguide.

Apart from the TE_{10} mode, in the dominant mode frequency range, all TE_{mn} modes for which $(1 < m < \infty, 0 < n < \infty)$ and TM_{mn} modes for which $(0 < m < \infty, 0 < n < \infty, m = n \neq 0)$ are cut-off and decay exponentially from the electromagnetic wave source. These modes are referred to as evanescent modes and carry no power. They account for stored energy which accumulates around discontinuities in the waveguide in order that the total field structure satisfies the boundary conditions there. Furthermore, this doubly infinite set of orthogonal TE and TM modes is mathematically 'complete' and enables full wave modelling of waveguide discontinuities such as slot radiators, as we shall see in Chaps. 4 and 5. There we develop a moment method model of the slot discontinuity problem. Texts 4 and 9 in the Bibliography, provide full and comprehensive treatments of EM modelling techniques.

1.5 Waveguide Slot Radiators

Arguably, waveguide fed slot radiators have provided a very significant source of ideas for recent developments in compact antenna technology. However, early embodiments, such as those depicted in Fig. 1.11, were impelled by large antenna arrays for military radars [6–8]. Rapid developments in the 1940s were driven by

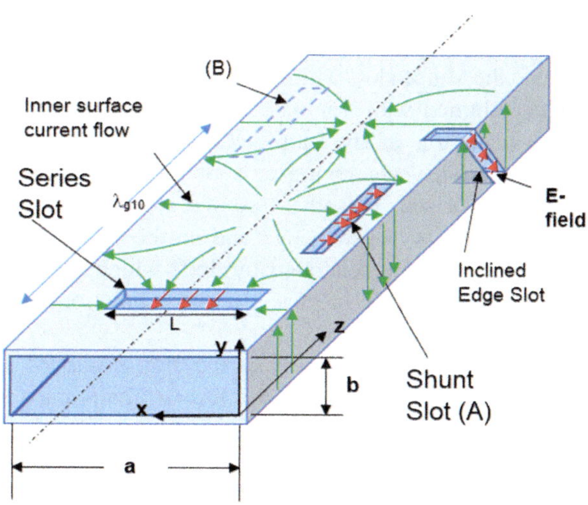

Fig. 1.11 A hollow rectangular TE_{10} mode waveguide showing some possible slot locations required to secure radiation. One guide wavelength (λ_{g10}) of the modal current pattern is depicted

the demands of World War II for long range radio detection systems. The frequencies used were generally below or at the low end of the microwave range, so the waveguide and the resultant antennas were inevitably large. Nevertheless, war-time advances on slotted waveguide array antennas laid important and reliable foundations on the electromagnetic issues associated with waveguide fed slots and waveguide slot arrays [8]. This early work continues to inform current developments, as hinted at in Chaps. 4, 5, 6 and 7.

1.5.1 Conventional Waveguide Slot Radiators

Traditional slot array antennas generally employ either longitudinally directed offset 'shunt' slots, or transverse centrally located 'series' slots, in the broadwall of the rectangular waveguide feed. Inclined slots located in the sidewall of waveguide are also common (see Fig. 1.11). Once the TE_{10} mode wall current flows have been established, as shown in the Fig. 1.9, the slot excitation mechanism is not too different from that of the TEM mode examples. Current flow interruption by the slot induces charge build up on its long edges resulting in an instantaneous E-field formation across the narrow dimension of the slot. This slot field changes in sympathy with the modal pattern as it passes through the waveguide at the phase velocity given by Eq. (1.7) (or Eq. B.15 in Appendix B). At a given incident power level in the propagating mode, the slot radiation is maximised by operating it near its resonant length, namely $L = \lambda_o/2$.

Shunt slot in waveguide

Figure 1.11 also reveals that the shunt slot is weakly excited close the waveguide centre line where the current flow is largely axially directed, and strong near the side wall where the current flow is almost entirely transverse. This behaviour is used to control radiation strength. The scattering nature of waveguide fed slot radiators in explored in detail in Chap. 5.

For the shunt slot, which is z-directed in Fig. 1.11, the induced electric field is clearly aligned with $\pm x$ axis. Therefore, in the direction of the pattern maximum, this orientation of the slot dictates that the radiation from it is polarised in a direction transverse to the axis of the waveguide—usually referred to as transversely polarised. This also true of a slot at position (B) which is separated axially from the slot at (A) by $\lambda_{g10}/2$ and on the opposite side of the centre line. In phase terms the slots are separated by 360° which can be interpreted as 0°. This means that the slots radiate in phase in the y-direction. The importance of this feature is outlined in Chap. 7.

The radiation pattern, when the slot is part of a planar array for example, is predominantly directed into the forward half-space (y > b) since the array surface presents a large conducting ground plane to the slot. It can be viewed as a 'half-dough-ring' in shape with nulls in the axial direction ($\pm$ z-direction), and

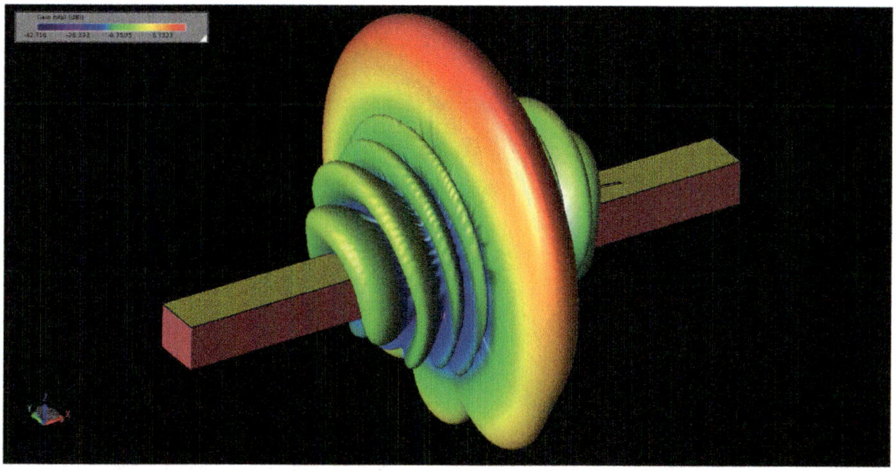

Fig. 1.12 Computed radiation pattern for a six element shunt slot waveguide fed array (with acknowledgement to www.remcom.com)

displaying omni-directional power density distribution in the radial direction in the x-y plane. These features can be discerned from Fig. 1.12 which displays a computed radiation pattern for a linear array of six shunt slots.

The structure of a typical shunt slot array, in planar form, is illustrated in Fig. 1.13. If fabricated on a programmed lathe, the top plate and base plate are generally created separately. The sizes, shapes and locations of the slots forming the radiating slot array, are determined from simulation on an electromagnetic solver, and these dimensional details are programmed for automatic machining into a metallic top plate of predetermined thickness. The base plate, on the other hand, carries the network of feed waveguides, again with dimensions predetermined by simulation, enabling computer controlled machining on a thick metallic (usually aluminium) work piece. In the example shown the eight parallel grooves forming the waveguide feed structures are shorted at each end forming resonant cavities. This arrangement is narrow band (see Chap. 7) but very cost efficient. Power is leaked into the resonant cavities from the transverse waveguide by narrow apertures in their end walls. The coupling apertures are separated by $\lambda_{g10}/2$ in the transverse member which means that the feed waveguides are excited in antiphase as Fig. 1.13b indicates. To ensure that the radiating slots are all radiating in-phase at the operating frequency the wall current directions are exploited to ensure that the slot staggering takes account of this. The transverse member is, in turn, powered through a T-junction formed by introducing an input waveguide from behind the array feed block. The 'choke', which is included in Fig. 1.13, comprises a quarter wavelength deep groove machined into the base plate approximately a quarter wavelength from the waveguide recess. It ensures that radio frequency leakage between the base plate and the slotted lid is negligible.

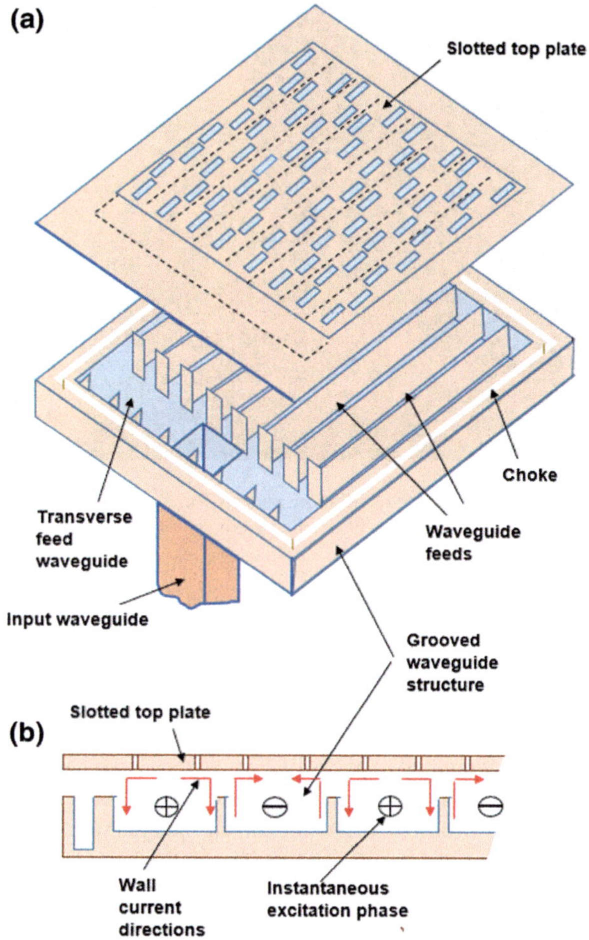

Fig. 1.13 Planar shunt slot resonant waveguide array, **a** isometric view, **b** end sectional view

Such an array can be surprisingly compact in high frequency sensor applications. For example, at mm-wave frequencies (circa 100 GHz) and employing (say) WG27 (2.4 mm × 1.2 mm) a sensor array could exhibit dimensions typically in the range 2.8 cm wide, 3.8 cm long and 2.5–3 mm thick. The radiating slots would be of length ~1.5mm and ~0.2 mm wide. Conventional slot arrays of the type illustrated in Fig. 1.13 are examined fully in Chap. 6.

Series slot in waveguide

The transverse slot in Fig. 1.11 is noticeably less versatile, and hence less practical, than the shunt slot. Since the slot is now excited by the longitudinal currents flows in the broadwall of the waveguide, which replicates the axial current flow along the wires of a two wire transmission line, the slot exhibits the behaviour of a lumped electrical component inserted in series with the line. For a slot, which is most commonly operated close to resonance ($L = \lambda_o/2$), transverse positional movements in order to control radiation strength are severely limited (L is not much smaller than a). Furthermore, from an array perspective broadside radiation requires the z-polarised transverse slots to be separated by the guide wavelength λ_{g10}, to procure in-phase excitation. If the waveguide is air filled this magnitude of slot spacing has significant and undesirable radiation pattern consequences. These are thoroughly examined in Chaps. 6 and 7. The solution is usually to fill the waveguide with a dielectric, or load it with a periodic surface, so that λ_{g10} is reduced in length toward $\lambda_o/2$. In a closely related application, as outlined in Chap. 8, series slots are ideal elements for the implementation of waveguide fed leaky-wave array antennas. This is because transverse broadwall slots, when tightly spaced axially, create periodic ladder structures which readily support surface waves and leaky waves.

The radiation pattern for a single slot is omnidirectional in the y-z plane while in the x-z plane nulls occur in the $\pm x$ direction. Apart from polarisation direction, the pattern differs from that of the shunt slot only in radiation distribution at wide-angles. This is attributable to divergences in the electromagnetic wave scattering into the backward half-space.

Edge slot in waveguide

Planar slotted-waveguide array antennas which employ inclined edge slots have been quite frequently reported upon in the early literature with a focus on antennas for radar systems operating at microwave frequencies [6–8]. On the sidewall of rectangular waveguide the TE_{10} mode surface current flows are directed transversely to the waveguide axis, that is y-directed in Fig. 1.11. Such currents could obviously be interrupted by z-directed slots. Strong excitation would result, but such slots offer little control of the radiation strength if half wavelength long slots are preferred. The usual practice therefore is to incline the slots which introduces radiation strength control using the tilt angle (the greater the tilt the stronger is the coupling to the TE_{10} mode).

Needless to say the tilted slots exhibit disadvantages and advantages in slot array applications. Firstly, because of the tilt alone the electric field induced in the slot is neither aligned with, nor transverse to, the waveguide axis. Consequently in order to avoid the linearly polarised radiation being in a ‘squint’ relationship to the waveguide axis adjacent slots in a linear array are inclined in opposite directions. It is not difficult to deduce from Fig. 1.11 that the resultant polarisation for such an array is axial. While the cross-polar (y-directed) components of the slot radiations cancel in the broadside direction, strong cross-polar sidelobes of the array appear at wide radiating angles. This represents a major disadvantage of inclined edge slot arrays. A schematic illustration of such an array is provided in Fig. 1.14 showing clearly the oppositely inclined slot implementation scheme. For this array, if the waveguides are parallel to the ground, the antenna provides a horizontally and linearly polarised radiating beam in the forward broadside direction.

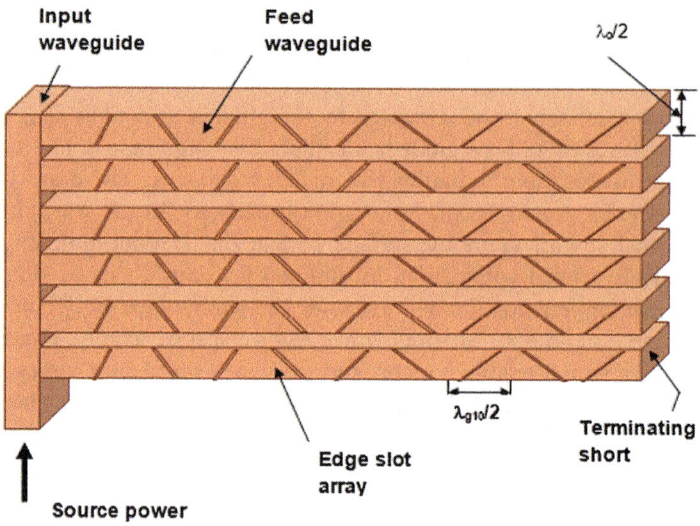

Fig. 1.14 Horizontally polarised resonant waveguide planar array antenna formed from edge-slots

In principle, vertical stacking of an edge slot array should be simpler than for broadwall slots since the waveguide height (*b*) is generally considerably smaller that $\lambda_o/2$, the critical distance for the avoidance of grating lobes (see Chaps. 6 and 7). However this advantage is somewhat negated by the encroachment of resonant length edge slots into the broadwall (see Fig. 1.11), so that electromagnetic wave leakage into the inter-guide gaps occurs. The result is radiation pattern unpredictability and additional cross-polar deficiencies. With each of the slotted waveguides in Fig. 1.14 depicted as terminating in a short circuit usually $\lambda_{g10}/4$ from the last slot and the slots separated by exactly $\lambda_{g10}/2$, the sketch is representative of a 'resonant' array (see Chap. 7). It produces a precise high gain principal beam in the broadside direction at the design frequency. However, high gain delivery is available over only a very narrow bandwidth.

1.5.2 *Unconventional Waveguide Slot Radiators*

Omni-directional slot doublet

The heuristic approach to evaluating, or comprehending, the radiation mechanisms developed by slots in transmission systems, can quite successfully be applied to more complex geometries. Below we investigate several examples which are germane to later chapters in this text.

Perhaps, the simplest unconventional slot pattern in waveguide is the slot doublet which has applications in medicine when embedded in a surgical implement or

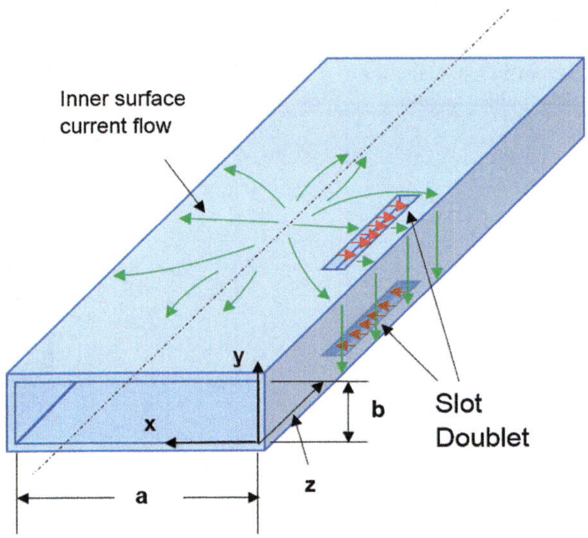

Fig. 1.15 Omni-directional slot-doublet in rectangular waveguide

as a compact, robust search and rescue transponder [9, 10]. In much the same way as is expounded in the previous section the mode of operation of the slot doublet in waveguide (Fig. 1.15) can be explained by knowledge of the TE_{10} mode wall currents and the disruption caused to these currents by the presence of the slots. Instantaneous charge build along the long edges of the slot is the source of electric field in the slot which always traverses the shortest distance within it (see Fig. 1.15). For the TE_{10} mode, at the instant shown where the top wall currents flow outward to the sidewalls they flow inwards on the bottom wall. This means that the electric field in the top wall slot is polarised in the opposite direction to the bottom slot, and this remains true for all instances in time as the mode propagates through the waveguide. Electromagnetic wave scattering by the upper slot produces $-x$ polarised radiation into the upper half-space while the lower slot radiates into the lower half-space with $+x$ polarised radiation. A comprehensive simulation of the slot doublet geometry has been formulated [9, 10] using the moment method (Chaps. 4 and 5). It exhibits good omni-directional radiation in the transverse plane of the waveguide (Fig. 1.16) for a low height waveguide (25% less than conventional waveguide). For the pattern in the figure the ratio of the maximum to minimum radiation intensity (at constant radius) is of the order of 1 dB. Perhaps not surprising, for a full height waveguide it is not nearly so good, at approximately 10 dB. Unfortunately lowering the waveguide height severely restricts power handling capability.

Polarisation agile waveguide slots [11]

An example of a significantly more complex arrangement of radiating slots in waveguide has been proposed for a polarisation agile 3-D antenna for a long range

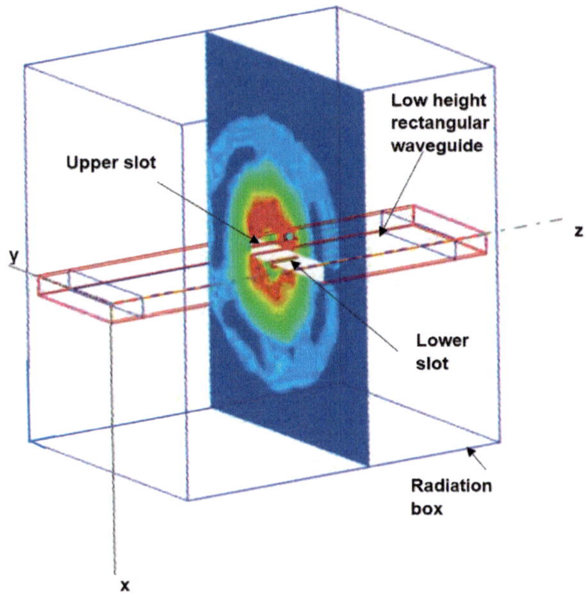

Fig. 1.16 Simulated omni-directional radiation pattern in x-y plane for slot-doublet radiator

ground radar [11, 12]. A typical system of this kind usually comprises a parabolic reflector providing azimuth gain and a front fed linear array providing elevation gain and frequency scanning (see Chap. 8). To bring about polarisation agility it is necessary to create a means of altering the current flows in the walls of the waveguide feed structure in order to switch TE_{10} mode excitation from transversely polarised slots to axially polarised slots. One way of doing this is to employ bifurcated waveguide as suggested in Fig. 1.17.

This particular bifurcated waveguide format can be excited from a conventional four port 'magic-T' (see Bibliography: D.M. Pozar). The two output ports dovetail with the bifurcated waveguide feed while of the two input ports one sets up the even mode of propagation, the other the odd mode alternative. Switching power between the two input ports represents routine microwave technology. Consequently, polarisation switching requires a broadwall slot pattern with orthogonally directed slots which are independently excited by the even and odd propagation modes. The slot pattern is dictated by the TE_{10} mode surface current flows as illustrated in Fig. 1.17.

Given that for effective switching slot radiators must be immersed in the even mode as a whole, or the odd mode as a whole, there are two possibilities—a transverse slot straddling the central wall, or bifurcation, and a longitudinal slot running along the bifurcation. The former is excited by axial wall current flows as we have already seen. In this case only the even mode is effective in exciting the transverse slot with an uni-directional axially polarised electric field (see Fig. 1.17a). For the odd mode the induced electric field in the transverse slot is weak and in anti-phase across the centre line. The resultant radiation from this slot

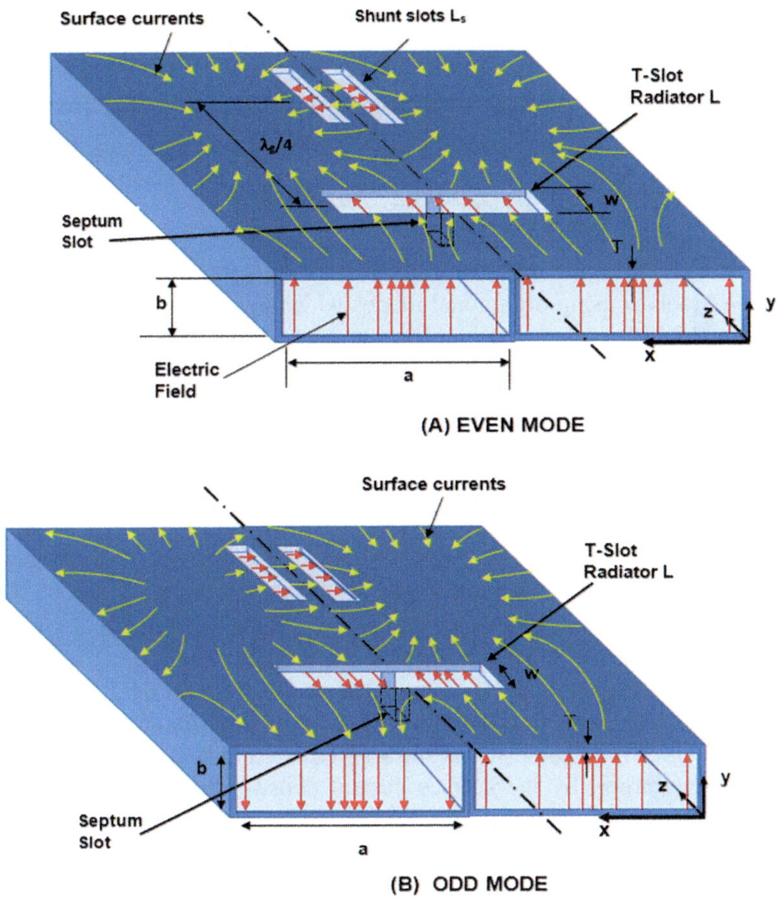

Fig. 1.17 Dual-mode bifurcated rectangular waveguide structure showing even mode to odd mode polarisation switching mechanism

is relatively insignificant. Note the inclusion of a clearance slot, or septum slot, in the bifurcation in order to minimise suppression of the slot field by the metal wall.

A longitudinally directed slot located above the common wall of the dual mode waveguide can be excited only by the odd mode. This mode generates wall current flows which cross the centre line and can be intercepted by an axially directed slot (Fig. 1.17b). However, an axial slot positioned over the common wall is impractical because of the damping effect, on the slot E-field, of the metal wall beneath it. To overcome this problem, the simple answer is to split the axial slot into two elements which straddle the bifurcation, as shown in the figure.

If a bifurcated waveguide array antenna, formed from slots of this description, were mounted vertically, relative to the ground, even-mode excitation would generate vertical polarisation, while the odd-mode would produce a horizontally polarised radiation pattern. If the inputs to the 'magic-T' were fed in quadrature phase through a 3 dB hybrid, for example, the radiation pattern would be circularly polarised.

1.5.3 Waveguide Cavity Backed Slot Radiators

A not unrelated, but subtly different, route toward a polarisation agile slot pattern in waveguide involves using orthogonal modes which are formed in a square waveguide cavity [13, 14], rather than the even and odd mode scheme outlined above. Rectangular waveguide is readily convertible to a high-Q cavity resonator by terminating a section of waveguide with a short circuit. Transmission line theory then informs us that this zero impedance load is replicated at every half-wavelength ($\lambda_g/2$) as we withdraw from the load toward the source (see Bibliography: J.D. Kraus). If the propagating mode in the original waveguide is the TE_{10} mode then by locating a second short circuit at the nearest E-field null to the load traps a TE_{101} cavity mode, while the next one traps a TE_{102} cavity mode—and so on. For a square cavity of dimensions $a \times a \times b$ the resonant frequency for the cavity mode can be deduced from equation (B.15). It is fixed and given by $f_0 = c/\sqrt{2}a$ for TE_{101} and $f_0 = \sqrt{2}c/a$ for TE_{102}. For example for WG16 for which a = 22.86 mm a TE_{101} square cavity resonates at 9.3 GHz, while a TE_{102} cavity resonates at 18.6 GHz.

Appropriate broadwall slot locations, commensurate with strong radiation, are indicated in Fig. 1.18. The positions shown are determined, as with the waveguide slots discussed earlier, by the surface current density flows (J) in the upper (or lower) inner square faces of the cavity. For the TE_{101} mode depicted in the left hand diagram of Fig. 1.18, electromagnetic coupling to the cavity fields occurs equally strongly at four possible slot locations as shown. Four slots in a balanced or

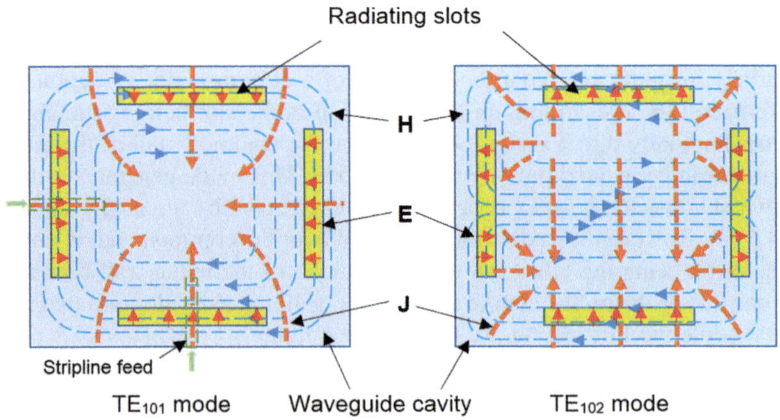

Fig. 1.18 Polarisation agile waveguide cavity backed slot antenna

symmetrical arrangement are chosen to extract orthogonal linear polarisations. However, in the broadside direction (normal to the square surface) this slot pattern actually produces negligible radiation for TE_{101} mode excitation. The instantaneous current flow directions lead to slot electric fields, in the horizontal slot pair and in the vertical slot pair, which are opposed in phase resulting in E-field cancellation, and hence power suppression, in the desired radiation direction. The figure suggests that the TE_{101} mode can be excited from stripline tabs located beneath one or more of the top-wall slots [14], but other mechanisms are equally possible.

While the TE_{101} cavity mode does not lead to a slot based antenna with meaningful polarisation agility, this is not true for a cavity tuned to the TE_{102} mode, as is illustrated in the right hand diagram of Fig. 1.18. The same balanced and symmetrical slot pattern is employed and it is assumed that the higher frequency TE_{102} mode can be excited by similar stripline tabs to those described above for the TE_{101} mode. If we again follow the empirical process of examining the relationship between the modal wall current flows and the disruptively placed slots it is not difficult to appreciate that the slot electric field alignments are as shown in the figure. These current flows for the TE_{102} mode dictate the vertical slot pair do not radiate, since the E-field reverses within the length of each slot. On the other hand the horizontal pair radiate strongly and in phase producing a vertically polarised radiation pattern with a maximum at broadside. This mode and this polarisation would be produced by a stripline feed entering the cavity either below the upper slot or the lower slot, or both. Feeding the cavity from the side, under either the left hand or the right hand slot sets up the TE_{201} mode, for which the magnetic field and current density patterns are rotated by 90° relative to the TE_{102} mode depicted in Fig. 1.18. In this case the horizontal slots are weakly excited while the vertical slots radiate strongly, and in phase, generating a horizontally polarised broadside beam.

An antenna of this description can therefore be switched between orthogonal linear polarised emissions by electronically switching the input signal between the two stripline ports. Furthermore, if the ports are fed simultaneously, and in quadrature phase, circularly polarised radiations can be generated. In microwave circuits this is generally achieved by incorporating a 3 dB hybrid coupler between the antenna and the source. A photograph of an experimental prototype [14] of an antenna with this functionality is presented in Fig. 1.19. It is fabricated from two copper clad substrates which have been combined to create a triplate sandwich. In the cavity region the inner copper cladding is cleared leaving only the stripline feed tabs. The outer cladding forms the top and bottom walls of the square cavity which is completed by the metal plated vias clearly shown in the photograph. The area between the cavity and the coaxial connectors supports a stripline feed network which comprises a 3 dB hybrid coupler to generate circularly polarised radiation from the slot pattern. The presence of the dielectric within the cavity lowers its frequency and its Q-level, as does the via constructed sidewalls. However, measurements demonstrate levels of Q which are more than high enough to maintain modal purity [14].

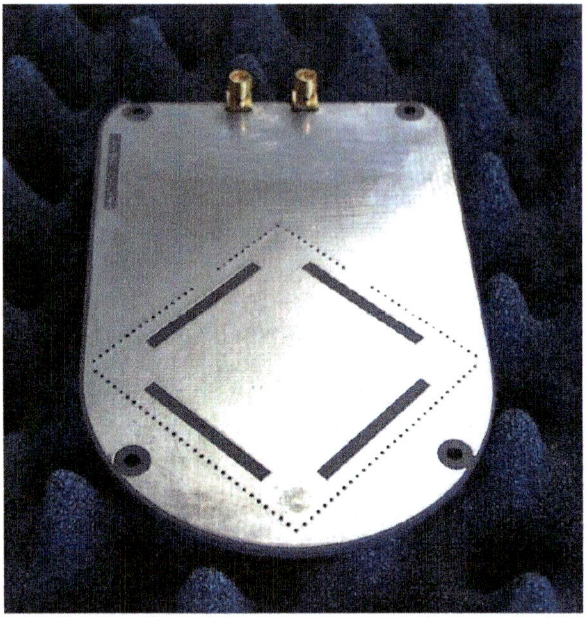

Fig. 1.19 Dual-polarised cavity resonator antenna for RFID scanning applications at 2.45 GHz ($\sim$4 cm wide by 7 cm long)

1.6 Planar Reflect-Arrays for Compact Antennas

Pressure from communication systems developers for access to compact alternatives to traditional antenna types, such as the mainstay parabolic reflector antenna, has increasingly been driven by the inevitable drift toward higher and higher frequencies by expanding demand for broadband systems and from growing requirements for mobile platforms which can provide direct access to satellites. At below microwave frequencies this role has traditionally been filled by versatile reflector antennas which provide high efficiency, high gain, high bandwidth and can readily deliver tracking using well established motion control techniques. Nevertheless, the conventional parabolic reflector antenna presents a primary disadvantage in applications such as mobile platforms capable of tracking low earth orbit (LEO) satellites, namely in its relatively bulky shape and in its weight. Incorporating such an antenna, while preserving low wind resistance, into a fast moving vehicle, which is itself streamlined, is difficult and costly. Furthermore, tracking a satellite from a fast moving vehicle severely stretches the capability of mechanical control systems. Solutions based on simply 'flattening' the reflector, while preserving the focusing property of the original antenna, are being explored.

The electromagnetic processes which require to be followed to 'flatten' a parabolic reflector antenna, without sacrificing 'focusing', are developed in detail in Chap. 9. However, it is not too difficult to illustrate the process with a few simple 2-D sketches as shown in Fig. 1.20. The 'flattening' argument applies equally to 3-D developments. In Fig. 1.20a the parabolic reflector surface (shown as a brown

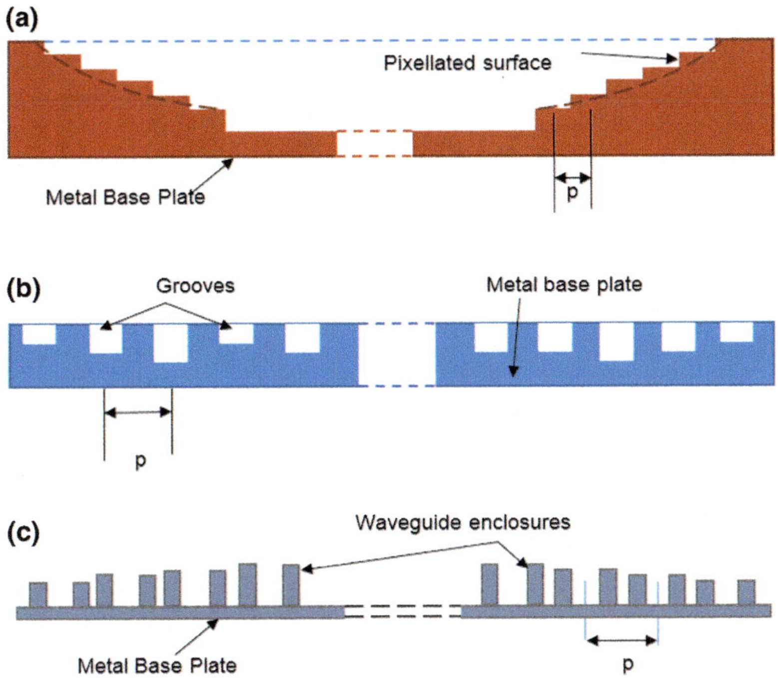

Fig. 1.20 Evolution of 'flat' focused reflectors for incorporation into compact reflectarray antennas with applications in communication systems

dashed line) is presumed to be replicable by a stepped, or pixelated, surface which is evidently possible if the pixellation steps (p) are infinitesimally small. In practice *p* is set at about 10% of the operating wavelength. Ideally at the reflector aperture (blue dashed line) the normal reflected wave from the centre of each step should match that of the reflection from a similar point on the original reflector, to procure 'equivalence' between the parabolic surface and the pixelated replacement.

To 'flatten' the reflector as suggested in Fig. 1.20b, its upper surface has to replicate the reflection phase of the original reflector at the aperture plane (blue dashed line) in Fig. 1.20a. This is achieved by arranging that at the pixelated step locations a 'shorted' groove is provided which reflects the incident electromagnetic wave so that at the 'flat' surface the reflected wave exhibits the phase distribution of the original reflector at its aperture plane. In three dimensions the grooves can be viewed as shorted waveguides of a depth which generates the requisite phase. An alternative approach is depicted in Fig. 1.20c where thin walled 'waveguides' are raised from a flat backing plate. The upper surface in this case is not exactly flat but the reflector is lighter than the case (b) alternative. In Chap. 9 more sophisticated solutions are addressed. These are based on the use of patch, or slot, arrays etched into flat copper clad substrates.

1.7 Compact Cavity Resonator Antennas

As we have observed in earlier sections of this chapter the manipulation of electromagnetic wave radiation patterns, in the context of compact antennas, can be procured and controlled by the judicious disruption of surface current flows in suitable feed systems such as striplines and waveguides. Furthermore, if bandwidth is not an issue, it is suggested that additional forms of pattern control can be devised by cleverly amalgamating slot patterns and orientations with carefully selected cavity modes. But if antenna compactness, ease of fabrication and cost effectiveness are primary developmental goals, then the implementation of cavity antennas in conventional copper clad substrate form, is generally preferred and becomes a high priority. Unfortunately, as Fig. 1.19 shows, stripline or microstrip technology does not easily lend itself to high-Q cavity formation without resorting to the use of non-standard techniques such as multiple metallised vias to realise sidewalls. This represents a distinct disadvantage for cavity based antenna initiatives where fabrication costs can be critical.

Needless to say solutions have been, and are being, sought by for example employing resonance conditions in parallel plate waveguide, and in particular the Fabry-Perot resonance to circumvent stripline fabrication issues. The Fabry-Perot scheme is shown diagrammatically in Fig. 1.21a. It can be viewed as a rectangular waveguide (see Fig. 1.9), with width a replaced by separation distance d, lying on its side with the top and bottom walls removed. The Fabry-Perot resonance equates to cut-off for the TE_{10} mode. Note that this mode is not influenced by the absence of walls at $y = -b/2$ and $y = +b/2$ (see Fig. 1.9b). The surface current flows in the walls at $z = 0$ and $z = d$ will be y-directed and unvarying for all y, so the Fabry-Perot resonator can act as an antenna for x-directed slots in the $z = d$ wall, assuming excitation of the cavity is through the $z = 0$ wall. For two dimensions the nature of this radiation, for an array of slots forming a partially reflective surface (PRS) at $z = d$, is not difficult to explain. The radiating elements of the PRS are excited in phase and consequently a high directivity principal beam is formed at broadside. It is analysed and discussed fully in Chap. 10.

In three dimensions, provided the resonator is excited by a non-directional feed such as a z-aligned coaxial probe, orthogonal TE_{10} modes (with x-directed or y-directed E-field) are possible, both of which behave as described above. The resultant broadside radiated pencil beam will be polarised at 45° to x or y. However, in the planar Fabry-Perot cavity antenna, each mode exhibits additional features which are much less straightforward. When the structure is viewed on the x-z plane, as shown in Fig. 1.21b, a more complicated field pattern for the same TE_{10} mode is revealed. The surface wall current flows remain y-directed and continue to excite x-directed slots. But now, for slots distributed in the $\pm x$ directions, in-phase excitation is only guaranteed at cut-off where $\lambda_{g10} \Rightarrow \infty$. In this case a high gain 'pencil' beam is generated by the 2-D Fabry-Perot cavity resonator antenna for this mode

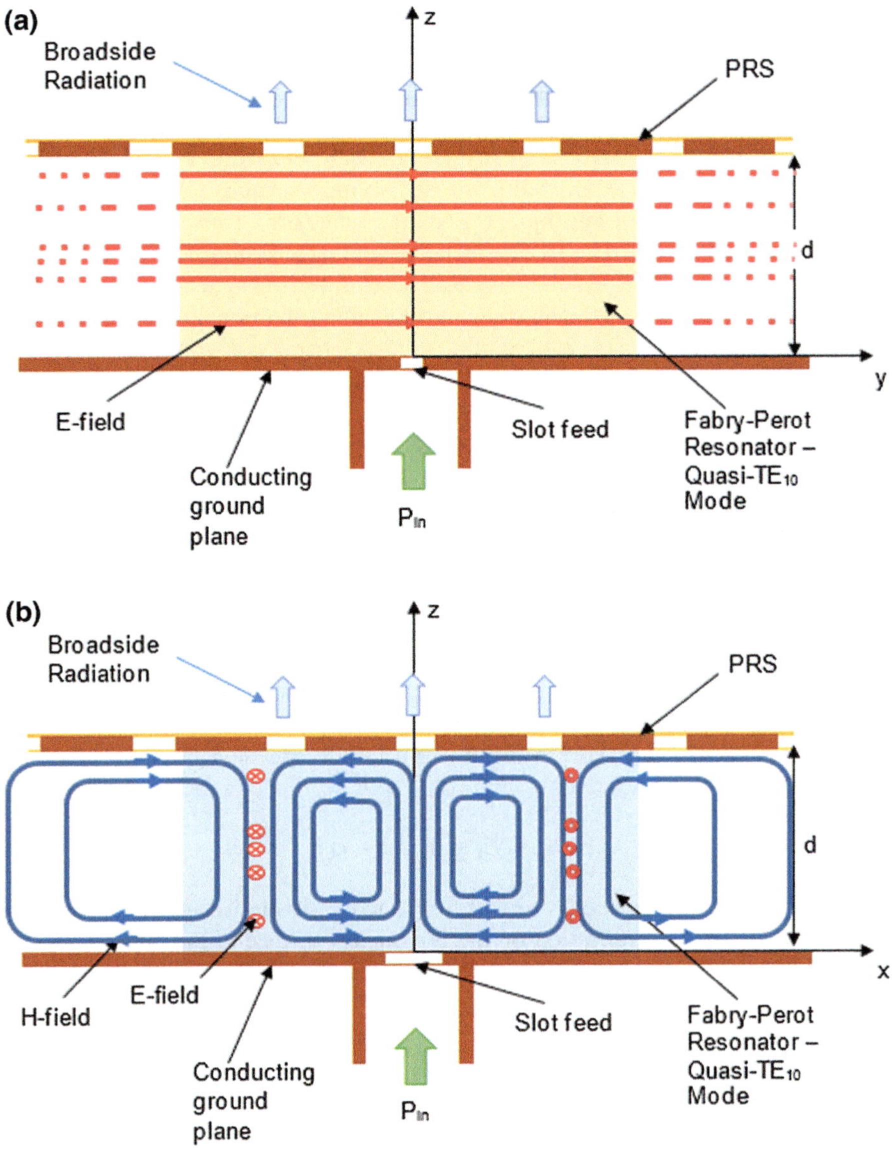

Fig. 1.21 Schematic representation of a Fabry-Perot resonator antenna, **a** end view of TE mode structure, **b** top view of TE mode structure

and the orthogonal mode. However, less helpfully, at frequencies above cut-off, waves begin to propagate in the $\pm x$ directions (or $\pm y$) within the parallel plate structure leading to frequency sensitive and scanning leaky-wave beams. This operating condition is examined in detail in Chap. 10.

1.8 Retro-Directive Array Antennas

The principle behind the retro-directive array, which is addressed in Chap. 11, has been known about since the 1960s. Retro-direction, whereby an incoming radar signal, on striking a target reflector is returned back in the direction from which it came, usefully boosts the 'visibility' of the original scatterer. This outcome has historically been achieved by employing 'corner reflectors'. These are often seen on the masts of yachts [15]. The device increases the radar cross-section of the target thereby enhancing its detectability against severe electromagnetic scattering emanating from the sea. A schematic drawing illustrating the principle of operation of the corner reflector is presented in Fig. 1.22. Generally it comprises three flat conducting metal surfaces positioned orthogonally to each other—for example the inner faces of a cube as suggested in Fig. 1.22a. In practice the corner reflector is made less bulky and hence lighter than the 'open cube' by refashioning it into a trihedral shape as shown in Fig. 1.22b.

The retrodirective nature of the corner reflector can be illustrated relatively easily by expressing the ray directions in Fig. 1.22a in vectorial terms. For example the incident portion of ray 1 can be defined as travelling in direction

$$\mathbf{r}_1 = \mathbf{a}_x x_1 + \mathbf{a}_y y_1 + \mathbf{a}_z z_1 \tag{1.11}$$

where $\mathbf{a}_x$, $\mathbf{a}_y$, $\mathbf{a}_z$ are the unit vectors in the x, y and z directions respectively. If this ray represents a plane electromagnetic wave at frequency ω propagating towards the corner reflector, then (see Appendix A) the electric field component can be expressed as:

$$\mathbf{E}_1 = \mathbf{a}_\perp E_1 \exp j(\omega t - k_0 \mathbf{r}_1) \tag{1.12}$$

where $\mathbf{a}_\perp$ defines a unit vector normal to $\mathbf{r}_1$, and k_0 is the free-space propagation coefficient. On the first reflection on the bottom x-y surface of the corner reflector, the components of $\mathbf{r}_1$ in the x and y directions are unchanged while the z-component

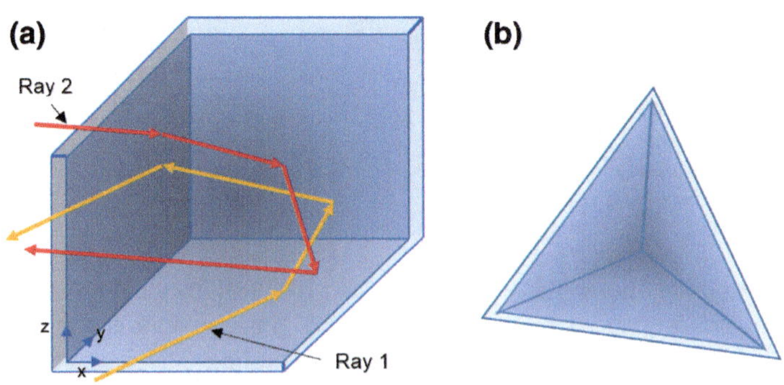

Fig. 1.22 Traditional corner reflectors showing retrodirective operational mode

is reversed by Snell's law, to $-\mathbf{a}_z z_1$. The resultant reflected ray direction can be expressed vectorially as:

$$\mathbf{r}_1' = \mathbf{a}_x x_1 + \mathbf{a}_y y_1 - \mathbf{a}_z z_1 \tag{1.13}$$

At the next reflection on the rear *x-z* plane (Fig. 1.22a), the x and z components of $\mathbf{r}_1$ are unchanged while it is the y component that is now reversed in phase. Hence

$$\mathbf{r}_1'' = \mathbf{a}_x x_1 - \mathbf{a}_y y_1 - \mathbf{a}_z z_1 \tag{1.14}$$

Finally, at the third reflection at the y-z surface, only the *x* component is reversed in phase, ending up with a ray travelling in the following direction:

$$\mathbf{r}_1''' = -\mathbf{a}_x x_1 - \mathbf{a}_y y_1 - \mathbf{a}_z z_1 \tag{1.15}$$

This is a precise reversal of the incident wave direction (it is said to be retro-directed) resulting in an E-field representation of the retro-wave which can be expressed mathematically as

$$\mathbf{E}_1 = \mathbf{a}_\perp E_1 \exp j(\omega t + k_0 \mathbf{r}_1) \tag{1.16}$$

For perfect conductors Eqs. (1.12) and (1.16) are essentially identical except for the sign of the second term within the exponential argument. The spatial component of the reflected wave is phase conjugated relative to the incident wave. Note that this is generally true of all incident plane waves which illuminate the corner reflector.

In mobile communication terms, and particularly in communications with satellites, the ability to retain a communications link by retro-directing the received signal back to its source has considerable potential, if it can be achieved with a less limited and bulky antenna which the corner reflector represents. The key feature of retro-directive action is the complex conjugate relationship between the received and transmitted wave phases. It is shown in Chap. 11 that retro-directivity can be procured using elegant antenna array formats with each element backed by a phase conjugating circuit. Not surprisingly, retro-directive antennas, in array format and incorporating varying levels of technical sophistication, have been evolving since pioneering work in the 1960s [16, 17] and arguably the planar retro-directive array offers the most cost effective option for satellite tracking. Recently, a planar retro-directive array design has been reported for potential use in terminals accessing the Inmarsat BGAN system [18].

References

1. M. Nosrati, N. Tavassolian, Miniaturised circularly polarised square slot antenna with enhanced axial ratio bandwidth using an anti-podal Y-strip. IEEE Antennas Wirel. Propag. Lett. **16**, 817–820 (2016)
2. Y. Seo, H. Lee, Y. Lim, Design of a compact dual-meander antenna for WLAN operation. Microwave Opt. Technol. Lett. **54**(7), 1595–1599 (2012)
3. C.-Z. Li, C.-M. Tong, W. Shaoqing, L.-H. Qi, Compact slot Yagi-Uda like antenna design with directional pattern, in *IEEE International Conference on Communications Software & Networks*, Chengdu, China, 6–7 June 2015. https://doi.org/10.1109/iccsn.2015.7296143
4. K.R. Carver, J.W. Mink, Microstrip antenna technology. IEEE Trans. Antennas Propag. **29** (1), 2–24 (1981)
5. A.J. Sangster, P. Smith, A polarisation agile T-shaped slot antenna fed from balanced microstripline, in *26th European Microwave Conference Proceedings*, vol. 1, pp. 288–292, Prague, 9–12 Sept 1996. ISBN 1 899919 08 2
6. S. Silver, *Microwave Antenna Theory & Design*, vol. 12 (Radiation Laboratory Series, 1949)
7. R.S. Elliott, *Antenna Theory and Design* (Prentice-Hall Ltd., Englewood Cliffs, 2006)
8. R.C. Johnson, *Antenna Engineering Handbook*, 3rd edn. (McGraw-Hill Inc., New York, 1993)
9. H.Y. Wang, S.Q. Zhang, A.J. Sangster, Numerical modelling of omni-directional slotted waveguide antennas. Microwave Opt. Tech. Lett. **31**(5), 374–377 (2001)
10. A.J. Sangster, H. Wang, Omni-directional blade antenna for microwave hyperthermia treatment. J. Electromagn. Waves Appl. **19**(14), 1935–1948 (2005)
11. K. Milne, The combination of pulse compression with frequency scanning for three dimensional radars, in *Proceeding of Symposium on Signal Processing in Radar and Sonar Direction Systems*, Birmingham, 6–9 July 1964
12. A.W. Rudge, K. Milne, A.D. Olver, P. Knight (eds.) *The Handbook of Antenna Design: Volumes I and II* (Peter Peregrinus Ltd., London, 1986)
13. A.J. Sangster, H. Wang, P. Smith, R. Jacobs, Moment method solutions for polarisation agile slots in bifurcated waveguide. Electromagnetics **19**(1/2), 91–107 (1999)
14. A.J. Sangster, P. Smith, E. McErlean, K. Sinclair, R. Jacobs, A dual-polarised stripline fed slot antenna incorporating signal cancellation. IEE Proc. (MAP) **148**(6), 357–362 (2001)
15. L.G. Van Atta, Electromagnetic Reflector. US Patent 2,908,002, 6 Oct 1959
16. C.Y. Pon, Retrodirective array using heterodyne technique. IEEE Trans. Antennas Propag. **AP-12**, 176–180 (1966)
17. R.N. Ghose, Electronically adaptive antenna systems. IEEE Trans. Antennas Propag. **AP-12** (2), 161–169 (1964)
18. N. Buchanan, V.F. Fusco, M. Van Der Vorst, N. Williams, C. Winter, New Retro-directive antenna for mobile terminal applications, in *32nd Antenna Workshop*, ESA/ESTEC, Noordwijk, Netherlands, 5–8 Oct 2010

Bibliography

1. R. Ferrari, *An Introduction to Electromagnetic Fields* (Van Nostrand Reinhold Co., Ltd., New York, 1975)
2. C.T.A. Johnk, *Engineering Electromagnetic Fields and Waves* (Wiley, New York, 1988)
3. R.P. Feynman, R.B. Leighton, M. Sands, *Lectures in Physics*, vol. II (Addison-Wesley Publishing Co., London, 1972)
4. P.M. Morse, H. Feshbach, *Methods of Theoretical Physics* (McGraw-Hill Book Co., New York, 1953)
5. J.D. Kraus, *Electromagnetics* (McGraw-Hill Book Co., London, 1984)

6. P. Lorrain, D. Corson, *Electromagnetic Field and Waves* (W.H. Freeman & Co., San Francisco, 1962)
7. A.J. Baden-Fuller, *Engineering Electromagnetism* (Wiley, New York, 1993)
8. P. Hammond, *Applied Electromagnetism* (Pergamon Press Ltd., Oxford, 1971)
9. J.A. Stratton, *Electromagnetic Theory* (Wiley, New Jersey, 2007)
10. D.M. Pozar, *Microwave Engineering*, 4th edn. (Wiley, USA, 2011). ISBN: 0470631554

Chapter 2
Fundamentals of Electromagnetic Radiation

2.1 Theoretical Formulation of Radiation from a Slot

Today, our deep understanding of the nature of antennas, and in particular slot antennas, originated with the study of what is now a classical problem in electromagnetics—namely that of the aperture in a plane conducting screen of infinite extent illuminated by a plane electromagnetic wave onto one face (see Fig. 2.1). The earliest significant examination of this, at the time intriguing, problem of how to calculate the electromagnetic fields, at arbitrary location P(x, y, z), in the empty half-space to the right of the screen, is attributed to Lord Rayleigh in 1897 [1].

The essence of a solution lies in creating a credible representation of the induced electric field or magnetic field within the aperture which becomes the source of the scattered fields. This is not a trivial problem and called for some considerable ingenuity on the part of early investigators at a time when the available mathematical tools were strictly analogue in character and while, contemporaneously, only closed form solutions were practicable or workable.

Antenna radiation (or reception) is classified as a field theory problem of the boundary value type not too unlike the electromagnetic propagation problems outlined in Appendix A. If the field region is remote from the source, field theory, as we have seen, identifies transverse electromagnetic waves (TEM waves) in free-space, or trapped electromagnetic waves in confined or bounded spaces. In this case the boundary value problem is generally termed ‘source-free’. For such problems Maxwell’s equations, in the source free differential form presented below, are detailed in Appendix A.

$$\nabla \times \mathbf{E} = -\mathrm{j}\omega\mu\mathbf{H} \tag{2.1}$$

$$\nabla \times \mathbf{H} = \mathrm{j}\omega\varepsilon\mathbf{E} \tag{2.2}$$

$$\nabla \cdot \mathbf{D} = 0 \tag{2.3}$$

A. J. Sangster, *Compact Slot Array Antennas for Wireless Communications*, Signals and Communication Technology, https://doi.org/10.1007/978-3-030-01753-8_2

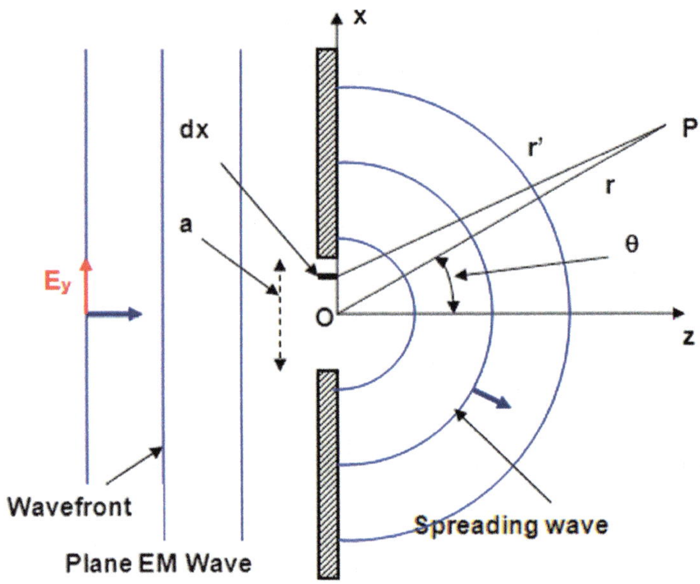

Fig. 2.1 Original problem of electromagnetic scattering by an aperture in a conductive plane screen of infinite extent

$$\nabla \cdot \mathbf{B} = 0 \tag{2.4}$$

In radiation problems on the other hand, the source in no longer remote and becomes an integral part of the field problem; radiation from a slot in a conducting surface is just such a problem. It should, perhaps, be noted that the normal convention, in formulating the mathematical equations associated with radiation, is to presume that the waves emanate from the 'source'. However, the transmit/receive process is entirely reciprocal as indicated in Fig. 2.2, so we could equally evolve the equations on the presumption that the focus of the waves is a 'sink' and the mathematics would not change.

In these source/sink present electromagnetic boundary value problems the most general form of Maxwell's equations are required. These are, in phasor form:

$$\nabla \times \mathbf{E} = -j\omega\mu\mathbf{H} \tag{2.5}$$

$$\nabla \times \mathbf{H} = j\omega\varepsilon\mathbf{E} + \mathbf{J} \tag{2.6}$$

$$\nabla \cdot \mathbf{D} = \rho \tag{2.7}$$

$$\nabla \cdot \mathbf{B} = 0 \tag{2.8}$$

where **J** and ρ are the *impressed* 'electric' current density and 'electric' charge density respectively, which form the source terms for the differential equations.

In radiation problems where the electromagnetic fields emanate from an aperture (e.g. horn and slot radiators) the basic source of the radiation is by analogy a

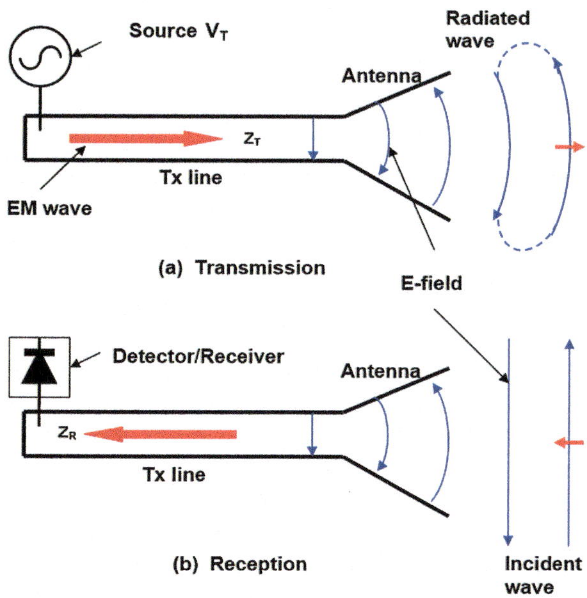

Fig. 2.2 **a** Radiation source and **b** radiation sink

'magnetic' current (electric field) rather than conventional 'electric' current (magnetic field), in which case it is more appropriate to use the following 'dual' forms for Maxwell's equations:

$$\nabla \times \mathbf{E} = -j\omega\mu\mathbf{H} - \mathbf{J}_m \tag{2.9}$$

$$\nabla \times \mathbf{H} = j\omega\varepsilon\mathbf{E} \tag{2.10}$$

$$\nabla \cdot \mathbf{D} = 0 \tag{2.11}$$

$$\nabla \cdot \mathbf{B} = \rho_m \tag{2.12}$$

where $\mathbf{J}_m$ and ρ_m are the impressed 'magnetic' current density and 'magnetic' charge density respectively representing 'aperture-source' terms for the differential equations. These forms are pursued in Chap. 3.

It is important, particularly in the 'electric' case to distinguish between the impressed currents and charges, and currents and charges which arise because of the presence of the fields in a medium having finite conductivity. Conduction current density $\mathbf{J}$ is proportional to the electric field and is given by $\sigma\mathbf{E}$, where σ is the conductivity of the material. This current component can readily be accounted for, in the Maxwell equations, by replacing the permittivity ε by the complex form $\varepsilon(1 - j\tan\delta) = \varepsilon(1 - j\sigma/\omega\varepsilon)$. The density of free charge, apart from that associated with impressed currents, may be presumed to be zero. Thus, in subsequent discussion, the charge density ρ and the current density $\mathbf{J}$ appearing in the field equations can be taken as comprising purely impressed charges and currents.

Any other currents, if they arise, will be accounted for by the complex electric permittivity which will be written simply as ε for convenience.

2.2 Reciprocity for Macroscopic Antennas

The reciprocity hinted at in Fig. 2.2 can be established as follows. The two antennas one in transmit mode (a) and one in receive mode (b) exhibit, respectively, input impedances Z_T and Z_R. It is assumed that the antennas exist is the same linear, isotropic space, which is not necessarily homogeneous, and that the electromagnetic power radiated from antenna (a) propagates toward (b). The equivalent circuit formations representing this antenna juxtaposition are presented in Fig. 2.3.

In the transmitter circuit (Fig. 2.3a) the generator internal impedance is represented by Z_g and to procure maximum power transfer to space it is set equal to the complex conjugate of the antenna impedance $Z_T\left(Z_g = Z_T^*\right)$. In the receiver circuit, on the other hand, optimum power transfer to the load (Z_L) requires that Z_L is set equal to the complex conjugate of the receive antenna Z_R; that is $Z_L = Z_R^*$. These impedance settings are convenient for the following reciprocity calculation, but are not essential.

The calculation proceeds by computing, firstly, the power (P_1) delivered to antenna (a) from the generator. This power is given by:

$$P_1 = \frac{1}{2}\mathrm{Re}(V_T I_T^*) = \frac{1}{2}\mathrm{Re}\left[\left(\frac{V_g Z_T}{Z_T + Z_g}\right)\left(\frac{V_g^*}{(Z_T + Z_g)^*}\right)\right] = \frac{|V_g|^2}{8R_T} \tag{2.13}$$

This power radiates into the surrounding space, and some of it is collected by the receive antenna and becomes absorbed by the load. This power P_2 can by computed if we ascribe a transfer admittance Y_{21} to the transmit/receive network which incorporates the generator impedance, the impedances of the antennas and the load impedance. Hence

$$P_2 = \frac{1}{2}\mathrm{Re}\left[Z_L(V_g Y_{21})(V_g Y_{21})^*\right] = \frac{1}{2}R_L|V_g|^2|Y_{21}|^2 \tag{2.14}$$

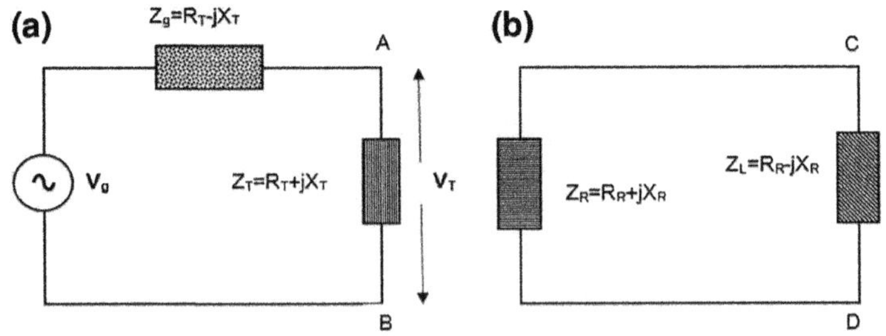

Fig. 2.3 Equivalent circuit representation of transmit/receive antenna combination

The division of Eq. (2.14) by Eq. (2.13) yields an expression for the receive-to-transmit power ratio for the antenna combination—namely:

$$\frac{P_2}{P_1} = 4R_L R_T |Y_{21}|^2 \tag{2.15}$$

If the transmit antenna and the receive antenna are interchanged with (b) transmitting power P_1 and (a) receiving power P_2 in its load, the transfer admittance is now Y_{12} and the power ratio becomes:

$$\frac{P_2}{P_1} = 4R_L R_T |Y_{12}|^2 \tag{2.16}$$

For linear circuits it is clear that $Y_{21} = Y_{12}$ and consequently the antenna transmit/receive combination displays the property of reciprocity.

2.2.1 Reciprocity Theorems for Electromagnetic Fields

The property of duality in electromagnetic theory as it evidently exists between electric current and magnetic current sources, is augmented by the concept of reciprocity which essentially dictates that sources can be sinks and *vice versa* as we have seen for macroscopic antennas occupying the same volume of space. Actually, familiarity with electrical circuit theory in general makes the idea of reciprocity not too surprising since it is quite natural to accept that in a linear circuit driven by a current source in one part of the circuit with a voltmeter monitoring the voltage in another, interchanging the current source and the voltmeter returns the same meter reading. The same is true of a voltage source and an ammeter in a linear circuit. Nevertheless, reciprocity is particularly appropriate to the development of antenna theory as we have illustrated above where it evidently enables transmitting antenna behaviour to be reinterpreted for the illumination of, or the better understanding of, receiving antenna characteristics. The reverse is also true.

To establish the reciprocity theorem at the microscopic field level we commence by assuming that electric current densities $\mathbf{J}_1$, $\mathbf{J}_2$, and magnetic current densities $\mathbf{J}_{m1}$, $\mathbf{J}_{m2}$ form two sets of sources in a linear isotropic medium which may, or may not, be homogeneous. We further assume that the sources radiate singly or together into the same medium at the same frequency and generate fields $\mathbf{E}_1$, $\mathbf{H}_1$ and $\mathbf{E}_2$, $\mathbf{H}_2$ respectively. Such sources and fields, which are governed by the Maxwell equations, are as a consequence forced to satisfy the Lorentz reciprocity theorem [1, 2], which in differential form yields:

$$-\nabla \cdot (\mathbf{E}_1 \times \mathbf{H}_2 - \mathbf{E}_2 \times \mathbf{H}_1) = \mathbf{E}_1 \cdot \mathbf{J}_2 + \mathbf{H}_2 \cdot \mathbf{J}_{m1} - \mathbf{E}_2 \cdot \mathbf{J}_1 - \mathbf{H}_1 \cdot \mathbf{J}_{m2} \tag{2.17}$$

Applying a volume integration over V to both sides of this equation, and subsequently enforcing the divergence theorem to the right hand side, leads to the arguably more useful integral form of the Lorentz reciprocity theorem—namely:

$$\oiint_S (\mathbf{E}_1 \times \mathbf{H}_2 - \mathbf{E}_2 \times \mathbf{H}_1) \cdot dS = \iiint_V (\mathbf{E}_1 \cdot \mathbf{J}_2 + \mathbf{H}_2 \cdot \mathbf{J}_{m1} - \mathbf{E}_2 \cdot \mathbf{J}_1 - \mathbf{H}_1 \cdot \mathbf{J}_{m2}) dV \tag{2.18}$$

Equation (2.18) leads to some useful relationships in certain limited, but relevant, conditions. In a source free region ($\mathbf{J}_1 = \mathbf{J}_2 = \mathbf{J}_{\mathrm{m}1} = \mathbf{J}_{\mathrm{m}2} = 0$) Eq. (2.18) evidently reduces to:

$$\iint_S (\mathbf{E}_1 \times \mathbf{H}_2 - \mathbf{E}_2 \times \mathbf{H}_1) \cdot dS = 0 \tag{2.19}$$

This equation is a special case of the Lorentz reciprocity theorem and must be satisfied on the enclosing surface S of a source free region containing the fields $\mathbf{E}_1$, $\mathbf{H}_1$ and $\mathbf{E}_2$, $\mathbf{H}_2$ simultaneously.

From an antenna perspective it is often the case that the sources and sinks (transmit and receive antennas) can be considered to reside in a sphere (free space) of infinite radius. At the surface of this sphere (S) at infinity, the fields generated by the finite sources within it ($\mathbf{J}_1$, $\mathbf{J}_2$, $\mathbf{J}_{\mathrm{m}1}$, $\mathbf{J}_{\mathrm{m}2}$) will have reduced to zero, which means that the left hand side of Eq. (2.18) is zero. Hence:

$$\iiint_V (\mathbf{E}_1 \cdot \mathbf{J}_2 + \mathbf{H}_2 \cdot \mathbf{J}_{m1} - \mathbf{E}_2 \cdot \mathbf{J}_1 - \mathbf{H}_1 \cdot \mathbf{J}_{m2}) dV = 0 \tag{2.20}$$

which can obviously be expressed in the form:

$$\iiint_V (\mathbf{E}_1 \cdot \mathbf{J}_2 + \mathbf{H}_2 \cdot \mathbf{J}_{m1}) dV = \iiint_V (\mathbf{E}_2 \cdot \mathbf{J}_1 - \mathbf{H}_1 \cdot \mathbf{J}_{m2}) dV \tag{2.21}$$

It is pertinent to note here that Eqs. (2.19) and (2.21) are not power or energy expressions because complex conjugation is absent. What these equations do represent is field/source coupling or reaction resulting in further field components [1]. In antenna terms, and in particular array antennas, the process gives rise to mutual coupling (see Chap. 6) which can present a major source of error, if handled incorrectly, when precision pattern control is sought in such arrays. In addition, the above reciprocity relations are particularly relevant to the electromagnetic modelling of elemental antennas, such as dipoles and slots embedded within compact antenna geometries, as we shall see in Chap. 5.

2.2.2 *Inhomogeneous Vector Helmholtz Equation*

The impressed currents and charges, expressed explicitly in the source-present form of Maxwell's equations (Eqs. 2.5–2.8, and 2.9–2.12), must be related to each other through the equation of continuity for current and charge. That is:

$$\nabla \cdot \mathbf{J} = -\frac{\partial \rho}{\partial t} \tag{2.22}$$

The dual relationship for 'magnetic current' and 'magnetic charge' is

$$\nabla \cdot \mathbf{J}_{\mathrm{m}} = -\frac{\partial \rho_{\mathrm{m}}}{\partial t} \tag{2.23}$$

Furthermore, to solve the Maxwell equations for E-field or H-field behaviour in a bounded region it is first necessary to form an equation for either E or H alone which is facilitated by employing Eq. (2.22) or (2.23) as appropriate. The standard procedure for achieving this conversion is to perform a curl operation on either the curl equation for E or the corresponding equation for H (see Appendix A). This process leads to second order differential equations in **E** and **H**. In the electric current case we obtain:

$$\begin{aligned} \nabla \times \nabla \times \mathbf{E} - k^2\mathbf{E} &= -\mathrm{j}\omega\mu\mathbf{J} \\ \nabla \times \nabla \times \mathbf{H} - k^2\mathbf{H} &= \nabla \times \mathbf{J} \end{aligned} \tag{2.24}$$

where ρ is electric charge density. Equation (2.24) are *inhomogeneous* vector Helmholtz equations in which the term on the right hand side of the equals sign forms a *source* function. By employing the equation of continuity for charge and current, the source function can be expressed in terms of the more familiar differential operators on the left of the equation, but at the expense of more complex source terms on the right. That is

$$\nabla^2\mathbf{E} + k^2\mathbf{E} = \mathrm{j}\omega\mu\mathbf{J} - \frac{\nabla\nabla \cdot \mathbf{J}}{\mathrm{j}\omega\varepsilon} \tag{2.25}$$

By an entirely analogous procedure, a corresponding inhomogeneous vector wave equation for **H** can be derived, which has the form:

$$\nabla^2\mathbf{H} + k^2\mathbf{H} = -\nabla \times \mathbf{J} \tag{2.26}$$

In deriving the above equations it has been presumed that μ and ε are invariant with time. In free space or vacuum, $\mu = \mu_0 = 4\pi \times 10^{-7}\,\mathrm{H/m}$ and $\varepsilon = \varepsilon_0 = \frac{1}{36\pi} \times 10^{-9}\,\mathrm{F/m}$.

In a source free region, for which the impressed current $\mathbf{J}$ is non-existent, Eqs. (2.25) and (2.26) reduce, as they should, to the well known homogeneous vector Helmholtz equations, namely:

$$\begin{aligned} \nabla^2\mathbf{E} + k^2\mathbf{E} &= 0 \\ \text{and} \quad \nabla^2\mathbf{H} + k^2\mathbf{H} &= 0 \end{aligned} \tag{2.27}$$

These are the phasor forms of the wave equations.

When a radiating source of the magnetic current type ($\mathbf{J}_\text{m}$—e.g. slot radiator) exists within the region of interest, a similar derivation when applied to Eqs. (2.25) and (2.26) leads to the following inhomogeneous vector differential equation

$$\begin{aligned} \nabla \times \nabla \times \mathbf{H} - k^2\mathbf{H} &= +\text{j}\omega\varepsilon\mathbf{J}_m \\ \nabla \times \nabla \times \mathbf{E} - k^2\mathbf{E} &= -\nabla \times \mathbf{J}_m \end{aligned} \tag{2.28}$$

or equivalently:

$$\begin{aligned} \nabla^2\mathbf{E} + k^2\mathbf{E} &= \nabla \times \mathbf{J}_m \\ \text{and} \quad \nabla^2\mathbf{H} + k^2\mathbf{H} &= -\text{j}\omega\varepsilon\mathbf{J}_m + \frac{\nabla\nabla \cdot \mathbf{J}_m}{\text{j}\omega\mu} \end{aligned} \tag{2.29}$$

These magnetic current sources, and the antenna forms which they generate, are treated in depth in Chap. 3.

It is clear on examining their forms that the impressed current density vectors $\mathbf{J}$ and $\mathbf{J}_m$ enter into the inhomogeneous differential Eqs. (2.25), (2.26) and (2.29) in a particularly complicated manner, making their direct solution extremely difficult, although with modern computer based electromagnetic solvers employing finite difference or finite element techniques—not impossible. Nevertheless, this difficulty is normally circumvented by resorting to the use of *auxiliary potential functions,* which generate more easily solvable wave equation forms. These auxiliary potential functions may, or may not, represent clearly definable physical entities (especially is this so in the absence of sources), and so it is usual to adopt the view point that these potentials are merely useful functions from which the electromagnetic fields can be derived. The two most common potential functions which arise in radiation problems are discussed in the next section.

2.3 Auxiliary Potential Functions

In formulating auxiliary potential functions, with the aim of facilitating the solution of the source present second order differential equations as represented by Eqs. (2.25), (2.26), (2.28) and (2.29), it is usually advisable to choose forms that satisfy at least one or more of the basic Maxwell equations. For the 'electric' current Eqs. (2.25), (2.26) such a choice is:

$$\mathbf{B} = \nabla \times \mathbf{A} \tag{2.30}$$

which immediately satisfies Eq. (2.8), since by definition $\nabla \cdot \nabla \times \mathbf{S} = 0$ for any vector $\mathbf{S}$. In addition the choice:

$$\mathbf{E} = -\frac{\partial \mathbf{A}}{\partial \mathbf{t}} + \nabla \phi \tag{2.31}$$

satisfies Maxwell Eq. (2.5) since $\nabla \times \nabla \varphi = 0$ for any scalar function. The auxiliary vector $\mathbf{A}$ is usually termed the magnetic vector potential, which has relevance in magnetostatics, while auxiliary scalar ϕ is termed the electric potential (sometimes denoted by V) and has its origins in electrostatics. Consequently, if we can develop solvable source present second order differential equations for $\mathbf{A}$ and ϕ, then the associated electric and magnetic field forms can be deduced subsequently from the above two equations.

To develop differential forms in $\mathbf{A}$ and ϕ we again enrol the Maxwell equations as expressed in Eqs. (2.5)–(2.8). On substituting $\mathbf{D}$ (= $\varepsilon_0 \mathbf{E}$ for vacuum) in Eq. (2.7) with the expression for $\mathbf{E}$ given in Eq. (2.31) we obtain:

$$-\nabla^2 \phi - \frac{\partial}{\partial t}(\nabla \cdot \mathbf{A}) = \frac{\rho}{\varepsilon_0} \tag{2.32}$$

which hardly seems an advance, in simplification terms, on Eqs. (2.25) and (2.26). However, Eqs. (2.30) and (2.31) only partially define $\mathbf{A}$ and ϕ. Equation (2.30) delineates only a rotational component of $\mathbf{A}$ ($\mathbf{A}_r$ say) while Eq. (2.31) defines a part of ϕ which has finite gradient in the space domain. Therefore, we can add to $\mathbf{A}$ any component which has a zero curl (termed a lamellar component) without altering Eq. (2.30), and we can add to ϕ any function whose gradient is zero, leaving Eq. (2.31) unchanged. Provided the Maxwell relationships are not disrupted, these arbitrary introductions to $\mathbf{A}$ and ϕ can be chosen in a manner which simplifies Eq. (2.32). Such a choice is the Lorentz condition whereby:

$$\nabla \cdot \mathbf{A} = -\frac{1}{c^2}\frac{\partial \phi}{\partial t} \tag{2.33}$$

Actually this choice is not totally arbitrary. The Lorentz condition can be identified with the Kirchhoff law which relates conduction current to rate of change of charge (Eq. 2.22). It also brings the resultant radiation equations into line with the requirements of special relativity [3].

On substituting Eq. (2.33) into Eq. (2.32) a second order differential in ϕ alone is formed, namely:

$$\nabla^2\phi - \frac{1}{c^2}\frac{\partial^2\phi}{\partial t^2} = -\frac{\rho}{\varepsilon_0} \tag{2.34}$$

or in phasor form:

$$\nabla^2\phi + k_0^2\phi = -\frac{\rho}{\varepsilon_0} \tag{2.35}$$

where

$$k_0 = \frac{\omega}{c} = \omega\sqrt{\mu_0\varepsilon_0} \tag{2.36}$$

Recalling that $\mathbf{H} = \mathbf{B}/\mu_0$ and $\varepsilon = \varepsilon_0$ in free space then substituting Eqs. (2.30) and (2.31), together with the Lorentz condition, into Eq. (2.6), yields an equivalent second order differential equation for **A**. In phasor form, this equation is termed the vector Helmholtz equation:

$$\nabla^2\mathbf{A} + k_0^2\mathbf{A} = -\mu_0\mathbf{J} \tag{2.37}$$

Equations (2.35) and (2.37) are clearly wave equations but with source functions on the right of the equals sign which are much less convoluted than those in Eqs. (2.25) and (2.26). These source functions have now been simplified to the impressed charge density (divided by a constant ε_o) in the region of interest in Eq. (2.35), and impressed current density (multiplied by a constant μ_o) in Eq. (2.37). If ρ or **J** are known in space and time, on an antenna located in a volume of space (say), then in principle the above equations can be solved directly for ϕ or **A** respectively. In practice, only one of the two wave equations need be solved since ϕ and **A** are related through the Lorentz condition. However, rather than demonstrate the direct route to a solution, it is actually more instructive to construct a solution by employing electrostatic and magnetostatic analogies.

2.4 Electrostatics Analogy

If Eq. (2.7) is applied to a small spherical volume of stationary and isolated free charge (q), located in a magnetic field *free* region, then symmetry and the circulation law for E (Eq. 2.5) dictate that the electric vector **D** will be everywhere radial and unvarying over an enveloping spherical surface (S) of radius r (Fig. 2.4), enclosing a volume V. Under these circumstances volume integration on both sides of Eq. (2.7) yields:

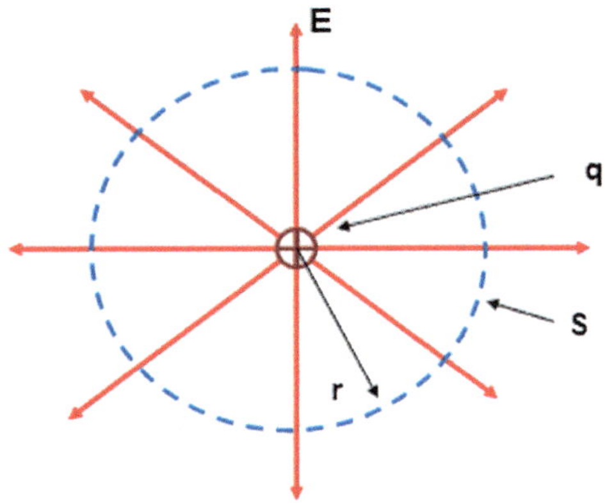

Fig. 2.4 Electric field flux through a spherical surface containing a point charge q

$$\iiint_V \nabla \cdot \mathbf{D} dV = \oiint_S \mathbf{D} \cdot \mathbf{dA} = 4\pi r^2 \mathrm{D_r} = q \tag{2.38}$$

Clearly Maxwell's first equation reduces to the inverse square law, or Gauss' law which is usually written:

$$E_r = \frac{q}{4\pi r^2 \varepsilon_0} \tag{2.39}$$

provided that, for air, or vacuum:

$$\mathbf{D} = \varepsilon_0 \mathbf{E} \tag{2.40}$$

In the m.k.s. system, **D** has the dimension C/m^2, while the constant of proportionality ε_0 is termed the absolute permittivity of free space. Measurement reveals that it has the value 8.84×10^{-12} F/m, where in the m.k.s. system a Farad in a Coulomb/volt. More generally, $\mathbf{D} = \boldsymbol{\varepsilon}_r \varepsilon_0 \mathbf{E}$, where $\boldsymbol{\varepsilon}_r$ is termed the relative permittivity of the material concerned. It need not be a constant, and could be a complex tensor to represent a general non-linear, anisotropic, inhomogeneous medium.

In conjunction with the Lorentz force equation:

$$F = qE \,\text{Newtons}$$

Equation (2.39) yields the Coulomb force law which can be expressed as:

$$F = \frac{qQ}{4\pi r^2 \varepsilon_0} \,\text{Newtons} \tag{2.41}$$

for isolated point charges q and Q (in coulombs) distance r metres apart in a vacuum.

Equation (2.39) really only applies if q is concentrated at a point, and so the situation in Fig. 2.5 where a distributed charge of density ρ is assumed to surround the origin of the Cartesian coordinate system (shaded region). To determine **E** in this case we must first dice the shaded volume up into elemental volumes such a $dV'(= dx'dy'dz')$ at r'. For such a volume the charge enclosed is $q = \rho(x, y, z)dV'$ coulombs and if it can be considered to reside at a point in space, then given that potential $\phi = -Edr$ where $dr = |r - r'|$, the potential at P due to this element is:

$$d\phi = \frac{\rho(x, y, z)dV'}{4\pi\varepsilon_0|r - r'|} \tag{2.42}$$

The potential at P due to the entire charge volume (V) at O, is the summation of all contributions $\rho(x, y, z)dV'$ at all possible positions O′ throughout the charge volume. In the limit this summation becomes an integral, giving:

$$\phi = \frac{1}{4\pi\varepsilon_0}\iiint\limits_V \frac{\rho dV'}{|r - r'|} \tag{2.43}$$

While this equation exhibits the correct spatial form for electric potential at distance from an arbitrary charge distribution, it is applicable only to time invariant charge concentrations. The discussion in relation to Fig. 3.1 suggests that if the charge alters with time then Eq. (2.43) could justifiably be applied to fields in close proximity to the charge, but at more distant locations as a time retardation mechanism is required. So can we build time delay into Eq. (2.43)? It turns out that we can.

If the charge distribution at O′ in Fig. 2.5 alters with time, the information that this change has occurred can be felt at point P only after a period of time $|r - r'|/c$, which is the time it takes for a TEM wave to travel distance $|r - r'|$. Hence, it is not

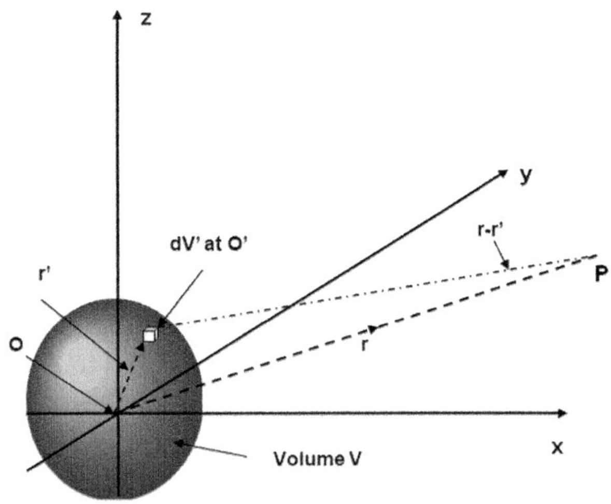

Fig. 2.5 Geometry for construction of retarded potentials

implausible to suggest that at time t and at position **r** the contribution to $\phi(\mathbf{r}, t)$ at P can be calculated using Eq. (2.43) if we insert the charge distribution at time $t - |r - r'|/c$. This is true for all volume elements dV' within V. Consequently, the desired time varying solution to Eq. (2.35) can be formulated by simply incorporating the time retardation into Eq. (2.43), to give:

$$\phi(\mathbf{r}) = \frac{1}{4\pi\varepsilon_0} \iiint\limits_V \frac{\rho(\mathbf{r}', t - |\mathbf{r} - \mathbf{r}'|/c)dV'}{|\mathbf{r} - \mathbf{r}'|} \tag{2.44}$$

That Eq. (2.44) is a solution to Eq. (2.35) can be established by substitution. The solution to the phasor form of Eq. (2.35) simplifies to:

$$\phi(\mathbf{r}) = \frac{1}{4\pi\varepsilon_0} \iiint\limits_V \frac{\exp(-jk_o|\mathbf{r} - \mathbf{r}'|\rho(\mathbf{r}')dV'}{|\mathbf{r} - \mathbf{r}'|} \tag{2.45}$$

with time expressed through the common factor $exp(j\omega t)$ which can be suppressed.

2.5 Magnetostatics Analogy

The fundamental measurement based laws of magnetostatics, as deduced by the early pioneers in electrical science can be constructed in similar manner from Eqs. (2.10) and (2.12). Equation (2.12), which is usually referred to as the flux law for magnetic field, simply states that however generated—by a permanent magnet, by the Earth's core, by current on a wire, by a current carrying coil—magnet flux density forms closed loops. At the macroscopic level there are no magnetic field sources performing the same role as charge in electrostatics. Hence, if we apply Eq. (2.10) to a steady (DC) current (I) on a long straight wire, we can confidently say that the magnetic field which the law predicts must form closed circular loops centred on the wire—assuming the wire has a cylindrical cross-section. We therefore obtain:

$$\begin{aligned} \oint\limits_C \mathbf{H} \cdot \mathbf{d}\ell &= 2\pi r H_t = I \\ H_t &= \frac{I}{2\pi r} \\ B_t &= \frac{\mu_0 I}{2\pi r} \end{aligned} \tag{2.46}$$

Equation (2.46) is Ampere's law, and since I is expressed in amps (A) in m.k.s. system, the dimension of magnetic field strength H is A/m. With H in A/m and B in Tesla consistency of the magnetic field laws requires that for vacuum:

$$\mathbf{B} = \mu_0 \mathbf{H} \tag{2.47}$$

where the constant μ_0 is termed the absolute permeability of free space and has the magnitude $4\pi \times 10^{-7}$ H/m (Henry = T · m^2/A). As with electrostatics we can say that more generally, $\mathbf{B} = \mu_r\mu_0\mathbf{H}$, where μ_r is termed the relative permeability of the material concerned. It need not be a constant, being a function of **H** in a ferrite material, and it could be a complex tensor to represent a general, anisotropic, inhomogeneous medium. Application of the circuital law for H to a current element $Id\ell$ rather than to a current on a long straight wire, leads to the Biot and Savart law (see Balanis in Bibliography). Referring to Fig. 2.6 where a current element Idl' is located at O' a distance r' from the origin at O. In determining the magnetic flux density at P distance r from O, we can assume from Ampere's law that the flux forms circular loops centred on the current I. The contribution **dB** at P must be proportional to I in magnitude and inversely proportional to the square of the distance between O' and P, namely $|\mathbf{r} - \mathbf{r}'|$. Directionally the vector **dB** must be orthogonal to both Idl' and the vector $(\mathbf{r} - \mathbf{r}')$. So mathematical reasoning suggests that we can express **dB** in the following form:

$$\mathbf{dB} = \frac{\mu_0 I}{4\pi} \frac{\mathbf{dl}' \times (\mathbf{r} - \mathbf{r}')}{|\mathbf{r} - \mathbf{r}'|^3} \tag{2.48}$$

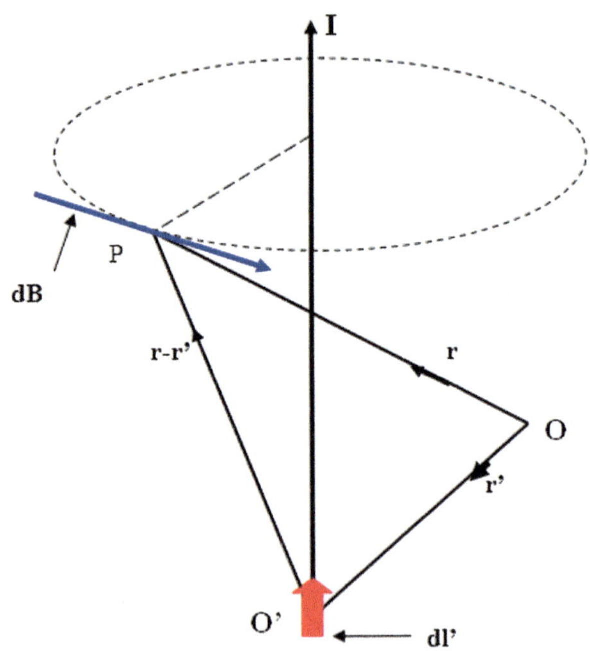

Fig. 2.6 Magnetic field at P due to current element Idl'

This the Biot-Savart law which is most commonly expressed in the form:

$$dB = \frac{\mu_0 I d\ell}{4\pi r^2} \tag{2.49}$$

for a current element at the centre of the magnetic field loop. If the straight portion of wire carrying I is actually part of a complete electrical circuit s, then on summing all the Idl' contributions around s leads, in the limit of infinitesimally small dl', to an integration giving:

$$\mathbf{B}(\mathbf{r}) = \frac{\mu_0 I}{4\pi} \oint_s \frac{\mathbf{dl}' \times (\mathbf{r} - \mathbf{r}')}{|\mathbf{r} - \mathbf{r}'|^3} \tag{2.50}$$

Remembering that we have chosen to define the vector magnetic potential through Eq. (2.30) and noting that [3]:

$$\nabla \times \frac{\mathbf{dl}'}{|\mathbf{r} - \mathbf{r}'|} = -\frac{(\mathbf{r} - \mathbf{r}')}{|\mathbf{r} - \mathbf{r}'|^3} \times \mathbf{dl}'$$

thence, combining Eqs. (2.30) and (2.50) yields the following:

$$\mathbf{A}(\mathbf{r}) = \frac{\mu_0 I}{4\pi} \oint_s \frac{\mathbf{dl}'}{|\mathbf{r} - \mathbf{r}'|} \tag{2.51}$$

The time retarded form for **A** can be deduced in much the same way as we have already done for ϕ with the additional observation that current and current density are related through:

$$I\mathbf{dl}' = \mathbf{J}dV' \tag{2.52}$$

Hence we obtain:

$$\mathbf{A}(\mathbf{r}, t) = \frac{\mu_0}{4\pi} \iiint_V \frac{\mathbf{J}(\mathbf{r}', t - |\mathbf{r} - \mathbf{r}'|/c)}{|\mathbf{r} - \mathbf{r}'|} dV' \tag{2.53}$$

or in phasor form:

$$\mathbf{A}(\mathbf{r}) = \frac{\mu_0}{4\pi} \iiint_V \frac{\exp(-jk_o|\mathbf{r} - \mathbf{r}'|)\mathbf{J}(\mathbf{r}')}{|\mathbf{r} - \mathbf{r}'|} dV' \tag{2.54}$$

The retarded potentials **A** and ϕ represent powerful tools in the search for solutions to complex electromagnetic radiation and reception problems, and are widely used in various simulation packages by providing the ‘building blocks’, such as the ‘short current element’, to construct field patterns for distributed current sources, which are known in space and time. This is particularly true of those packages employing moment methods. The short current element, or Hertzian dipole, is examined in the next section.

Needless to say, equivalent forms to **A** and ϕ also exist, which provide expedient routes to the solutions to Eqs. (2.28) and (2.29). These are usually defined by the parameters $\mathbf{A}_{\mathrm{m}}$—an auxiliary potential function associated with 'magnetic current' $\mathbf{J}_{\mathrm{m}}$, and ϕ_{m}—an auxiliary potential linked to 'magnetic charge' ρ_{m}. These forms will be considered in more detail in Chap. 3.

2.6 Hertzian Dipole

The Hertzian dipole and the 'short current element' are essentially synonymous. The concept is presented schematically in Fig. 2.7. The short current element is viewed as a filament of current *Idl* which exhibits constant magnitude along its length *dl*. As such it is physically impossible to realise, but remains a helpful mathematical construct. It functions as an exceptionally useful building block in antenna modelling which enables the field patterns of practical antenna structures to be determined by integration. Consequently it is instructive to examine the field behaviour of this current filament, and this is implemented by applying the appropriate potential function to the example shown in Fig. 2.7.

The current element is assumed to be located at the origin of a Cartesian coordinate system and that it takes the form a constant z-directed current varying in time at frequency ω rad/sec—i.e. the current $I = I_0 \cos \omega t$. If time retardation is incorporated into the definition of current then this becomes:

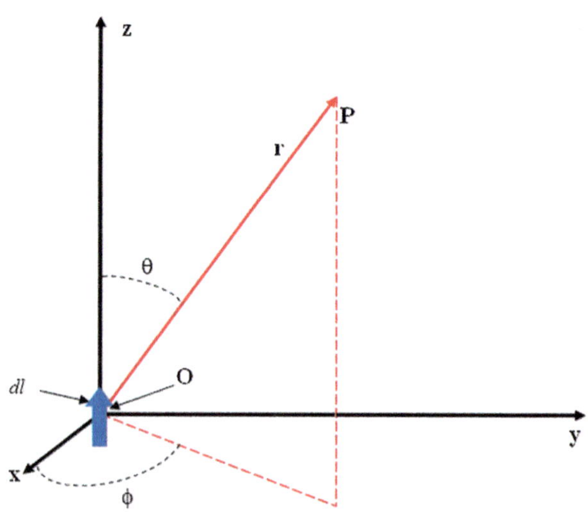

Fig. 2.7 Short current element at origin of Cartesian coordinate system

$$I = I_0 \cos\left(\omega(t - \frac{r}{c})\right) = I_0 \cos(\omega t - k_0 r) \tag{2.55}$$

In complex notation this can be written:

$$I = \mathrm{Re}[I_0 \exp(j(\omega t - k_0 r))]$$

In these two equations the phase coefficient $k_0 = \omega/c = \omega\sqrt{\mu_0 \varepsilon_0}$ for free space. So, for this one dimensional element of length dl in the z direction Eq. (2.54) becomes:

$$\mathbf{A} = \frac{\mu_0 I_0 \mathbf{dl}}{4\pi r} \exp(j(\omega t - k_0 r)) \tag{2.56}$$

If we adopt the unit vector notation, for the Cartesian system, namely $(\hat{\mathbf{a}}_\mathbf{x}, \hat{\mathbf{a}}_\mathbf{y}, \hat{\mathbf{a}}_z)$ then:

$$\mathbf{A} = A_z \hat{\mathbf{a}}_\mathbf{z} = \frac{\mu_0 I_0 dl \hat{\mathbf{a}}_\mathbf{z}}{4\pi r} \exp(j(\omega t - k_0 r)) \tag{2.57}$$

It is evident from Eq. (2.57) that at distant point P (Fig. 2.7) the magnetic vector potential is z-directed. However, we know from our earlier discussion on the radiation mechanism that the important field components are orthogonal to the direction vector **r** and are therefore best represented by spherical components. Consequently we let:

$$\mathbf{A} = \hat{\mathbf{a}}_\mathbf{r} A_r + \hat{\mathbf{a}}_\theta A_\theta + \hat{\mathbf{a}}_\phi A_\phi \tag{2.58}$$

where $A_r = A_z \cos\theta$, $A_\theta = -A_z \sin\theta$ and $A_\phi = 0$. Hence the magnetic field components of the radiated wave at P can be derived using $\mathbf{B} = \mu_0 \mathbf{H} = \nabla \times \mathbf{A}$ resulting in [6]:

$$H_\phi = \frac{I_0 dl}{4\pi}\left[\frac{jk_0}{r} + \frac{1}{r^2}\right] \sin\theta \exp(-jk_0 r) \tag{2.59}$$

$$H_\theta = H_r = 0 \tag{2.60}$$

The electric field components of the same radiated wave can be constructed by applying the Maxwell Eq. (2.2) to generate:

$$E_\theta = \sqrt{\frac{\mu_0}{\varepsilon_0}} \frac{I_0 dl}{4\pi}\left[\frac{jk_0}{r} + \frac{1}{r^2} - \frac{j}{k_0 r^3}\right] \sin\theta \exp(-jk_0 r) \tag{2.61}$$

$$E_r = \sqrt{\frac{\mu_0}{\varepsilon_0}} \frac{I_0 dl}{2\pi}\left[\frac{1}{r^2} - \frac{j}{k_0 r^3}\right] \cos\theta \exp(-jk_0 r) \tag{2.62}$$

$$E_\phi = 0 \tag{2.63}$$

From the discussion in Sect. 2.1 the fields radiating away from the short current element are expected to be normal to the direction vector r, and to each other. They are also predicted to diminish in magnitude at a rate which is inversely proportion to distance r from the source. The only field components which meet this requirement are the first term of H_ϕ in Eq. (2.59) and the first term of E_θ (Eq. 2.61). So in the far field, generally defined as occupying the region given by $k_0r > 1$:

$$H_\phi = \frac{jk_0 I_0 dl}{4\pi r} \sin\theta \exp(-jk_0 r) \tag{2.64}$$

$$E_\theta = \sqrt{\frac{\mu_0}{\varepsilon_0}} H_\phi = \eta_0 H_\phi \tag{2.65}$$

A pictorial representation of the E_θ field distribution is shown in Fig. 2.8. Note that the pattern is circularly symmetric in the 'horizontal' plane (x-y plane in Fig. 2.7) which slices through the current element, but it is not 'isotropic'—radiating equally in all directions. The 'solid' pattern is 'dough-ring' shaped with electric (and magnetic) field nulls along the z-axis (θ = 0°, 180°) of the source current. The H_ϕ field is everywhere normal to the E-field, and normal to the plane of the page containing Fig. 2.8.

So, what is the significance of the other field components, which diminish with distance as $1/r^2$ and $1/r^3$ in Eqs. (2.61) and (2.62)? The $1/r^3$ term is reminiscent of the electrostatic field behaviour for an electric dipole [6]. It arises because the current I must, in accordance with Kirchoff's law, be associated with positive and negative charge build up $\left(I = -\frac{dq}{dt}\right)$ at its extremities, so forming, at any given

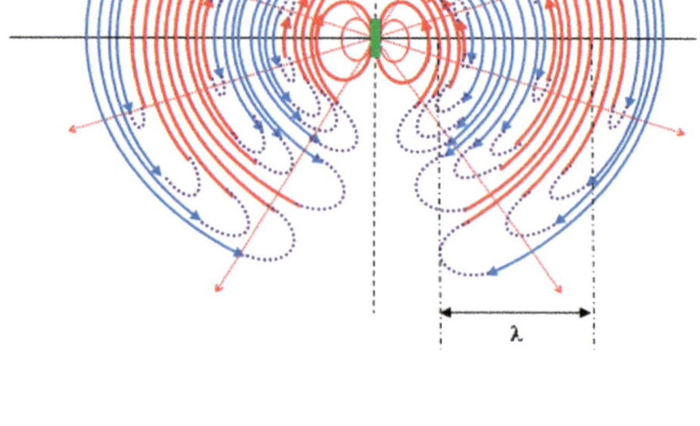

Fig. 2.8 Electric field contours for radiation from a short current element at a specific instant in time (red = peaks; blue = troughs in the wave)

instant in time, an electric dipole. The field exists only in the 'near' field region ($k_o r < 1$) of the source, being negligible in the far field. It supports electrostatic or capacitive energy storage.

The $1/r^2$ which exists in both the electric and magnetic field formulations (Eqs. 2.59, 2.61 and 2.62) exhibits the characteristics of the stored inductive magnetic field in the vicinity of a current carrying wire. In magnetostatics it is derived using the Biot-Savart law. The electric field components arise because the antenna current, and hence this inductive magnetic field, is time varying. The fields are derivable using the Faraday law (Eq. 2.5). Again these fields diminish rapidly with distance from the source and are essentially zero in the far field.

The nature of these field components can be further illuminated by examining complex power flow in the near field of the source. Complex power flow density in W/m^2 is generally expressed in the form:

$$\mathbf{p} = \frac{1}{2}\mathbf{E} \times \mathbf{H}^* \ \mathrm{W/m^2} \tag{2.66}$$

where $\mathbf{p}$ is a vector directed normal to $\mathbf{E}$ and $\mathbf{H}$. Following convention, $\mathbf{H}$ is conjugated to suppress the exponential term, in much the same way as in electrical circuit theory, complex power is expressed as $P = \frac{1}{2}VI^*$. Hence expressing $\mathbf{E}$ and $\mathbf{H}$ in their r, θ and ϕ components we obtain:

$$\mathbf{p} = \frac{1}{2}(\hat{\mathbf{a}}_\phi E_r H_\theta^* - \hat{\mathbf{a}}_\theta E_r H_\phi^* - \hat{\mathbf{a}}_\phi E_\theta H_r^* + \hat{\mathbf{a}}_\mathbf{r} E_\theta H_\phi^* + \hat{\mathbf{a}}_\theta E_\phi H_r^* - \hat{\mathbf{a}}_\mathbf{r} E_\phi H_\theta^*) \tag{2.67}$$

On introducing Eqs. (2.59) to (2.63) into Eq. (2.67) the following components to the complex power density $\mathbf{p}$ for the short current element are derived:

$$\begin{aligned}\mathbf{p} &= \hat{\mathbf{a}}_\theta \frac{-jk_0 I_0^2 dl^2}{16\pi^2}\sqrt{\frac{\mu_0}{\varepsilon_0}}\left[\frac{1}{r^3} + \frac{1}{k_0^2 r^5}\right]\sin\theta\cos\theta \\ &+ \hat{\mathbf{a}}_\mathbf{r} \frac{k_0 I_0^2 dl^2}{32\pi^2}\sqrt{\frac{\mu_0}{\varepsilon_0}}\left[\frac{k_0}{r^2} - \frac{j}{k_0^2 r^5}\right]\sin^2\theta\end{aligned} \tag{2.68}$$

The real power flow density radiating in the r-direction away from the source is given by $p_r = \mathrm{Re}\,\mathbf{p} = \frac{1}{2}\mathrm{Re}(\mathbf{E} \times \mathbf{H}^*)$ and has the form:

$$p_r = \frac{k_0^2 I_0^2 dl^2}{32\pi^2}\sqrt{\frac{\mu_0}{\varepsilon_0}}\frac{\sin^2\theta}{r^2} \tag{2.69}$$

The remaining components of Eq. (2.68) are imaginary in character and represent stored energy in the near field on the source. These components can be extracted by expressing $p_i = \mathrm{Im}\,\mathbf{p}$. This leads to an r-directed component:

$$p_i|_r = \frac{-jI_0^2 dl^2}{32k_0\pi^2}\sqrt{\frac{\mu_0}{\varepsilon_0}}\frac{\sin^2\theta}{r^5} \tag{2.70}$$

and a θ-directed component:

$$p_i|_\theta = \frac{-jk_0 I_0^2 dl^2}{16\pi^2}\sqrt{\frac{\mu_0}{\varepsilon_0}}\left[\frac{1}{r^3}+\frac{1}{k_0^2 r^5}\right]\sin\theta\cos\theta \tag{2.71}$$

The real power flow density (Eq. 2.69) diminishes, as expected, in proportion to the inverse radius squared (inverse square law), while the sine squared dependence of the equation denotes that the radiating lobe maximises in the x-y plane (at $\theta = 90°$) for all values of ϕ. The stored energy density term $p_i|_r$ (Eq. 2.70), which decays in inverse proportion to radius to the power of five, also exists mainly in the x-y plane. This electric dipole energy essentially oscillates at the frequency of the source, between electric and magnetic stored forms, with the instantaneous power flow density moving in a radially direction. On the other hand, the stored energy density represented by Eq. (2.71), which displays θ-directed instantaneous power flow, is concentrate in 'energy lobes' located at $\theta = \pm 45°$, for all values of ϕ (see Fig. 2.9). Both electric dipole ($1/r^3$) and inductive ($1/r^5$) energy form are represented.

To complete the picture of the radiation behaviour of the short current element it is instructive to examine to total power delivered by the element for a given current. To determine the total power it is necessary to integrate the power density over a spherical surface centred on the element and enclosing it. That is:

$$\begin{aligned} P_{rad} &= \iint_S p_r\hat{\mathbf{a}}_{\mathbf{r}}\cdot dS = \int_0^{2\pi}\int_0^{\pi}\frac{k_0^2 I_0^2 dl^2}{32\pi^2}\sqrt{\frac{\mu_0}{\varepsilon_0}}\frac{\sin^2\theta}{r^2}r^2\sin\theta\, d\theta\, d\phi \\ &= \frac{k_0^2 I_0^2 dl^2}{32\pi^2}\sqrt{\frac{\mu_0}{\varepsilon_0}}2\pi\int_0^{\pi}\sin^3\theta\, d\theta \end{aligned} \tag{2.72}$$

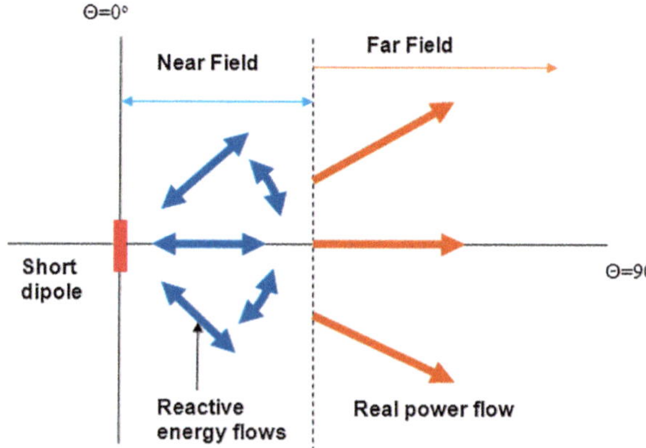

Fig. 2.9 Power and energy flows in near and far field regions for short current element

But $\int_0^\pi \sin^3\theta\, d\theta = 4/3$ and $k_0 = \frac{2\pi}{\lambda_0}$ where λ_0 is the free space wavelength. Hence:

$$P_{rad} = \frac{\pi\eta_0 I_0^2}{3}\left(\frac{dl}{\lambda_0}\right)^2 \tag{2.73}$$

This can be further reduced to a now traditional form:

$$P_{rad} = 40\pi^2\left(\frac{dl}{\lambda_0}\right)^2 I_0^2 \quad \text{Watts} \tag{2.74}$$

by using the relation $\eta_0 = 120\pi$. Not unexpectedly, the power radiated is proportional to the square of the current applied, but more importantly it is proportional to the square of the element length normalised to the free space wavelength. This ratio occurs repeatedly in antenna theory and underpins the unmistakeable trend that antennas largely diminish in size as frequency rises.

In electrical circuits it is useful to express power in terms of a resistance, and so it is convenient to make the relationship:

$$P_{rad} = I_{rms}^2 R_{rad} \tag{2.75}$$

where I_{rms} is the root mean squared current in the source, to yield:

$$P_{rad} = \frac{1}{2} I_0^2 R_{rad} \tag{2.76}$$

R_{rad} is termed the radiation resistance of the source. On combining Eqs. (2.76) and (2.74), the following classical equation is obtained:

$$R_{rad} = 80\pi^2\left(\frac{dl}{\lambda_0}\right)^2 \quad \text{Ohms} \tag{2.77}$$

If an antenna is viewed as a device which provides a transitioning mechanism from a guided wave in a feed line supplied from a transmitter to a free space (TEM) wave, then radiation resistance represents the coupling between the antenna and free space through its radiation pattern. It should be understood that R_{rad} has nothing to do with ohmic loss in the antenna structure. For optimum coupling to free space the output impedance of the transmitter, the characteristic impedance of the feed line and the radiation resistance should be matched. For the short current element, however, this is not possible. The assumptions incorporated into the development of the above equations require $dl/\lambda_0 < 0.1$, which gives $R_{rad} < 8\ \Omega$. An antenna feed line typically exhibits a characteristic impedance between 50 and 100 Ω—obviously very different from R_{rad}. Clearly techniques are required to embed and match practical antennas into real transmission systems and these will be explored in Chaps. 6, 7 and 8.

2.7 Chapter Summary

The role of the Maxwell equations in the formulation of a comprehensive theory of electromagnetic radiation is explored. It is shown that in the presence of an active source of radiation the solution of the Maxwell's equations leads to the formulation of second order vector differential equations with decidedly complex source functions. Substitution techniques are required to simplify the equations and thus effect a solution. It is demonstrated that the most common and effective substitution entails the introduction of a magnetic vector potential **A** and a scalar electric potential ϕ. The functional forms of these intermediary terms are constructed from the physical nature of both electric and magnetic fields at the static level. The full time varying forms for ϕ and **A** are deduced from the fundamental nature of electromagnetic wave propagation.

While the mathematical niceties of the Hertzian dipole are not too relevant to antennas evolved from slots (Chap. 3), the well established representations of radiated field distributions, of power density distributions, of radiation pattern fundamentals, of radiated power and stored energy, of radiation resistance and impedance relations, provide an invaluable insight into the nature of radiation and antennas.

Bibliography

1. C.A. Balanis, *Antenna Theory* (Wiley, New York, 1982)
2. R.E. Collin, F.J. Zucker, *Antenna Theory* (McGraw-Hill Book Co., New York, 1969)
3. I.S. Grant, W.R. Phillips, *Electromagnetism* (Wiley, Chichester, 1975)
4. F.H. Read, *Electromagnetic Radiation* (Wiley, Chichester, 1980)
5. F.T. Ulaby, *Fundamentals of Applied Electromagnetics* (Prentice Hall Inc., New Jersey, 1997)
6. J.D. Kraus, *Electromagnetics* (McGraw-Hill Book Co., Singapore, 1984)

Chapter 3
Compact Aperture Antennas

3.1 Aperture Antenna Models

The auxiliary potential functions introduced in Chap. 2, help to facilitate the analysis and modelling of wire, or current carrying, antennas, including simple reflectors as used in a wide range of applications. Their successful adoption is dependent on knowing the surface current density everywhere on the metallic structure. This is very often not possible as with lens antennas and dielectric rod antennas for which there is no identifiable conduction current flow, while it is very difficult to arrange, for antennas such as horns. These antenna types (see Fig. 3.1) are generally referred to as aperture antennas. Those represented by sketches a, b and c in Fig. 3.1 are horn types in which an empty rectangular waveguide opens out into a rectangular or square aperture depending on the application. In sketches d and e the slots cut into the broad and side walls of the waveguide form the aperture radiators, while f represents a leaky-wave dielectric rod radiator, where the aperture is the surface of the dielectric rod. The classification also includes reflector antennas (g), although these are also amenable to the techniques described in Chap. 2.

To determine the fields radiated by an aperture antenna it is usually necessary to employ a 'field equivalence principle' [1–3]. By adopting such a procedure, the antenna structure of interest, which would represent a largely intractable mathematical problem in its original manifestation, is typically reconfigured to a much simpler more amenable geometrical form, while remaining in essence electromagnetically equivalent to the original. While this modified form may no longer lead to an exact solution on analysis, it can be modelled more readily, and if constructed insightfully and intelligently, generally yields more than acceptable results. The basic concept is illustrated in Fig. 3.2a. In the left hand diagram input electric current and magnetic current densities $\mathbf{J}$ and $\mathbf{J_m}$ (e.g. on and inside a horn antenna) sets up a radiation field $\mathbf{E}$, $\mathbf{H}$ in a large area of space (V_2). Within this area an arbitrary surface S of volume V_1 (say) is inserted, which is invisible to the field distribution $\mathbf{E}$, $\mathbf{H}$. The equivalence theorem states that if on the *surface* S surface

A. J. Sangster, *Compact Slot Array Antennas for Wireless Communications*, Signals and Communication Technology, https://doi.org/10.1007/978-3-030-01753-8_3

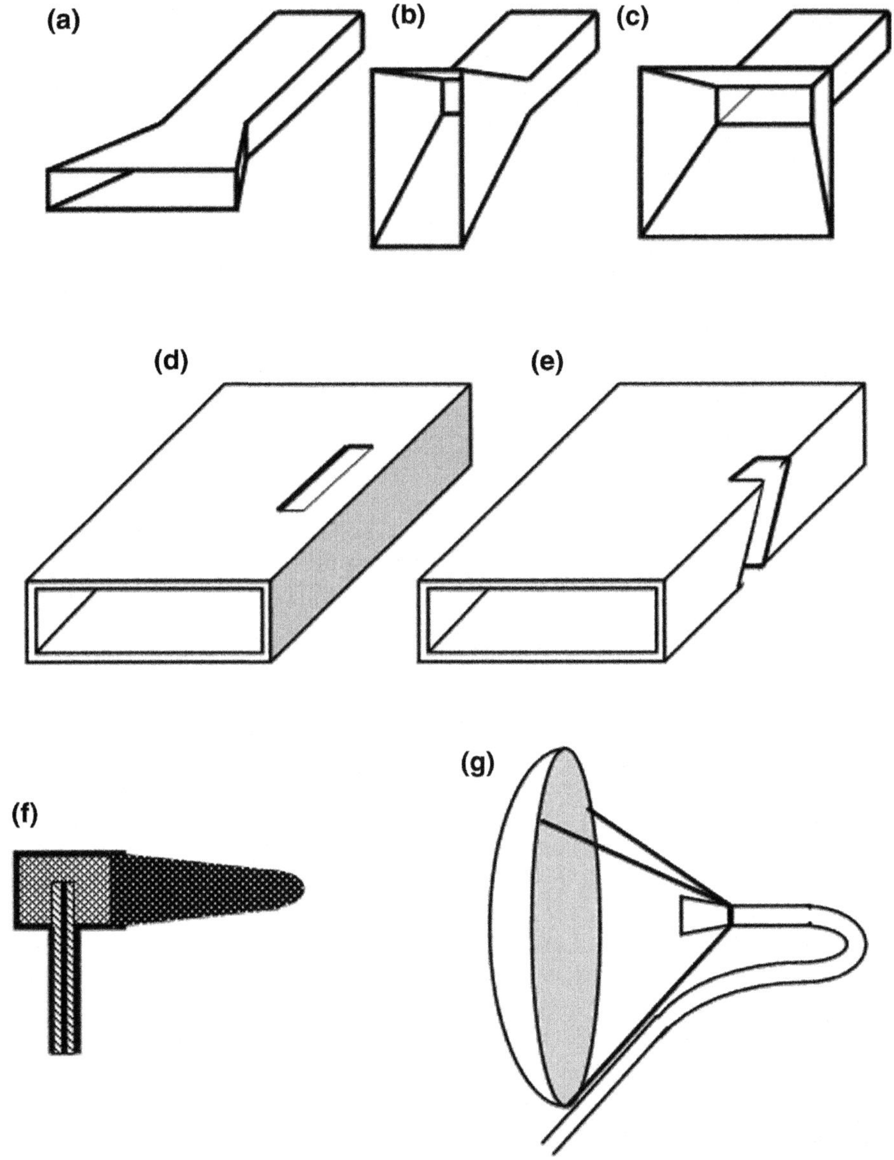

Fig. 3.1 Examples of aperture antennas: **a** H-plane horn, **b** E-plane horn, **c** sectoral horn, **d** shunt-slot in waveguide, **e** edge-slot in waveguide, **f** dielectric rod, **g** horn fed parabolic antenna

currents $\mathbf{J_s}$ and $\mathbf{J_{ms}}$ are imposed such that the external field $\mathbf{E}$, $\mathbf{H}$ is unaltered, then the introduced currents are equivalent sources for the region V_2. Therefore on S, $\mathbf{J_s} = \hat{\mathbf{n}} \times (\mathbf{H} - \mathbf{H_i})$ and $\mathbf{J_{ms}} = -\hat{\mathbf{n}} \times (\mathbf{E} - \mathbf{E_i})$ where the interior of S is presumed to contain a field distribution $\mathbf{E_i}$, $\mathbf{H_i}$ which is specified only on S by means of the above boundary relations. Since the interior volume V_1 is not of interest in setting up the equivalent sources, the interior fields $\mathbf{E_i}$, $\mathbf{H_i}$ can be arbitrarily chosen. The most common choice is $\mathbf{E_i} = \mathbf{H_i} = 0$ which is termed Love's equivalence principle [3]. In this case, on S, $\mathbf{J_s} = \hat{\mathbf{n}} \times \mathbf{H}$ and $\mathbf{J_{ms}} = -\hat{\mathbf{n}} \times \mathbf{E}$. Furthermore this means that the medium within S can be chosen arbitrarily. For example, a conductor can be judiciously placed there with no electromagnetic effect on the arrangement.

In Fig. 3.2a it should be noted that both electric current density $\mathbf{J_s}$ and magnetic current density $\mathbf{J_{ms}}$ have been inserted onto the surface S. However, according to the uniqueness theorem [3] we need only specify $\mathbf{J_s}$ *or* $\mathbf{J_{ms}}$ on S, not both, to ensure that the resultant field solution is unique. For example, consider the conducting body S with an aperture A as shown in Fig. 3.2b for which the field in the aperture ($\mathbf{E}_a \times \hat{\mathbf{n}}$) is known. The radiated field in the exterior region is unaffected if the aperture in the conducting body is closed by a conducting wall as depicted in Fig. 3.2b (right hand diagram) and replaced by the magnetic current $\mathbf{J_{ms}} = \mathbf{E}_a \times \hat{\mathbf{n}}$ placed closely adjacent to the position of the original aperture—essentially on the surface of the aperture wall.

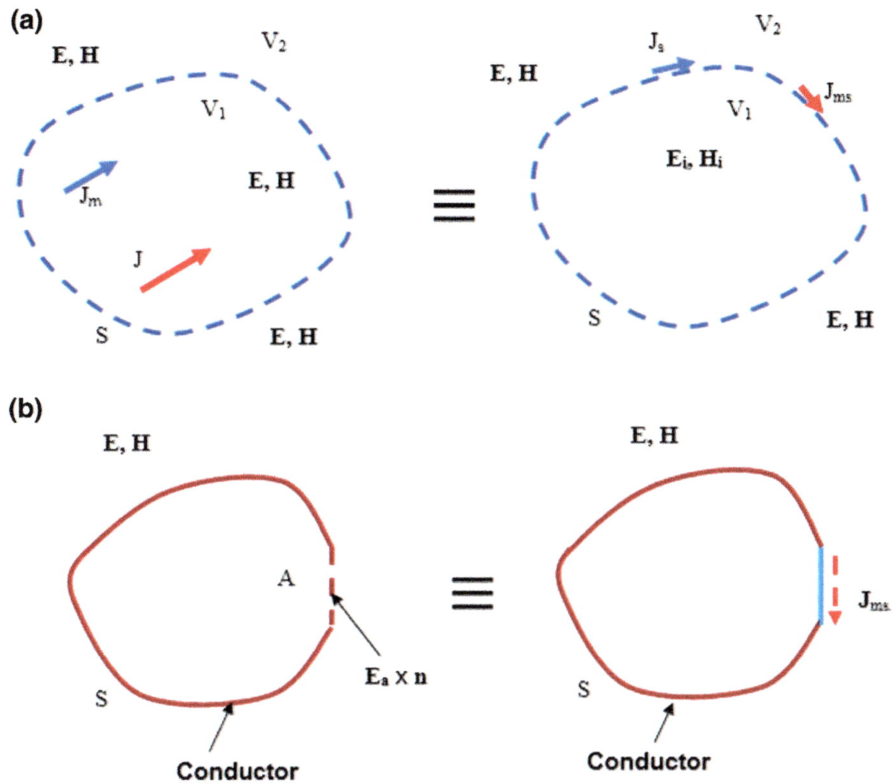

Fig. 3.2 Equivalence principles **a** for arbitrary surface S **b** for conducting body S with aperture A

Let us consider applying this equivalence technique to the horn antenna (Fig. 3.3). Firstly, the horn—the electromagnetic wave source—is enclosed (Fig. 3.3b) by an arbitrary surface which both accommodates the 'radiating aperture' of the source and completely encases it. It is commonly referred to as the Huygen's surface. A field free region is introduced within the surface S if surface currents are introduced to form fields external to S equivalent to those generated by the original source. This entails placing on S a distribution of electric and magnetic surface current densities $\mathbf{J_s}$ and $\mathbf{J_{ms}}$. If selected appropriately these current replacements generate the original external field. Given that on a defined surface the electromagnetic boundary conditions dictate that $\mathbf{J}_s = \hat{\mathbf{n}} \times \mathbf{H}$ and $\mathbf{J}_{\mathbf{m}s} = -\hat{\mathbf{n}} \times \mathbf{E}$, where $\hat{\mathbf{n}}$ denotes the unit normal to the surface, the application of the equivalence principle requires precise knowledge of both **E** and **H** over the surface S. These field definitions must obviously be dictated by the original source geometry. Actually, because of the relationship between **E** and **H**, built into the Maxwell equations, it is not necessary to define both field quantities over S. It is usually enough to define either **E** or **H** alone over the entire surface, or more commonly **E** alone over part of S and **H** alone over the rest. The resultant external field solution is unique, provided that on no part of S is surface current density unspecified. This means that S can be viewed as a perfect conductor, with $\mathbf{J_{ms}}$ specified over an area adjacent to the position of the horn aperture where E_a is finite (Fig. 3.3b). In principle this is a simpler problem to solve using auxiliary potential functions, particularly as we are relatively free in the way we choose the shape and form of S [3]. The E-field is imaged in the conducting surface to satisfy the boundary conditions so that $\mathbf{J_{ms}} = -2\hat{\mathbf{n}} \times E_a$. For example, if the portion of the surface which intercepts the aperture is presumed to be a plane perfectly conducting surface of infinite extent, then the equivalent model now comprises two half-spaces separated by an infinitesimally thin conducting plane, with the horn aperture field replaced by a magnetic current sheet adjacent to the conducting plane as suggested in Fig. 3.3c. This is clearly a much simpler radiation problem to solve than the raw original. To solve the problem all we now need is a knowledge of the E-field, and hence $\mathbf{J_{ms}}$, in the horn aperture, which is generally available from waveguide theory. Hence using an appropriate auxiliary function the radiated field in the right hand half space can be determined (see Sect. 3.2). The disadvantage of the technique is that only field patterns in the forward half space of the antenna are generated—back radiation is ignored. However, this is seldom a serious limitation since it is the forward pattern that is of primary concern in the majority of antenna applications whether operated in transmission or reception mode.

3.2 Auxiliary Potentials

In very much the same way as was expounded in Chap. 2, if we resort to an auxiliary potential function to simplify the electromagnetic boundary value problem created by a radiation source, the choice should satisfy at least one of the

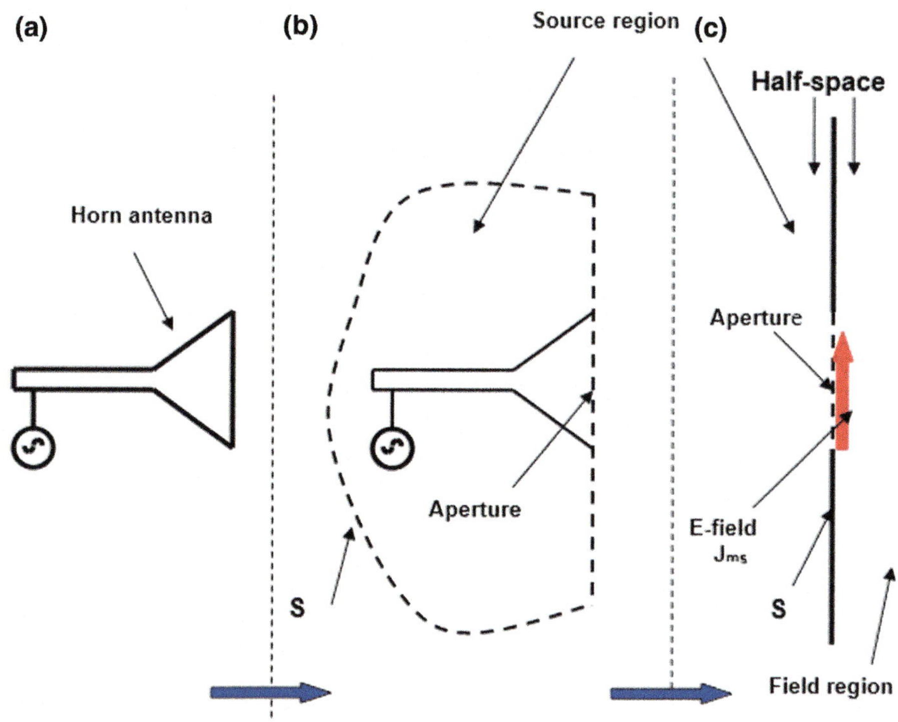

Fig. 3.3 Application of equivalence principle **a** original horn structure **b** enclosing surface S **c** equivalent aperture in infinite screen

source-free Maxwell equations in an electric charge-free space. Here we are concerned with magnetic current (electric field) sources and consequently an apt choice is to set (see [3]):

$$\mathbf{E} = -\frac{\mathbf{1}}{\varepsilon_0}\nabla \times \mathbf{A_m} \tag{3.1}$$

where $\mathbf{A_m}$ is the auxiliary potential function. Unlike the magnetic potential function **A** in Chap. 2, $\mathbf{A_m}$ has little electrical significance, other than as a useful mathematical generator of electric field, and hence magnetic field.

By duality, the Maxwell equations which are applicable to the problem of radiation from a magnetic current source of density $\mathbf{J}_\mathrm{m}$ and magnetic charge density ρ_m are:

$$\nabla \times \mathbf{E} = -\mathrm{j}\omega\mu\mathbf{H} - \mathbf{J}_\mathrm{m} \tag{3.2}$$

$$\nabla \times \mathbf{H} = j\omega\varepsilon\mathbf{E} \tag{3.3}$$

$$\nabla \cdot \mathbf{D} = 0 \tag{3.4}$$

$$\nabla \cdot \mathbf{B} = \rho_{\mathrm{m}} \tag{3.5}$$

The magnetic current density $\mathbf{J}_{\mathrm{m}}$ and magnetic charge density ρ_{m} are linked through the equation:

$$\nabla \cdot \mathbf{J}_{\mathrm{m}} = -j\omega\rho_m \tag{3.6}$$

which is suggestive of the Kirchoff relationship between electric current and charge (see Chap. 2).

Following a derivation paralleling Chap. 2, these lead to two second order vector differential equations for electric and magnetic fields, namely:

$$\nabla^2\mathbf{E} + k^2\mathbf{E} = \nabla \times \mathbf{J}_m \tag{3.7}$$

and

$$\nabla^2\mathbf{H} + k^2\mathbf{H} = \mathrm{j}\omega\varepsilon\mathbf{J}_m - \frac{\nabla\nabla \cdot \mathbf{J}_m}{\mathrm{j}\omega\mu} \tag{3.8}$$

Clearly the right hand side source functions in these equations are unappealingly complex. However the auxiliary potential function choice delineated above, when substituted into the derivation process as outline in Chap. 2, generates a much more amenable second order vector differential equation, as follows:

$$\nabla^2\mathbf{A}_m + k_0^2\mathbf{A}_m = -\varepsilon_0\mathbf{J}_m \tag{3.9}$$

where $k = k_0 = \omega\sqrt{\mu_0\varepsilon_0}$ for radiation into free space or vacuum. By borrowing an obvious parallel with the construction of the solution to the analogous equation in Chap. 2 [Eqs. (2.28) and (2.45)], it is not difficult to assert that the solution to Eq. (3.9) must have the phasor form.

$$\mathbf{A}_m(\mathbf{r}) = \frac{\varepsilon_0}{4\pi}\iiint\limits_V \frac{\exp(-jk|\mathbf{r}-\mathbf{r}'|)\mathbf{J}_m(\mathbf{r}')}{|\mathbf{r}-\mathbf{r}'|}\mathrm{d}V' \tag{3.10}$$

That Eq. (3.10) is a solution to Eq. (3.9) can be established by substitution. Although the functional forms of Eqs. (2.45) and (3.10) are essentially identical the volumes V are not the same. While the current element $\mathbf{J}$ exists in isolation in a 'whole' space (V an infinite volume), $\mathbf{J}_{\mathrm{m}}$ is presumed to reside on an infinite conducting screen (Fig. 3.3) and radiates into a half-space (V).

The evident mathematical duality that exists between radiation solutions emanating from surface conduction current densities ($\mathbf{J} \equiv \mathbf{J}_\mathbf{s}$) on 'wire' structures, and magnetic current densities ($\mathbf{J}_\mathbf{m} \equiv \mathbf{J}_\mathbf{ms}$) representing field sources in apertures, is highlighted in Table 3.1.

Table 3.1 Duality of equations associated with electric and magnetic current sources

Electric sources ($\mathbf{J} \neq 0$, $\mathbf{J}_m = 0$)	Magnetic sources ($\mathbf{J}_m \neq 0$, $\mathbf{J} = 0$)
$\nabla \times \mathbf{E} = -j\omega\mu_0\mathbf{H}$	$\nabla \times \mathbf{H} = -j\omega\varepsilon_0\mathbf{E}$
$\nabla \times \mathbf{H} = \mathbf{J} + j\omega\varepsilon_0\mathbf{E}$	$-\nabla \times \mathbf{E} = \mathbf{J_m} + j\omega\mu_0\mathbf{H}$
$\mathbf{H} = \frac{1}{\mu_0}\nabla \times \mathbf{A}$	$\mathbf{E} = -\frac{1}{\varepsilon_0}\nabla \times \mathbf{A_m}$
$\mathbf{E} = -j\omega\mathbf{A} - j\frac{1}{\omega\mu_0\varepsilon_0}\nabla(\nabla \cdot \mathbf{A})$	$\mathbf{H} = -j\omega\mathbf{A}_m - j\frac{1}{\omega\mu_0\varepsilon_0}\nabla(\nabla \cdot \mathbf{A}_m)$
$\nabla^2\mathbf{A} + k_0^2\mathbf{A} = -\mu_0\mathbf{J}$	$\nabla^2\mathbf{A}_m + k_0^2\mathbf{A}_m = -\varepsilon_0\mathbf{J}_m$
$\mathbf{A}(\mathbf{r}, t) = \frac{\mu_0}{4\pi}\iiint_V \frac{\mathbf{J}(\mathbf{r}', t)}{\lvert\mathbf{r}-\mathbf{r}'\rvert}dV'$	$\mathbf{A}_m(\mathbf{r}, t) = \frac{\varepsilon_0}{4\pi}\iiint_V \frac{\mathbf{J}_m(\mathbf{r}', t)}{\lvert\mathbf{r}-\mathbf{r}'\rvert}dV'$

3.3 Radiation from a Linear Aperture

Radiation mechanisms associated with antennas which fall into the category of ‘aperture antennas’ are attributable to diffraction from the electric field distribution formed across the aperture at any given instant in time. The field distribution from an aperture can be modelled in many ways (as we shall see in Chap. 4) but the radiation mechanism can be understood most clearly and directly by examining the electromagnetic penetration of a narrow one-dimensional slit as depicted in Fig. 3.4. The arrangement comprises a perfectly flat conducting screen of infinite extent in the x-y plane, containing a slit of width a in the x-direction and extending to infinity at $\pm y$. A plane electromagnetic wave in the left hand half-space is incident on the screen and hence the slit and sets up, in sympathy with the x-polarised plane wave, an x-directed electric field E_0 in the aperture, which we can presume to be constant across the narrow aperture.

3.3.1 Huygen’s Principle and Equivalent Sources

At every point on the E-field in the aperture it can be presumed, referring to Fig. 3.4, that there exists a source of spreading waves into the half-space to the right of the aperture. In the left hand half-space the back scattered waves from the slot, if it is small are negligible by comparison with waves reflected by the screen and can therefore be ignored at this stage. This approximate technique is termed the Huygen’s principle. In two dimensions these waves spread with cylindrical wave-fronts into the half-space with $1/\sqrt{r}$ dependence on distance from the source. That this is so, is evident from the nature of Eq. (3.10). In three dimensions they would spread out spherically with $1/r$ dependence for the fields.

So let us consider an element of the slit of width dx at x in the x-direction (Fig. 3.4). We can assume that the fields everywhere are invariant in the y-direction for a slit of constant width for all y. In accordance with Huygen’s principle the field within the element can be considered to be a source of cylindrical waves in the right-hand medium ($z > 0$). The strength of this ‘secondary’ source will obviously

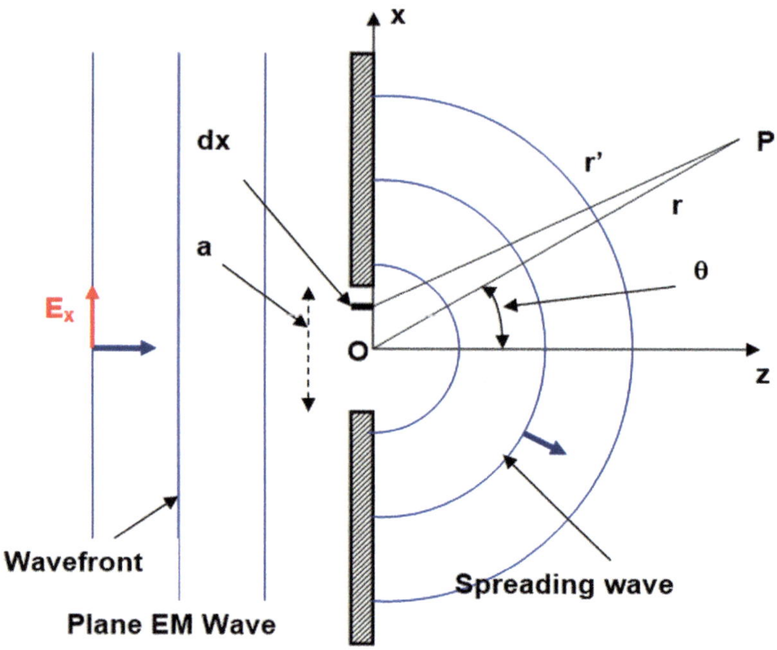

Fig. 3.4 Electromagnetic wave diffraction by a parallel sided slit in a perfectly conducting membrane (Huygen's mechanism)

be proportional to E_0 and also to the width dx of the element. Hence the field at the remote point P in Fig. 3.4, which is located at distance r' from the source at (x, 0, 0) and distance r from the origin at O, can logically be represent by a function of the following form:

$$\mathrm{d}E_P = \frac{AE_0\mathrm{d}x}{\sqrt{r'}}\exp(-jk_0r') \tag{3.11}$$

where A is an amplitude constant, and k_O is the propagation coefficient for free space electromagnetic waves in the right-hand region. If P is very remote from the aperture such that r $\gg$ a, then we can reasonably presume that $r' \approx r$ in the denominator of Eq. (3.11). However, for the much more sensitive phase shift we use $r' = r - x\sin\theta$ and so Eq. (3.11) becomes:

$$\mathrm{d}E_P = \frac{AE_0\mathrm{d}x}{\sqrt{r}}\exp(-jk_0(r - x\sin\theta)) \tag{3.12}$$

where θ is the angle between OP and the z-axis in Fig. 3.4.

The total diffracted field at P due to the field E_0 in the aperture can now be determined by summing all of the dx contributions at P for all values of x between $-a/2$ and $+a/2$. This results in the following integral equation:

$$E_P(r,\theta) = \frac{AE_0}{\sqrt{r}} \exp(-jk_0 r) \int\limits_{-a/2}^{+a/2} \exp(jk_0 x \sin\theta) \mathrm{d}x \tag{3.13}$$

which, on performing the integration, reduces to:

$$E_P(r,\theta) = \frac{AE_0 a}{\sqrt{r}} \exp(-jk_0 r) \frac{\sin(\pi\frac{a}{\lambda_0}\sin\theta)}{\pi\frac{a}{\lambda_0}\sin\theta} \tag{3.14}$$

Significantly, the form of this equation reveals that the diffracted wave at P from the superposition of cylindrical waves, emanating from all parts of the aperture containing the field E_0, combine to form an 'interference pattern' which in this case has a characteristic sin X/X distribution. Distributions of this nature are ubiquitous in aperture antenna theory, and they will be examined in more detail in the following section.

3.3.2 Plane Wave Spectrum

In several branches of electromagnetism, the wave solutions to the electromagnetic boundary value problem are too complex to recruit directly into device development. Fortunately however, such solutions can usually be resolved into constituent plane waves. A good example [2] is rectangular, or circular cylindrical, waveguide where the behaviours of the TE or TM modes can best be implemented by expressing them in plane wave terms (Appendix B). This approach can also be used to advantage in certain radiation problems, particularly those of the aperture type. Mathematically, as we shall see, the plane wave spectrum technique, as it is termed, is not difficult to implement for antennas with known aperture field shapes, and can lead to quick and reliable predictions of an antenna's radiation characteristics.

The plane wave spectrum of a radiating aperture is a fundamentally more rigorous means of estimating the diffraction pattern than the Huygen's method. For example, a point on the cylindrical wave front depicted in Fig. 3.4 can locally be viewed as comprising a 'plane' wave within an elemental sector dθ, particularly for large r values. This idea is illustrated in Fig. 3.5. The combination of plane waves representing all sectors dθ between $\theta = -\pi/2$ and $\theta = +\pi/2$ mathematically reconstructs the original cylindrical wave front. This combination of plane waves is usually termed the angular spectrum, or spectrum function ($F(\theta)$), and for the case of a very narrow slit it has a constant magnitude. Actually this function is conventionally expressed as $F(\sin\,\theta)$ rather than $F(\theta)$.

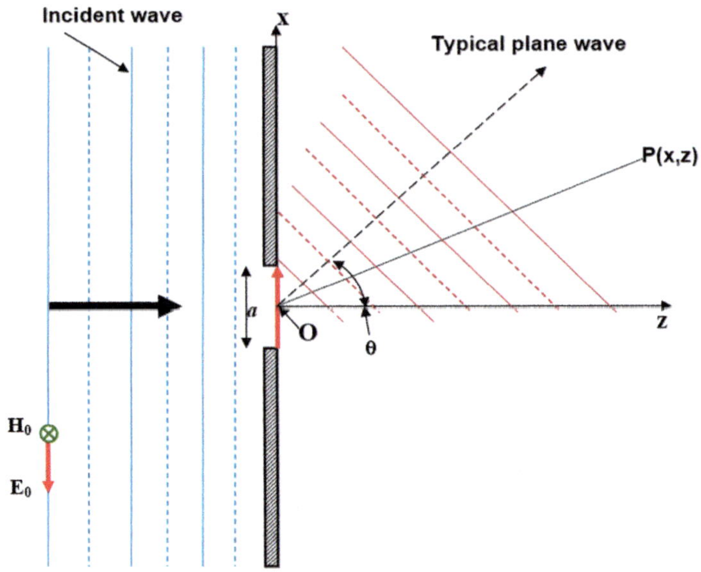

Fig. 3.5 Plane wave diffraction at a narrow slit in an infinite perfectly conducting screen—illustrating the diffraction pattern is a spectrum of plane waves

On the other hand for a radiating slit of finite width the spectrum function is by no means constant. In this more general case, it can be evaluated be reference to Fig. 3.5, where the slit has a width a in the x-direction. As in the Huygens example the slit is irradiated from the left by a single plane wave of E-field amplitude E_0. The figure shows a typical plane wave component of the radiated power in the right hand half-space. It propagates in the direction $\mathbf{k}_0$ which is at an angle θ to the z-axis. If the right hand region is free-space the magnitude of $\mathbf{k}_0$ is given by $k_0 = \omega\sqrt{\mu_0\varepsilon_0}$, and the direction is given by that of the constituent plane wave. The point P(x, z) (Fig. 3.6) is expressed in the cylindrical coordinate frame by the vector $\mathbf{r}$. The plane wave propagating in the direction θ subtends angle dθ at the origin, and logically the wave magnitude will be proportional to both $F(\sin\theta)$ and dsin θ, on using sin θ rather than θ to denote angle. Hence the E-field at P due to this representative plane wave can be expressed as:

$$\begin{aligned} \mathrm{d}E_P(x,z) &= F(\sin\theta)\mathrm{d}(\sin\theta)\exp(-j\mathbf{k_0}\cdot\mathbf{r}) \\ &= F(\sin\theta)\mathrm{d}(\sin\theta)\exp(-jk_0(x\,\sin\theta + z\,\cos\theta)) \end{aligned} \tag{3.15}$$

and on collecting and adding all possible contributions from all typical plane waves the following integral function is created:

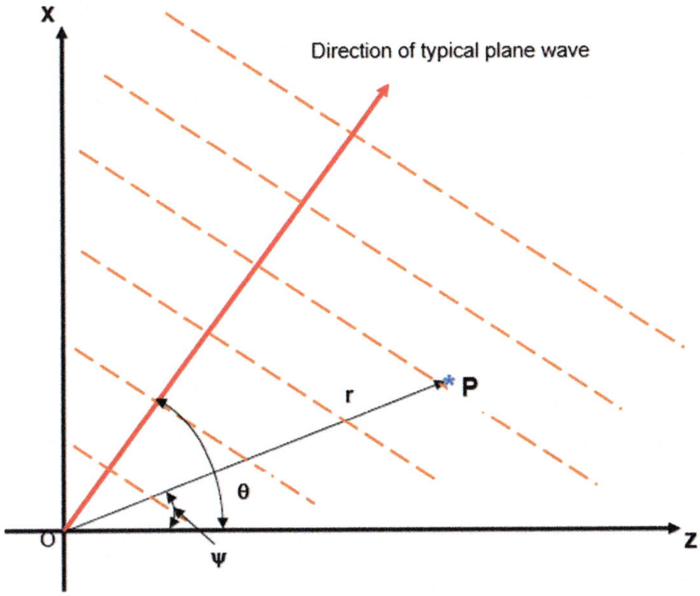

Fig. 3.6 Cylindrical coordinates (r, ψ) of a field point P

$$E_P(x,z) = \int_{-\infty}^{\infty} F(\sin\theta)\exp(-jk_0(x\sin\theta + z\cos\theta))\mathrm{d}(\sin\theta) \tag{3.16}$$

The integration limits are taken from $-\infty$ to $+\infty$ to ensure that all possible contributions to the diffracted field are included—both real and imaginary.

If the field in the aperture is expressed functionally as:

$$f(x) = E_P(x,\ 0) \tag{3.17}$$

then Eq. (3.16) becomes:

$$f(x) = \int_{-\infty}^{\infty} F(\sin\theta)\exp(-jk_0 x\sin\theta)\mathrm{d}(\sin\theta) \tag{3.18}$$

In electrical circuit theory Fourier transforms are regularly used to relate frequency to time and vice versa through the equations:

$$f(t) = \frac{1}{2\pi}\int_{-\infty}^{\infty} F(\omega)\exp(j\omega t)\mathrm{d}\omega \tag{3.19}$$

$$F(\omega) = \int_{-\infty}^{\infty} f(t) \exp(-j\omega t) \mathrm{d}t \tag{3.20}$$

So by analogy with this time/frequency transform pair, it is reasonable to suggest that Eq. (3.18) has the transform partner:

$$F(\sin \theta) = \frac{1}{\lambda_0} \int_{-\infty}^{\infty} f(x) \exp(jk_0 x \sin \theta) \mathrm{d}x \tag{3.21}$$

This transform relationship between the angular spectrum and aperture field, admittedly expressed in a limited two dimensional form here, in Eqs. (3.18) and (3.21), is a very important one in antenna theory and will be considered in more detail below.

In the elementary diffraction example presented in Fig. 3.5, the electric field, and hence $f(x)$, in the range of x from $-\infty$ to $+\infty$, is non-zero only in the slit: therefore,

$$f(x) = \begin{cases} E_0 & \text{for } |x| \leq a/2 \\ 0 & \text{for } |x| > a/2 \end{cases} \tag{3.22}$$

On substituting this form for $f(x)$ into Eq. (3.21) we obtain:

$$F(\sin \theta) = \frac{E_0}{\lambda_0} \int_{-\infty}^{\infty} \exp(jk_0 x \sin \theta) \mathrm{d}x$$

This integral is well known and exhibits the following solution:

$$F(\sin \theta) = \frac{E_0 a}{\lambda_0} \frac{\sin\left(\frac{\pi a \sin \theta}{\lambda_0}\right)}{\frac{\pi a \sin \theta}{\lambda_0}} \tag{3.23}$$

It has the same angular dependence in the far-field as the radiation pattern represented by Eq. (3.14) which was deduced from the Huygen's principle. That the spectrum function bears a direct relationship with the far-field pattern of an aperture antenna is more fully developed in many texts on antenna theory [3]. This powerful relationship remains true for all aperture antennas, not just the slit example examined here.

3.4 Spectrum Functions

The Fourier transform pair [Eqs. (3.18) and (3.21)] which expresses the relationship between spectrum function (far-field) and an aperture field function (antenna source field) provides an excellent base from which to examine the fundamental radiation characteristics of aperture antennas. They are of course limited to furnishing pattern information for the principal plane of the antenna only—that is the plane containing the major axis of the aperture and the normal to the aperture (the x-z plane in Fig. 3.5). Notwithstanding this limitation all the essential features of the antenna pattern can be demonstrated. To assess antenna behaviour in planes other than the principal plane the three dimensional forms of Eqs. (3.18) and (3.21) must be used. These are developed in Sect. 3.5.

In Fig. 3.7 spectrum functions for four commonly used antenna aperture field distributions are displayed in two-dimensional rectangular rather than polar form. The functions themselves are listed in Table 3.2. The rectangular form is more traditional but with the growing availability of sophisticated drafting tools within EM solvers the polar representation is becoming increasingly common place. Of course, for 'real' antennas radiation patterns are three dimensional, and modern drafting software can generate 3-D images. However 3-D representations on a 2-D page can lose subtle details of a given pattern, and so the 2-D depictions continue to

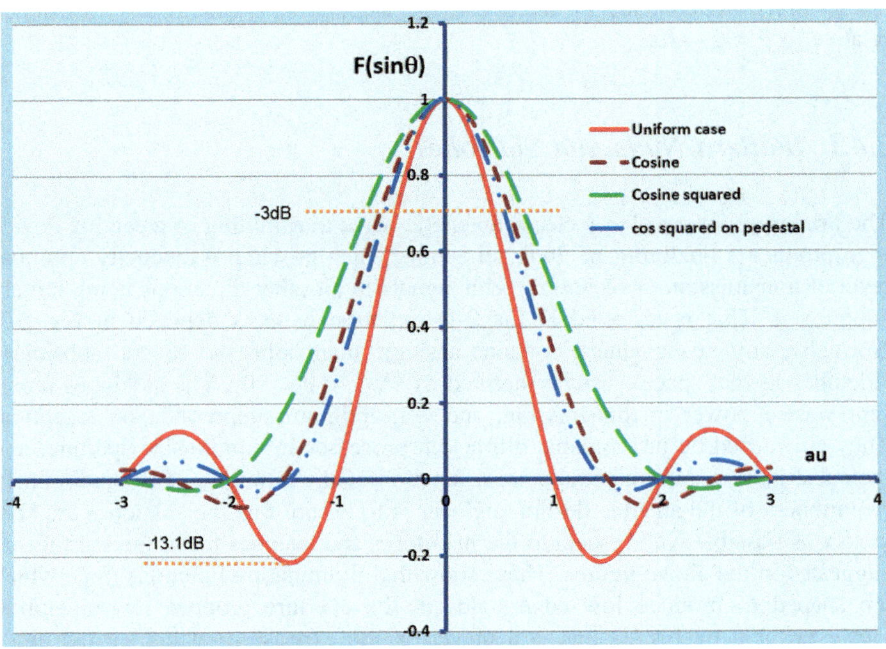

Fig. 3.7 Far-field patterns for apertures with the following field distributions—uniform, cosine, cosine squared, cosine squared on a 10 dB pedestal

Table 3.2 Commonly used spectrum functions

	Aperture distribution normalised to λ_0/a $f(x)$	Spectrum function $F(\sin\theta)$
1	Uniform $f(x) = \lambda_0/a$	$\frac{\sin(\pi au)}{\pi au}\left(u = \frac{\sin\theta}{\lambda_0}\right)$
2	Cosine $f(x) = \frac{\lambda_0}{a}\cos\left(\frac{\pi x}{a}\right)$	$\left(\frac{\pi}{2}\right)^2 \frac{\cos(\pi au)}{(\pi/2)^2 - (\pi au)^2}$
3	Cosine squared $f(x) = \frac{\lambda_0}{a}\cos^2\left(\frac{\pi x}{a}\right)$	$\frac{\sin(\pi au)}{\pi au}\frac{1}{1-(au)^2}$
4	Parabolic $f(x) = \frac{\lambda_0}{a}\left[1 - \left(\frac{2x}{a}\right)^2\right]$	$\frac{3}{(\pi au)^2}\left[\frac{\sin(\pi au)}{\pi au} - \cos(\pi au)\right]$
5	Parabolic squared $f(x) = \frac{\lambda_0}{a}\left[1 - \left(\frac{2x}{a}\right)^2\right]^2$	$\frac{15}{(\pi au)^2}\left[\frac{3}{(\pi au)^2}\left[\frac{\sin(\pi au)}{\pi au} - \cos(\pi au)\right] - \frac{\sin(\pi au)}{\pi au}\right]$
6	Cosine squared on 10 dB pedestal $f(x) = \frac{\lambda_0}{a}\left[0.3162 + 0.6838\cos^2\left(\frac{\pi x}{a}\right)\right]$	$\frac{\sin(\pi au)}{\pi au}\left[\frac{1-0.481(au)^2}{1-(au)^2}\right]$

be used to ensure that the pattern of interest is fully recorded. Normally this means that at least two planar patterns are required representing the major planes (or 'cuts') of the antenna. These are usually chosen to be orthogonal to each other. The traditional planes employed are predominantly the vertical and horizontal planes for ground mounted antennas. These are also termed the elevation and azimuth planes particularly in radar applications. The reference direction $au = 0$ (i.e. $\theta = 0°$) is generally assumed to be aligned with the normal to the surface of the aperture. These directions coincide in practice if the antenna is symmetrically distributed relative to $\theta = 0°$ axis.

3.4.1 Pattern Nulls and Sidelobes

The primary function of any electromagnetic wave transmitting or receiving device is to produce a predominant 'look' direction which provides a distinctly optimum level of transmission or reception, with signals in all other directions being largely suppressed. This is reflected in the 2-D radiation patterns depicted in Fig. 3.7. Normally, any given pattern contains a single main lobe and several subsidiary sidelobes in real space—usually defined as $-90° < \theta < 90°$. The sidelobes represent wasted power in transmission, and non-optimum suppression on reception. They are caused by unavoidable diffraction processes in a finite size aperture, and their distribution and magnitude are a function of the ratio a/λ_0. The predominant requirement of the antenna design engineer is to ensure that the sidelobes are kept as low as possible with respect to the mainlobe, and avenues for achieving this are suggested in the above figures. These show that illumination functions ($f(x)$) which are shaped to produce low edge fields in the aperture produce low sidelobes. However, at a fixed a/λ_0 this is achieved at the expense of widening the angle subtended by the mainlobe. On the other hand, the main beam width, which determines the directivity of the antenna, can be reduced by increasing a/λ_0. This stratagem introduces more nulls into the pattern, and hence moves the first null (nearest the mainlobe peak) toward au = 0, that is $\theta = 0$, thus 'squeezing down' the mainlobe width.

For example if $a = 0.3$ m and $\lambda_0 = 3$ cm the first null occurs at 5.7° for a uniform aperture illumination (Fig. 3.7). If a is now increased to 0.6 m the first null shifts to 2.85° thus narrowing the primary lobe. The magnitude of the first sidelobe relative to the peak of the mainlobe, as can be seen from Fig. 3.7, has the magnitude $0.22 = -13.1$ dB. For most applications this would normally be viewed as too high and aperture field shaping would be introduced to lower it.

3.4.2 *Mainlobe Beamwidth*

The angular width of the mainlobe of a directional antenna is arguably the parameter of primary concern in relation to achieving a design which meets a specific electromagnetic wave transmission or reception requirement. In two dimensions, the definition of this lobe-width or beamwidth could simply be the angular distance between first nulls as the spectrum patterns suggest. In three dimensions, as we shall see in Sect. 3.5, beamwidths in at least two orthogonal planes are usually required to define the lobe.

For the example considered above, of a 0.3 m wide uniformly illuminated slit radiator, the distance between first nulls can be deduced from Fig. 3.7 (solid red curve) and gives a null beamwidth of 11.4°. While in theory this is a perfectly acceptable definition, in practice this is not so, for three very sound reasons:

1. There may not be a null—a feature of some limited pattern types.
2. If the null exists, its precise location may be difficult to determine accurately.
3. The minimum detectable level of an EM signal in any practical measurement is well above zero field, as a result of noise and other signal perturbations.

As a consequence of these difficulties the convention, which has sensibly evolved, defines beamwidth as the angular separation in a given plane, between the points on either side of the mainlobe, which are 3 dB in power density below the density at the beam *maximum*. This is termed the half-power beamwidth (HPBW) and is identified on the spectrum pattern in Fig. 3.7. Note that the pattern is normalised to give a peak value of unity. Since the spectrum actually represents a field pattern, the 3 dB positions are located where the field is $1/\sqrt{2}$ below the peak value. The HPBW is determined by equating the spectrum function ($f(x)$) to 0.707. In the case of uniform illumination this gives: $\sin \pi au = 0.707\pi au$ with $\theta = \theta_{3\,\mathrm{dB}}$. For example, for a 0.3 m wide uniformly illuminated slit at 10 GHz ($\lambda_0 = 3.0$ cm) the beamwidth from Fig. 3.7 is approximately 5.1°, whereas for a cosine illumination it is 6.8°. Thus the improved sidelobe performance associated with the cosine illumination is obtained at the expense of some deterioration in beamwidth. It is assumed that a/λ_0 is the same in both cases.

3.4.3 *Pattern Gain*

A property of a directional antenna that is arguably of primary interest to users, whether in reception or transmission roles, is its gain. In any physically realisable antenna there will generally be one direction in which the radiated or collected power density is greatest. To give a quantitative value to this optimum operating condition, the power density at this pattern location is compared with some generally accepted reference level. This level is provided by estimating the power density in the same distance and direction produced by an isotropic source or sink delivering (or receiving) the same total power as the original antenna. As we have already noted, an isotropic antenna notionally produces an omni-directional or spherical radiation pattern. So, for a lossless antenna, or for a low loss antenna for which resistive and other losses can be ignored, the ratio of the peak power density delivered or received by the test antenna in the direction of the mainbeam to that produced by an equivalent isotropic antenna is termed the directive gain or directivity (D). On the other hand, when the antenna losses are not ignored, the resultant gain (G) is inevitably lower, and it is usually termed the power gain. Directivity and gain are related through the relation:

$$G = \eta D \tag{3.24}$$

where η (the efficiency) is a measure of the total power transmitted by the given antenna divided by the total power delivered to it, or in reception terms the total power delivered to the receiver divided by the total power received by the antenna. Directivity and gain will be examined further in Sect. 3.5.

3.5 Rectangular Aperture

While the examination of linear apertures has the advantage of simplifying the basic analysis process associated with the evolution of radiation patterns and with identifying the nature and properties of these patterns, we still need tools to investigate practical aperture antennas. These real antennas exhibit, of course, two dimensional apertures, and hence they generate three dimensional radiation patterns. Here it will be demonstrated that two dimensional apertures can be modelled by revisiting Eq. (3.10) and extending its applicability to two dimensions.

In the computational modelling of two dimensional aperture problems the main difficulty lies in performing the integrations embedded in Eq. (3.10). However, if interest is limited to the far-field, which is the normal scenario, the complexity of the integrations can be significantly reduced. It is helpful at this stage to make some simplifying definitions. Firstly in the far-field the following approximations can be made with little impact on accuracy:

$$R \approx |\mathbf{r}| = r \quad \text{(for magnitude changes)} \tag{3.25}$$

$$R = r + r' \cos \xi \quad \text{(for phase changes)} \tag{3.26}$$

Here ξ denotes the angle between the vectors **r** and **r′** which are depicted in Fig. 3.8.

For far-field computations $\xi \approx 0$ in which case **R** and **r** are essentially parallel. Hence applying Eq. (3.10) to a radiating aperture which can be represented by a surface S, we obtain:

$$\mathbf{A}_m = \frac{\varepsilon_0}{4\pi} \iint_S \mathbf{J_m} \frac{\exp(-jk_0 R)}{R} \mathrm{d}S' \tag{3.27}$$

If we now incorporate Eqs. (3.25) and (3.26) into Eq. (3.27), $\mathbf{A}_m$ can be written as:

$$\mathbf{A_m} \approx \frac{\varepsilon_0 \exp(-jk_0 r)}{4\pi r} \iint_S \mathbf{J_m} \exp(+jk_0 r' \cos \xi) \mathrm{d}S' \tag{3.28}$$

In the polar (r, θ, ϕ) coordinate system the vector $\mathbf{A}_\mathrm{m}$ has two components in the far-field, namely:

$$\mathbf{A_m} = \mathbf{a}_\theta A_{m\theta} + \mathbf{a}_\phi A_{m\phi} \tag{3.29}$$

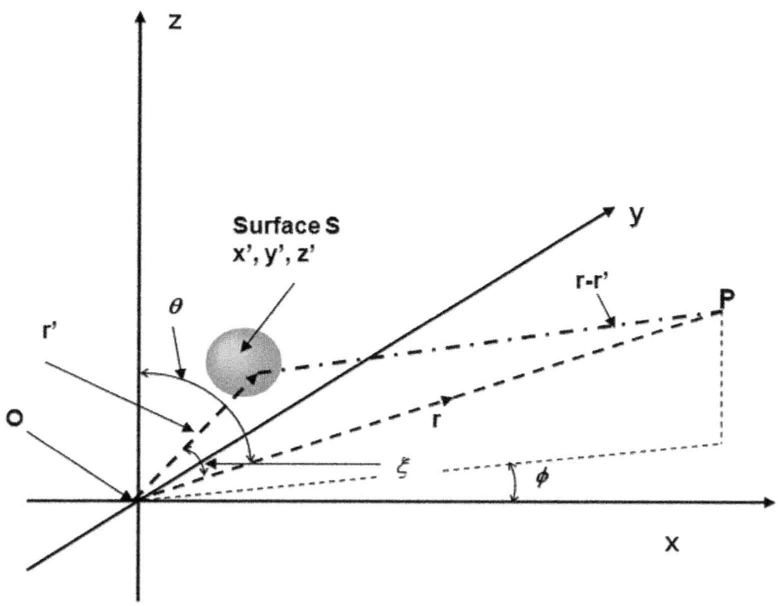

Fig. 3.8 Geometry for two-dimensional aperture antenna analysis

where $\mathbf{a}_\theta$ and $\mathbf{a}_\phi$ are unit vectors. There is no A_{mr} since in the far-field the radiation is transverse electromagnetic in character. At the position of the aperture surface at r' (x', y', z') the vector magnetic current can be expressed as:

$$\mathbf{J_m} = \mathbf{a_x}J_{mx} + \mathbf{a_y}J_{my} + \mathbf{a_z}J_{mz} \tag{3.30}$$

Hence combining Eqs. (3.28), (3.29), and (3.30) yields the following accessible forms:

$$A_{m\theta} = \frac{\varepsilon_0 \exp(-jk_0 r)}{4\pi} \iint_S \left[J_{mx} \cos\theta \cos\phi + J_{my} \cos 0 \sin\phi - J_{mz} \sin\theta\right] \exp(+jk_0 r' \cos\xi)\, dS' \tag{3.31}$$

$$A_{m\varphi} = \frac{\varepsilon_0 \exp(-jk_0 r)}{4\pi} \iint_S -\left[J_{mx} \sin\phi + J_{my} \cos\phi\right] \exp(+jk_0 r' \cos\xi)\, dS' \tag{3.32}$$

The electric and magnetic fields in the far-field can be generated using the appropriate equations in Table 3.1—noting that the divergence of $\mathbf{A}_\mathrm{m}$ is zero in the far-field.

Hence we obtain, finally, the far-field solution for the general two dimensional aperture antenna:

$$E_r \approx 0 \tag{3.33}$$

$$E_\theta \approx -\frac{jk_0 \exp(-jk_0 r)}{4\pi r} A_{m\varphi} \tag{3.34}$$

$$E_\phi \approx +\frac{jk_0 \exp(-jk_0 r)}{4\pi r} A_{m\theta} \tag{3.35}$$

$$H_r \approx 0 \tag{3.36}$$

$$H_\theta \approx \frac{jk_0 \exp(-jk_0 r)}{4\pi r} \left[-\frac{A_{m\theta}}{\eta_0}\right] \tag{3.37}$$

$$H_\phi \approx -\frac{jk_0 \exp(-jk_0 r)}{4\pi r} \left[\frac{A_{m\phi}}{\eta_0}\right] \tag{3.38}$$

3.5.1 Uniformly Illuminated Rectangular Aperture

It is useful to establish the effectiveness of these equations by examining a standard example which can be found in Balanis [3], Chap. 12. It comprises a rectangular aperture of width a and height b located within an infinite perfectly conducting

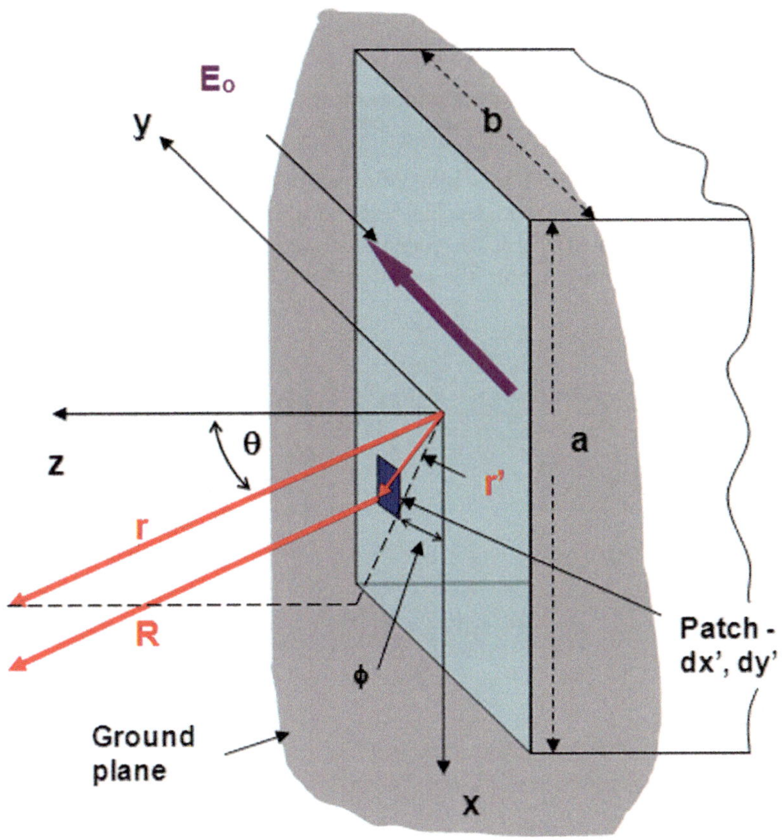

Fig. 3.9 Rectangular aperture antenna embedded in a conducting ground plane of infinite extent

ground plane (see Fig. 3.9). The electric field in the aperture is assumed to be constant in both the x- and y-directions and to be y-polarised. This is not a practical field distribution but it serves the purpose of the demonstration with minimal complexity. This field distribution can be expressed as:

$$\mathbf{E_a} = \mathbf{a_y} E_0 \tag{3.39}$$

existing in the range $-a/2 \leq x' \leq a/2$ and $-\mathrm{b}/2 \leq y' \leq \mathrm{b}/2$. E_0 is a constant in this case.

The equivalent solvable geometry comprises a 'closed' half-space bounded by a flat conducting screen extending over the entire x-y plane at $z = 0$. Hence applying the boundary conditions arising from Fig. 3.3, we obtain:

$$\mathbf{J_{ms}} = -2\hat{\mathbf{n}} \times \hat{\mathbf{a}}_{\mathbf{y}} E_a = 2\hat{\mathbf{a}}_{\mathbf{x}} E_0 \begin{cases} -\mathrm{a}/2 \le x' \le a/2 \\ -\mathrm{b}/2 \le y' \le b/2 \end{cases} \quad (3.40a)$$

$$\mathbf{J_{ms}} = 0 \quad \{\text{everywhere outside aperture} \quad (3.40b)$$

The far-fields radiated by this magnetic current distribution located in a rectangular aperture can be found by inserting definitions (3.40a) and (3.40b) into Eqs. (3.31) and (3.32) to evolve the potential function components, and hence into Eqs. (3.33) to (3.38) to generate the radiated fields. The result is:

$$E_r = 0 \quad (3.41)$$

$$E_\theta = \frac{jabk_0 E_0 \exp(-jk_0 r)}{2\pi r} \left\{ \sin\phi \left(\frac{\sin X}{X}\right)\left(\frac{\sin Y}{Y}\right) \right\} \quad (3.42)$$

$$E_\phi = \frac{jabk_0 E_0 \exp(-jk_0 r)}{2\pi r} \left\{ \cos\theta \ \cos\phi \left(\frac{\sin X}{X}\right)\left(\frac{\sin Y}{Y}\right) \right\} \quad (3.43)$$

$$H_r = 0 \quad (3.44)$$

$$H_\theta = -\frac{E_\phi}{\eta_0} \quad (3.45)$$

$$H_\phi = -\frac{E_\theta}{\eta_0} \quad (3.46)$$

X and Y in these equations are given by:

$$X = \frac{k_0 a}{2} \sin\theta \ \cos\phi \quad \text{and} \quad Y = \frac{k_0 b}{2} \sin\theta \ \sin\phi$$

The normalised E-field distributions typical of the above equations are plotted using standard mathematical software in Fig. 3.10 for a rectangular aperture with $a = 3\lambda_o$ and $b = 2\lambda_o$, and in Fig. 3.11 for a square aperture with $a = b = 3\lambda_o$. Note that in Fig. 3.10 the E-field is y-polarized (across the narrow dimension) in Fig. 3.11 it is x-polarized.

Figures 3.10 and 3.11 provide a good pictorial representation of the penetration of the aperture radiation pattern into the forward half space of the antenna, provided it is fully realised that the depicted lobe surfaces are actually constant magnitude power density contours. They could be viewed as 'iso-density' surfaces. Lower power density contours would push much further out into the half-space, while higher power densities produce surfaces closer to the antenna. Actually, experimentally this form of presentation is seldom used because the volume of measurement and the consequent time involved in generating such images is much too high to be cost effective. So, as we have intimated earlier the convention is to

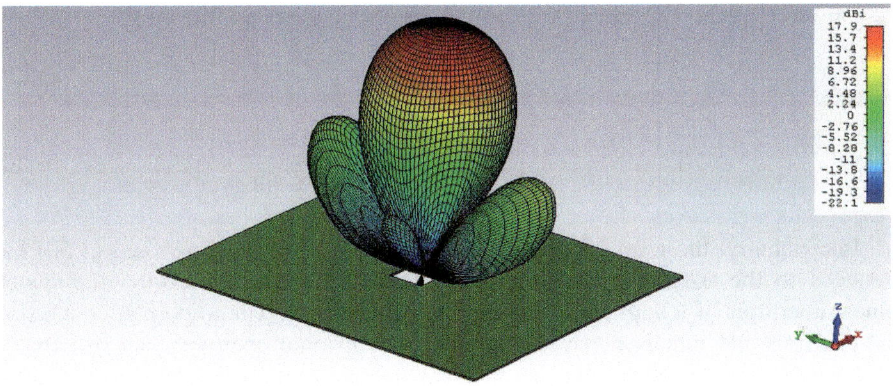

Fig. 3.10 Three-dimensional E-field pattern for a rectangular aperture ($3\lambda_o \times 2\lambda_o$) in a large conducting plane—constant E_y-directed case

Fig. 3.11 Three-dimensional E-field pattern for a square aperture ($3\lambda_o \times 3\lambda_o$) in a large conducting plane—constant E_x-directed case

present antenna patterns as two primary 'cuts'—generally the E-plane and the H-plane (see Fig. 3.10). As the figure demonstrates the E-plane for the aperture is the y-z plane or the $\phi = \pi/2$ plane. The E-field distribution on this plane is therefore:

$$E_r = E_\phi = 0 \tag{3.47}$$

$$E_\theta = \frac{jabk_0E_0\exp(-jk_0r)}{2\pi r}\left[\frac{\sin\left(\frac{k_0b}{2}\sin\theta\right)}{\frac{k_0b}{2}\sin\theta}\right] \tag{3.48}$$

On the H-plane (x-z plane or ϕ = 0 plane) the E-field has the form:

$$E_r = E_\theta = 0 \tag{3.49}$$

$$E_\varphi = \frac{jabk_0E_0 \exp(-jk_0r)}{2\pi r}\cos\theta\left[\frac{\sin\left(\frac{k_0a}{2}\sin\theta\right)}{\frac{k_0a}{2}\sin\theta}\right] \tag{3.50}$$

Interestingly, the term inside the square brackets in Eqs. (3.48) and (3.50) has reduced to the spectrum functions (see Table 3.2) of the uniformly illuminated linear apertures of length b and of length a respectively. The important conclusion here is that the radiation patterns of two dimensional apertures can usually be adequately represented on the two major planes, or 'cuts', by recruiting the spectrum functions. These functions generally provide good first approximations to mainlobe beamwidths, the locations of the sidelobe zeros and maxima, as well as the magnitudes of the sidelobes.

3.5.2 Directivity

As discussed in Sect. 3.4.2 the directivity of an antenna gives a measure of the enhanced power density at the peak of the mainlobe of an antenna of finite size relative to that of an isotropic antenna delivering the same power at the same far-field radius. If the antenna power density at the mainbeam peak at radius r is p_a (W/m^2) and the total power radiated is P_{rad} (Watts) the directivity of the antenna is defined as:

$$D = \frac{p_a}{P_{rad}/4\pi r^2} = \frac{4\pi r^2 p_a}{P_{rad}} \tag{3.51}$$

For an aperture antenna, as described above, if the electric and magnetic fields in the far-field have already been determined, as detailed in the previous section, then the radial power density flow at any angle in the forward half-space is given by:

$$\begin{aligned} U(\theta,\phi) &= \frac{1}{2}\mathrm{Re}\left[\left(\hat{\mathbf{a}}_\theta E_\theta + \hat{\mathbf{a}}_\phi E_\phi\right) \times \left(\hat{\mathbf{a}}_\theta H_\theta + \hat{\mathbf{a}}_\phi H_\phi\right)^*\right] \\ &= \frac{1}{2\eta_0}\left(\left|E_\theta\right|^2 + \left|E_\phi\right|^2\right) \end{aligned} \tag{3.52}$$

If at the 'nose' of the mainlobe $E_\theta = E_\theta^0$ and $E_\phi = E_\phi^0$ then the power flow density there we can computed from:

$$p_a = \frac{1}{2\eta_0}\left(\left|E_\theta^0\right|^2 + \left|E_\phi^0\right|^2\right) \text{ W/m}^2 \tag{3.53}$$

The total power P_{rad} is more difficult to evaluate. It involves a surface integration of the general power density term $U(\theta,\ \phi)$ over a hemisphere in the forward half-space of the radiating aperture (see for example Balanis [3]).

However, since the chosen aperture model is embedded in a plane conducting surface of infinite extent a simpler procedure exists to compute the total radiated power. If the electric field $(E_0(x', y'))$ in the aperture is available, which it usually is for known excitation methods, the power density in the aperture can be calculated. Thence the total power entering the forward half-space namely P_{rad}, can be computed by integrating the power density over the aperture area. That is:

$$P_{rad} = \frac{1}{2\eta_0}\iint\limits_S \left|E_0(x', y')\right|^2 \mathrm{d}x'\mathrm{d}y' \tag{3.54}$$

where S is the aperture surface. For the uniformly illuminated rectangular aperture in Fig. 3.9 this gives:

$$P_{rad} = ab\frac{|E_0|^2}{2\eta_0} \text{ W} \tag{3.55}$$

Also, for this aperture example, Eq. (3.53) in combination with Eqs. (3.48) and (3.50), evaluated at $\theta = 0$, yields:

$$p_a = \left(\frac{ab}{\lambda}\right)^2 \frac{|E_0|^2}{2\eta_0 r^2} \text{ W/m}^2 \tag{3.56}$$

Thus, for this particular case we obtain:

$$D = \frac{4\pi p_a}{P_{rad}} = \frac{4\pi ab}{\lambda_0^2} = \frac{4\pi A_{eff}}{\lambda_0^2} \tag{3.57}$$

where A_{eff} is the effective area of the aperture. It is worth stressing that, Eq. (3.57) is actually a very useful relationship for gauging the directivity (and hence gain) of any aperture antenna, not just for the cited example.

3.6 Chapter Summary

By taking advantage of well-established field equivalence theorems, it is demonstrated here that a duality exists between (a) time varying electric surface current density ($\mathbf{J}_s$) on a metal interface in air, which is the source electromagnetic wave radiation from 'wire' antennas, and (b) a magnetic surface current density ($\mathbf{J}_{ms}$) on a non-conducting interface, which is the source of radiation from an aperture. Consequently, it becomes possible to take advantage of a suite of analytical procedures, which have been evolved to explain and simulate wire antenna characteristics (see Chap. 2), and apply them to the related but separate area of investigation presented by aperture antennas, including slots.

From the perspective of electromagnetic theory the duality between electric current sources and magnetic current sources is emphasised in a table comparing and contrasting the fundamental equations. Furthermore, by employing Huygen's principle and plane wave spectrum techniques, the precise and uncompromising relationships between aperture size and shape, and radiation pattern characteristics, are exposed and explored. As we shall see these seemingly unyielding physical constraints on radiation pattern manipulation have major significance for compact slot based antennas.

Bibliography

1. R.E. Collin, F.J. Zucker, *Antenna Theory* (McGraw-Hill, New York, 1969)
2. R.F. Harrington, *Time Harmonic Electromagnetic Fields* (McGraw-Hill, New York, 1961)
3. C.A. Balanis, *Antenna Theory* (Wiley, New York, 1997)

Chapter 4
Computational Modelling Techniques for Slot Antennas

4.1 Introduction—Inhomogeneous Helmholtz Equation

In so far as the source present Maxwell equations (Eqs. 2.5–2.12) provide a rigorous theoretical model of electromagnetic radiation in a properly specified source present space, so the second order differential equations, repeated below, namely Eqs. (4.1) and (4.2) are rigorous representations expressed in terms of the vector potential functions $\mathbf{A}$ and $\mathbf{A}_\mathrm{m}$ defined in Chap. 2.

$$\nabla^2 \mathbf{A} + k_0^2 \mathbf{A} = -\mu_0 \mathbf{J} \tag{4.1}$$

$$\nabla^2 \mathbf{A}_m + k_0^2 \mathbf{A}_m = -\varepsilon_0 \mathbf{J}_m \tag{4.2}$$

The heuristically deduced solutions (see Chaps. 2 and 3) are:

$$\mathbf{A}(\mathbf{r}, t) = \frac{\mu_0}{4\pi} \iiint\limits_V \frac{\mathbf{J}(\mathbf{r}', t)}{|\mathbf{r} - \mathbf{r}'|} dV' \tag{4.3}$$

$$\mathbf{A}_m(\mathbf{r}, t) = \frac{\varepsilon_0}{4\pi} \iiint\limits_V \frac{\mathbf{J}_m(\mathbf{r}', t)}{|\mathbf{r} - \mathbf{r}'|} dV' \tag{4.4}$$

Although these solutions to the second order differential equations have been constructed by analogy with statics it should be stressed that the differential equations themselves remain rigorous provided that the parameters and the solution space are properly defined, as we have already emphasised.

However, for incorporation into computer simulation software, where wide applicability is a prime goal, a general approach to the solution of second order differential equations generated by electromagnetism (and other disciplines) is

A. J. Sangster, *Compact Slot Array Antennas for Wireless Communications*, Signals and Communication Technology, https://doi.org/10.1007/978-3-030-01753-8_4

required. The required versatility can be established by noting that Eqs. (4.1) and (4.2) can be rewritten in the forms:

$$[\nabla^2 + k_0^2]\mathbf{A} = -\mu_0 \mathbf{J} \tag{4.5}$$

$$[\nabla^2 + k_0^2]\mathbf{A}_m = -\varepsilon_0 \mathbf{J}_m \tag{4.6}$$

Equations of this type belong, generally, within the mathematics of linear spaces and operators (see Harrington [1]), where they take the form:

$$\mathsf{L}(\mathbf{f}) = \mathbf{g} \tag{4.7}$$

Here **L** denotes the scalar operator ($[\nabla^2 + k_0^2]$ in the above equations) while vector function **f** is the operated upon function and vector **g** represents the range of the operator domain [the source term in the inhomogeneous Eqs. (4.5) and (4.6)]. Equation (4.7) can be written in the form:

$$\mathsf{L}(\hat{\mathbf{u}}f) = \hat{\mathbf{u}}g \tag{4.8}$$

where $\hat{\mathbf{u}}$ denotes the unit vector, in the direction of **g**, which applies on both sides of the equation for a linear operator.

In Eq. (4.8), if f and g are functions of a continuous variable x, then evidently:

$$\mathsf{L}(f(x)) = g(x) \tag{4.9}$$

Also, if **L** is a linear operator, as we have already supposed, then it must [1] have an inverse $\mathbf{L}^{-1}$ which leads to an inverse form of Eq. (4.9) namely:

$$f(x) = \mathsf{L}^{-1}\, g(x) \tag{4.10}$$

For an operator **L** as defined by Eq. (4.9) it is possible to show that a function $\ell(x, x')$ termed the *kernel* function [2, 3] must exist (see Appendix C), such that:

$$g(x) = \int \ell(x, x') f(x, x') dx' \tag{4.11}$$

Of particular interest to electromagnetic modelling is the specific kernel $\delta(x, x')$ which exhibits the property that for any continuous function $e(x)$, the following very useful formulation becomes available:

$$e(x) = \int \delta(x, x') e(x') dx' \tag{4.12}$$

$\delta(x, x')$ is termed the Dirac delta function.

More importantly, this property of the Dirac function leads to one of its most significant applications, namely in Green's functions, which form a very powerful theoretical treatment of second order inhomogeneous differential equations (see Bibliography).

In general mathematical terms, Eqs. (4.9) and (4.10) respectively represent an inhomogeneous differential equation with operator **L** and its integral solution. Furthermore $g(x)$ is a continuous function of x, and x can be viewed as a vector with n components $x_1, x_2, x_3 \ldots, x_n$. The δ-function can now be employed to express the right hand side of Eq. (4.9) in the form:

$$g(x) = \int \delta(x, x')g(x')dx' \tag{4.13}$$

and recalling that L operates only on the non-primed x values we obtain in place of Eq. (4.10):

$$f(x) = \int g(x')\mathsf{L}^{-1}\delta(x' - x)dx' \tag{4.14}$$

Consequently, we can deduce that the solution to Eq. (4.9) takes the form:

$$f(x) = -\int g(x')\mathsf{G}(x, x')dx' \tag{4.15}$$

We can make this change provided the introduced term G(x, x′) is a solution of

$$\mathsf{G}(x, x') = \mathsf{L}^{-1}\delta(x' - x) \tag{4.16}$$

G(x, x′) is defined as the Green's function and is a solution to the second order inhomogeneous differential equation for which the source term is the δ-function. This may or may not be the whole solution. The right hand side of Eq. (4.16) could contain a non-singular term, for example $G_o(x, x')$, which is a solution only of the homogeneous equation, not unlike the 'constant' in conventional integration:

$$\mathsf{L}(\mathsf{G}_o) = 0 \tag{4.17}$$

Consequently Eq. (4.16) should more correctly be written:

$$\mathsf{G}(x, x') = \mathsf{L}^{-1}\delta(x' - x) + \mathsf{G}_o \tag{4.18}$$

Hence, we have:

$$\mathsf{L}\,\mathsf{G}(x, x') = \delta(x' - x) + \mathsf{L}\,\mathsf{G}_o(x, x') = \delta(x' - x) \tag{4.19}$$

The additional term $G_o(x, x')$ ensures that all the relevant boundary conditions, and initial conditions are satisfied, leading to a unique solution. In electromagnetic radiation terms this formulation can reasonably be viewed as the modelling of radiation from a 'point' emitter. For a defined space containing the point source of radio waves, G(x, x′) is the unique electromagnetic field solution which satisfies the boundary conditions.

In physical terms, Eq. (4.9) can be interpreted as expressing the relationship between the given source g and the field distribution which it generates. The Green's function expresses that part of the field at a distant point x (at some later instant t) which is generated by a unit intensity source located at the point x' (and earlier time t') within the source volume. The total field f can then be formed by integrating (see Eq. 4.10) over all source locations x' and over all time intervals over which it acts.

4.2 Common Green's Functions in Electromagnetic Radiation

In Chap. 2 it has been shown that in a bounded volume of space containing a source of radiation, the solution to Maxwell's equations leads to the formulation of an inhomogeneous scalar Helmholtz equation in the magnetic vector potentials $\mathbf{A}$, and a corresponding one $\mathbf{A}_\mathrm{m}$, with source terms representing electric current density $\mathbf{J}$ or magnetic current density $\mathbf{J}_\mathrm{m}$, respectively. It should be noted that the Helmholtz equation is not the only second order differential equation which emanates from Maxwell's equations. However, these other forms are encapsulated within the vector potential formulation. So we need examine only the Green's functions which are applicable to the Helmholtz equation. There is actually a range of such Green's functions for the solution of antenna related problems. These arise to accommodate changes in boundary conditions.

For the Helmholtz equation (Eqs. 4.5 and 4.6) we have demonstrated in the preceding section (see Eq. 4.19) that an appropriate Green's function is a solution of:

$$[\nabla^2 + k^2]\mathsf{G}(\mathbf{r}/\mathbf{r}') = -\delta(\mathbf{r}' - \mathbf{r}) \tag{4.20}$$

where $\mathbf{r}$ denotes the point (x, y, z) while $\mathbf{r}'$ denotes the point (x', y', z') as shown in Fig. 4.1.

In free space where the origin is naturally placed at zero (i.e. $\mathbf{r}' = 0$) the Green's form of the scalar Helmholtz equation becomes:

$$[\nabla^2 + k^2]\mathsf{G}(r) = -\delta(r) \tag{4.21}$$

This equation can be solved rigorously [4, 5] if the Sommerfeld radiation condition (phasor form) at infinity is applied. It is:

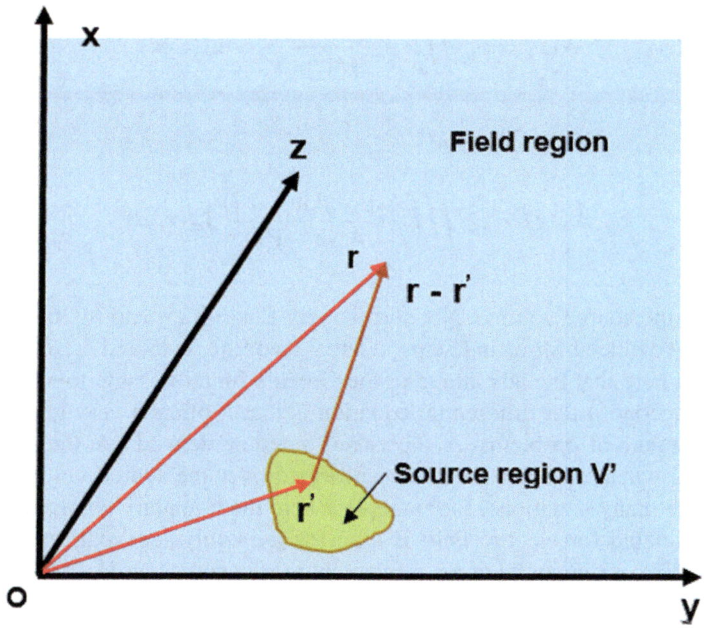

Fig. 4.1 Field and source coordinate system

$$\begin{aligned} &|r\mathsf{G}| > K \\ &\lim_{r\to\infty} r\left(\tfrac{\partial \mathsf{G}}{\partial r} + jk\mathsf{G}\right) = 0 \end{aligned} \tag{4.22}$$

The parameter K is an arbitrary constant. Actually the upper inequality is redundant since it must apply if the lower equation is enforced. The solution to Eq. (4.21) is then:

$$\mathsf{G}(r) = \frac{\exp(-jkr)}{4\pi r} \tag{4.23}$$

If the source is restored to the point $\mathbf{r}'$ in Fig. 4.1 then the scalar Green's function can be expressed more generally as:

$$\mathsf{G}(\mathbf{r}/\mathbf{r}') = \frac{\exp(-jk|\mathbf{r} - \mathbf{r}'|)}{4\pi|\mathbf{r} - \mathbf{r}'|} \tag{4.24}$$

In Eqs. (4.5) and (4.6) the right hand source term is $-\mu_o\mathbf{J}$, or $-\varepsilon_o\mathbf{J}_m$, respectively, rather than the delta function source as in Eq. (4.21). The solutions to Eqs. (4.5) and (4.6) can therefore be established by employing Eq. (4.15) to yield:

$$\mathbf{A}(\mathbf{r}) = \mu_0 \iiint_{V'} \frac{\exp(-jk|\mathbf{r}-\mathbf{r}'|)}{4\pi|\mathbf{r}-\mathbf{r}'|}\mathbf{J}(\mathbf{r}')dV' \tag{4.25}$$

or

$$\mathbf{A}_m(\mathbf{r}) = \varepsilon_0 \iiint_{V'} \frac{\exp(-jk|\mathbf{r}-\mathbf{r}'|)}{4\pi|\mathbf{r}-\mathbf{r}'|}\mathbf{J}_m(\mathbf{r}')dV' \tag{4.26}$$

where V' is the source volume. Not surprisingly Eqs. (4.25) and (4.26) replicate the heuristically deduced forms in Chaps. 2 and 3 (see Eqs. 2.45 and 3.10). It should be emphasised here that the advantage of the Green's function route toward a solution to a given second order differential equation is that it offers a very general method for a wide range of problems. A successful outcome depends on the nature of the integral and whether or not it can be performed over the source volume V'.

There are many situations, for example where the boundary conditions are more easily formulated for electric field **E** than for the equivalent magnetic vector potential. In this case it can be more convenient to proceed using Eq. (2.15) or (2.19). Equation (2.15) can be written in the form:

$$[\nabla \times \nabla \times -k^2]\mathbf{E}(\mathbf{r}) = -j\omega\mu\mathbf{J}(\mathbf{r}') \tag{4.27}$$

To solve this equation using the Green's function strategy we need an electric dyadic Green's function which is a solution of:

$$[\nabla \times \nabla \times -k^2]\overline{\mathsf{G}}_\mathsf{e}(\mathbf{r}/\mathbf{r}') = \overline{\mathsf{I}}\delta(\mathbf{r}-\mathbf{r}') \tag{4.28}$$

It is also chosen to satisfy the radiation condition. The quantity $\overline{\mathsf{I}}$ is the unit dyadic

$$\overline{\mathsf{I}} = \hat{\mathbf{a}}_\mathbf{x}\hat{\mathbf{a}}_\mathbf{x} + \hat{\mathbf{a}}_y\hat{\mathbf{a}}_y + \hat{\mathbf{a}}_z\hat{\mathbf{a}}_z \tag{4.29}$$

and $\hat{\mathbf{a}}_\mathbf{x}, \hat{\mathbf{a}}_y, \hat{\mathbf{a}}_z$, are the unit vectors in the x, y and z directions respectively.

Some standard manipulation within the rules of vector algebra [5] demonstrates that $\overline{\mathsf{G}}_\mathsf{e}$ is a solution of the vector Helmholtz equation:

$$[\nabla^2 + k^2]\overline{\mathsf{G}}_\mathsf{e}(\mathbf{r}/\mathbf{r}') = -\left(\overline{\mathsf{I}} + \frac{\nabla\nabla}{k^2}\right)\delta(\mathbf{r}-\mathbf{r}') \tag{4.30}$$

In other words, the electric dyadic Green's function $\overline{\mathsf{G}}_\mathsf{e}$ can be deduced from the scalar Green's function G, which satisfies Eq. (4.20) by operating on G with dyadic operator $\overline{\mathbf{I}} + \frac{\nabla\nabla}{k^2}$. Consequently, given that the scalar Green's function is already known (Eq. 4.24) then we must have:

$$\overline{\mathsf{G}}_{\mathsf{e}}(\mathbf{r}/\mathbf{r}') = \left(\bar{\mathsf{I}} + \frac{\nabla\nabla}{k^2}\right)\frac{\exp(-jk|\mathbf{r}-\mathbf{r}'|)}{4\pi|\mathbf{r}-\mathbf{r}'|} = \left(\bar{\mathsf{I}} + \frac{\nabla\nabla}{k^2}\right)\mathsf{G} \tag{4.31}$$

With the Green's function now known for the vector Helmholtz equation we can deduce, from Eq. (4.15), that the solution for **E** has the form:

$$\mathbf{E}(\mathbf{r}) = -j\omega\mu \iiint_{V'} \mathbf{J}(\mathbf{r}') \cdot \overline{\mathsf{G}}_{\mathsf{e}}(\mathbf{r}'/\mathbf{r})dV' \tag{4.32}$$

provided **r** is not within V'.

The inhomogeneous vector Helmholtz equation for **H** in Eq. (2.15) is similarly solvable if the requisite Green's function is available. It is a solution of:

$$[\nabla \times \nabla \times -k^2]\overline{\mathsf{G}}_{\mathsf{m}}(\mathbf{r}/\mathbf{r}') = \nabla \times [\bar{\mathsf{I}}\delta(\mathbf{r}-\mathbf{r}')] \tag{4.33}$$

where $\overline{\mathsf{G}}_{\mathsf{m}}(\mathbf{r}/\mathbf{r}')$ is termed the magnetic dyadic Green's function. Since $\overline{\mathsf{G}}_{\mathsf{e}}(\mathbf{r}/\mathbf{r}')$ and $\overline{\mathsf{G}}_{\mathsf{m}}(\mathbf{r}/\mathbf{r}')$ must generate electric and magnetic fields, respectively, which obey the Maxwell equation, it follows that

$$\nabla \times \overline{\mathsf{G}}_{\mathsf{e}} = \overline{\mathsf{G}}_{\mathsf{m}} \tag{4.34}$$

Hence, as with the electric case, we can express the free-space magnetic dyadic Greens function in terms of the scalar Green's function by combining Eqs. (4.31) and (4.34) to yield:

$$\overline{\mathsf{G}}_{\mathsf{m}} = [\nabla \mathrm{G}(\mathsf{r}/\mathbf{r}'] \times \bar{\mathsf{I}} \tag{4.35}$$

and, analogously to Eq. (4.32), **H(r)** can be determined from the equation:

$$\mathbf{H}(\mathbf{r}) = \iiint_{V'} \mathbf{J}(\mathbf{r}') \cdot \overline{\mathsf{G}}_{\mathsf{m}}(\mathbf{r}'/\mathbf{r})dV' \tag{4.36}$$

The duality between E and H in the Maxwell equations dictates that similar definitions exist for the magnetic current source $\mathbf{J}_{\mathrm{m}}$. In this case the Green's functions $\overline{\mathsf{G}}_{\mathrm{e}}$ and $\overline{\mathsf{G}}_{\mathrm{m}}$ are solutions of the equations:

$$[\nabla^2 + k^2]\overline{\mathsf{G}}_{m}(\mathbf{r}/\mathbf{r}') = -\left(\bar{\mathbf{I}} + \frac{\nabla\nabla}{k^2}\right)\delta(\mathbf{r}-\mathbf{r}') \tag{4.37}$$

$$[\nabla \times \nabla \times -k^2]\overline{\mathsf{G}}_{\mathsf{e}}(\mathbf{r}/\mathbf{r}') = \nabla \times [\bar{\mathsf{I}}\delta(\mathbf{r}-\mathbf{r}')] \tag{4.38}$$

and **E** and **H** can be deduced from:

$$\mathbf{H}(\mathbf{r}) = -j\omega\varepsilon \iiint\limits_{V'} \mathbf{J}_m(\mathbf{r}') \cdot \overline{\mathsf{G}}_m(\mathbf{r}'/\mathbf{r})dV' \tag{4.39}$$

$$\mathbf{E}(\mathbf{r}) = \iiint\limits_{V'} \mathbf{J}_m(\mathbf{r}') \cdot \overline{\mathsf{G}}_e(\mathbf{r}'/\mathbf{r})dV' \tag{4.40}$$

$\overline{\mathsf{G}}_e$ and $\overline{\mathsf{G}}_m$ can be expressed in terms of the scalar free space Green's function G, in much the same way as $\overline{\mathsf{G}}_e$ and $\overline{\mathsf{G}}_m$ in Eqs. (4.31) and (4.35). The results are:

$$\overline{\mathsf{G}}_m(\mathbf{r}/\mathbf{r}') = \left(\overline{\mathsf{I}} + \frac{\nabla\nabla}{k^2}\right)\frac{\exp(-jk|\mathbf{r}-\mathbf{r}'|)}{4\pi|\mathbf{r}-\mathbf{r}'|} = \left(\overline{\mathsf{I}} + \frac{\nabla\nabla}{k^2}\right)\mathsf{G} \tag{4.41}$$

$$\overline{\mathsf{G}}_e = [\nabla\mathsf{G}(\mathbf{r}/\mathbf{r}'] \times \overline{\mathsf{I}} \tag{4.42}$$

Table 4.1 demonstrates that for a source of electromagnetic radiation in free space Green's functions exist for solving the major inhomogeneous wave equations shown in the left hand column. For the scalar inhomogeneous Helmholtz equation the relevant Green's function is expressed in full in the top two entries in column 3. For dyadic equations listed in entries 4, 5, 6, and 7 of column 1 the appropriate Green's functions are formulated in terms of the scalar solution.

4.2.1 Green's Function Solution for Hertzian Dipole

As we have seen in Chap. 2, the essence of dipole radiation can be deduced by examining the radiation from a fictitious current element I of finite length dl (Fig. 4.2). The element is usually located at the origin of the Cartesian system and z-directed. If the current density (J A/m^2) is uniformly distributed through the element over a cross-sectional area A, then we can write:

$$Idl\hat{\mathbf{a}}_\mathbf{z} = J_z dV\hat{\mathbf{a}}_\mathbf{z} \tag{4.43}$$

where $dV = Adl$.

The relevant Helmholtz equations for this set up are presented in rows 1, 3 or 4 of Table 4.1. The most common approach is to first solve for the vector potential **A**. So applying Eq. (4.15) we obtain:

$$\mathbf{A}(\mathbf{r}) = \mu_0 \iiint\limits_{V'} \frac{\exp(-jk|\mathbf{r}-\mathbf{r}'|)}{4\pi|\mathbf{r}-\mathbf{r}'|}\mathbf{J}(\mathbf{r}')dV'$$

Table 4.1 Established Electromagnetic Green's Functions

Electromagnetic Helmholtz Equation	Green's Equation	Green's function
$\nabla^2\mathbf{A} + k_0^2\mathbf{A} = -\mu_0\mathbf{J}$	$[\nabla^2 + k^2]\mathsf{G}(\mathbf{r}/\mathbf{r}') = \delta(\mathbf{r}' - \mathbf{r})$	$\mathsf{G}(\mathbf{r}/\mathbf{r}') = \frac{\exp(-jk\|\mathbf{r}-\mathbf{r}'\|)}{4\pi\|\mathbf{r}-\mathbf{r}'\|}$
$\nabla^2\mathbf{A}_m + k_0^2\mathbf{A}_m = -\varepsilon_0\mathbf{J}_m$	$[\nabla^2 + k^2]\mathsf{G}(\mathbf{r}/\mathbf{r}') = \delta(\mathbf{r}' - \mathbf{r})$	$\mathsf{G}(\mathbf{r}/\mathbf{r}') = \frac{\exp(-jk\|\mathbf{r}-\mathbf{r}'\|)}{4\pi\|\mathbf{r}-\mathbf{r}'\|}$
$[\nabla \times \nabla \times -k^2]\mathbf{E}(\mathbf{r}) = -\mathrm{j}\omega\mu\,\mathbf{J}(\mathbf{r}')$	$[\nabla^2 + k^2]\overline{\mathsf{G}}_\mathsf{e}(\mathbf{r}/\mathbf{r}') = -(\bar{\mathsf{I}} + \frac{\nabla\nabla}{k^2})\delta(\mathbf{r} - \mathbf{r}')$	$\overline{\mathsf{G}}_\mathsf{e}(\mathbf{r}/\mathbf{r}') = (\bar{\mathsf{I}} + \frac{\nabla\nabla}{k^2})\mathsf{G}$
$[\nabla \times \nabla \times -k^2]\mathbf{H}(\mathbf{r}) = \nabla \times \mathbf{J}(\mathbf{r}')$	$[\nabla \times \nabla \times -k^2]\overline{\mathsf{G}}_\mathsf{m}(\mathbf{r}/\mathbf{r}') = \nabla \times [\bar{\mathsf{I}}\delta(\mathbf{r} - \mathbf{r}')]$	$\overline{\mathsf{G}}_\mathsf{m} = [\nabla G(\mathbf{r}/\mathbf{r}'] \times \bar{\mathsf{I}}$
$[\nabla \times \nabla \times -k^2]\mathbf{H}(\mathbf{r}) = +\mathrm{j}\omega\varepsilon\mathbf{J}_m(\mathbf{r}')$	$[\nabla^2 + k^2]\overline{\mathsf{G}}_m(\mathbf{r}/\mathbf{r}') = -(\bar{\mathsf{I}} + \frac{\nabla\nabla}{k^2})\delta(\mathbf{r} - \mathbf{r}')$	$\overline{\mathsf{G}}_m(\mathbf{r}/\mathbf{r}') = (\bar{\mathsf{I}} + \frac{\nabla\nabla}{k^2})\mathsf{G}$
$[\nabla \times \nabla \times -k^2]\mathbf{E}(\mathbf{r}) = -\nabla \times \mathbf{J}_m(\mathbf{r}')$	$[\nabla \times \nabla \times -k^2]\overline{\mathsf{G}}_\mathsf{e}(\mathbf{r}/\mathbf{r}') = \nabla \times [\bar{\mathsf{I}}\delta(\mathbf{r} - \mathbf{r}')]$	$\overline{\mathsf{G}}_\mathsf{e} = [\nabla\mathsf{G}(\mathbf{r}/\mathbf{r}'] \times \bar{\mathsf{I}}$

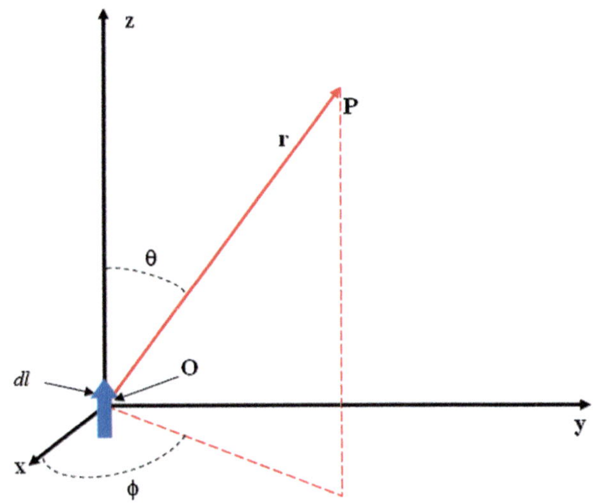

Fig. 4.2 Short current element *Idl* at origin of Cartesian system

For the short current element, with $V' \equiv dV'$ and $\mathbf{r}' = 0$, this becomes:

$$\mathbf{A}(\mathbf{r}) = \mu_0 Idl \frac{\exp(-jkr)}{4\pi r} \hat{\mathbf{a}}_\mathbf{z} \tag{4.44}$$

which compares, as it should, with the equivalent result in Chap. 2 (Eq. 2.48). Further detailed analysis of the distributions of electric and magnetic fields in the radiation pattern, including power and energy flows, would precisely parallel the discussion in Sect. 2.5 of Chap. 2, and need not be repeated here.

4.2.2 Modified Green's Functions for Electromagnetic Radiation

Clearly, an a priori knowledge of the current density distribution on the antenna of interest, as in the case of the Hertzian dipole above, (and other wire antennas radiating into free-space—see [5]), makes the Green's function method for the computation of radiated fields relatively straightforward. Needless to say, however, for real antennas in complex radiation environments solutions to the relevant inhomogeneous Helmholtz equation are generally more difficult to construct. In practice, radiation occurs in environments where conductive reflectors and scatterers can be present giving rise to imaged electric current distributions which in themselves are secondary sources of radiation. Real spaces can also contain dielectric interfaces on which secondary magnetic current densities can appear forming further sources of radiation.

Only in symmetrical geometries of largely simple layout is it possible to generate *modified* Green's functions capable of yielding direct solutions to the electromagnetic boundary value problem in a manner paralleling the Hertzian dipole example. Such modified Green's functions have been studied comprehensively in reference [4]. It encompasses, for example, the half-space, the parallel plate waveguide, the rectangular waveguide, the cylindrical waveguide, the circular cylinder in free-space, the elliptical cylinder in free-space, and the conducting sphere in free-space. Here we replicate the Green's function development for the half-space, which is arguably the most common radiation environment encountered in the electromagnetic simulation of antennas.

As we have seen above, the scalar inhomogeneous Helmholtz equation can be solved, for a free-space environment, using the Green's function G which has the form:

$$\mathsf{G}(\mathbf{r}/\mathbf{r}') = \frac{\exp(-jk|\mathbf{r}-\mathbf{r}'|)}{4\pi|\mathbf{r}-\mathbf{r}'|}.$$

In terms of G the electric dyadic Green's function can be written in the form (see Eq. (4.31)):

$$\overline{\mathsf{G}}_{\mathsf{e}}(\mathbf{r}/\mathbf{r}') = \left(\bar{\mathsf{I}} + \frac{\nabla\nabla}{k^2}\right)\mathsf{G}(\mathbf{r}/\mathbf{r}') \tag{4.45}$$

where, referring to the upper half of Fig. 4.3,

$$|\mathbf{r}-\mathbf{r}'| = \left[(x-x')^2 + (y-y')^2 + (z-z')^2\right]^{1/2} \tag{4.46}$$

Equation (4.45) applies to the upper half of Fig. 4.3, for a point source in V', as if the conducting sheet were non-existent. To accommodate its presence imaging of the source in the 'mirror', which the sheet represents, is the commonly applied technique [4]. This results in the imaged source region V_i' in the imaged Cartesian system x', y', z'. This imaged source also exhibits a Green's function contribution at the field point P, namely:

$$\overline{\mathsf{G}}_{\mathsf{e}\,i}(\mathbf{r}/\mathbf{r}_i') = \left[(-\hat{\mathbf{a}}_x\hat{\mathbf{a}}_x - \hat{\mathbf{a}}_y\hat{\mathbf{a}}_y + \hat{\mathbf{a}}_z\hat{\mathbf{a}}_z) + \frac{1}{k^2}\nabla\left(-\hat{\mathbf{a}}_x\frac{\partial}{\partial x} - \hat{\mathbf{a}}_y\frac{\partial}{\partial y} + \hat{\mathbf{a}}_z\frac{\partial}{\partial z}\right)\right]\mathsf{G}(\mathbf{r}/\mathbf{r}_i') \tag{4.47}$$

where

$$\mathsf{G}(\mathbf{r}/\mathbf{r}_i') = \frac{\exp(-jk|\mathbf{r}-\mathbf{r}_i'|)}{4\pi|\mathbf{r}-\mathbf{r}_i'|} \tag{4.48}$$

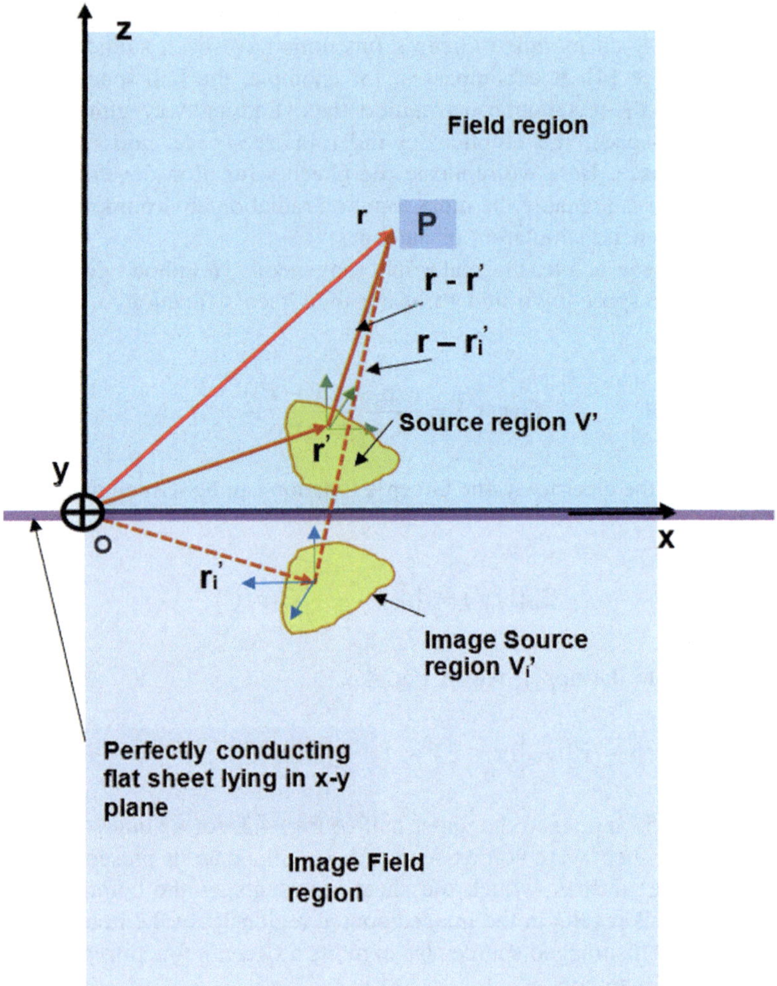

Fig. 4.3 Field and source regions imaged in an infinite and perfectly conducting ground plane

and

$$|\mathbf{r} - \mathbf{r}_i'| = \left[(x - x_i')^2 + (y - y_i')^2 + (z - z_i')^2\right]^{1/2} \tag{4.49}$$

Not unexpectedly, although it appears very different, Eq. (4.47) mirrors Eq. (4.45). The first term (in the round brackets) in Eq. (4.47) replaces the unit dyadic. Note that the green arrows in the upper half-space indicate the positive E-field directions due to the 'real' source, while the blue arrows in the lower half-space show the corresponding field directions associated with the image.

So, the elements in Eq. (4.47) have to be shown explicitly because of the sign differences associated with the imaged E-fields. In the second term of Eq. (4.47), a similar situation arises with the del operation on the field coordinates of the image-source.

Hence the Green's function for the half-space $\overline{\mathsf{G}}_{\mathsf{e}\,hs}(\mathbf{r}/\mathbf{r}')$ is merely the combination of $\overline{\mathsf{G}}_{\mathsf{e}}(\mathbf{r}/\mathbf{r}')$ and $\overline{\mathsf{G}}_{\mathsf{e}\,i}(\mathbf{r}/\mathbf{r}')$. That is:

$$\overline{\mathsf{G}}_{\mathsf{e}\,hs}(\mathbf{r}/\mathbf{r}') = \overline{\mathsf{G}}_{\mathsf{e}}(\mathbf{r}/\mathbf{r}') + \overline{\mathsf{G}}_{\mathsf{ei}}(\mathbf{r}/\mathbf{r}') \tag{4.50}$$

Finally, by resorting to the Maxwell equations, we can generate the magnetic dyadic Green's function for the half-space from:

$$\overline{\mathsf{G}}_{\mathsf{m}\,hs}(\mathbf{r}/\mathbf{r}') = \nabla \times \overline{\mathsf{G}}_{\mathsf{e}\,hs}(\mathbf{r}/\mathbf{r}') \tag{4.51}$$

The fields in the upper half-space of Fig. 4.3 can now be generated by applying Eq. (4.15), which for the electric field leads to:

$$\mathbf{E}(\mathbf{r}) = -j\omega\mu \iiint_{V'} \mathbf{J}(\mathbf{r}') \cdot \overline{\mathsf{G}}_{\mathsf{e}hs}(\mathbf{r}'/\mathbf{r})dV' \tag{4.52}$$

where $\mathbf{J}(\mathbf{r}')$ includes the imaged current density and V' incorporates V_i'. Consequently, on adopting Eq. (4.50) we obtain:

$$\mathbf{E}(\mathbf{r}) = -j\omega\mu \iiint_{V'} \mathbf{J}(\mathbf{r}') \cdot \overline{\mathsf{G}}_{\mathsf{e}}(\mathbf{r}'/\mathbf{r})dV' - j\omega\mu \iiint_{V_i'} \mathbf{J}(\mathbf{r}_i') \cdot \overline{\mathsf{G}}_{\mathsf{ei}}(\mathbf{r}_i'/\mathbf{r})dV' \tag{4.53}$$

4.2.3 *Hertzian Dipole Adjacent to Conducting Ground Plane*

If $\mathbf{J}$ were chosen to represent the current density for a z-directed short current element (current I_o and length l), located distance d in front of the perfectly conducting sheet, application of Eq. (4.53) leads, after some mathematical manipulation [4, 5], to the following scalar equation for the θ-component of the electric field at P:

$$E_\theta = \frac{j\omega\mu I_0 l}{4\pi}\left[\frac{e^{-jkr_1}}{r_1}\sin\theta_1 + \frac{e^{-jkr_2}}{r_2}\sin\theta_2\right] \tag{4.54}$$

where r_1, r_2, θ_1 and θ_2 are defined in Fig. 4.4. Similar scalar equations can be generated for E_r and E_ϕ, using:

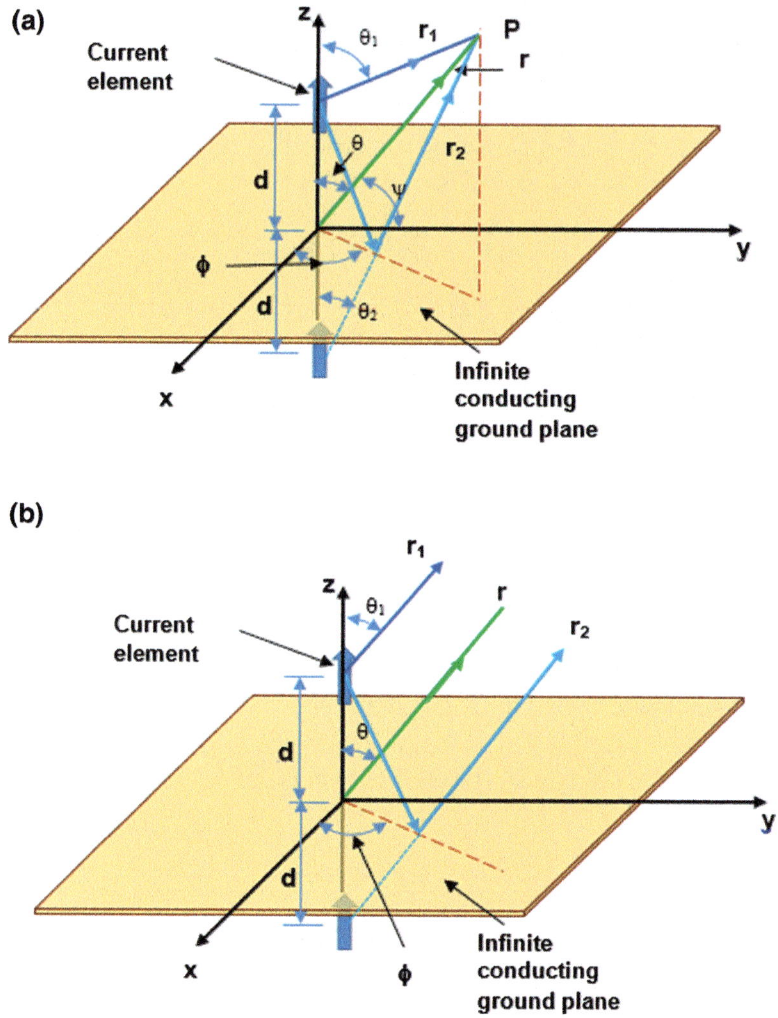

Fig. 4.4 Current element above a perfectly conducting ground plane and perpendicular to it **a** Converging rays at point P **b** rays converging at infinity

$$
\begin{aligned}
E_r &= \sin\,\theta\,\cos\,\varphi E_x \\
E_\theta &= \sin\,\theta\,\sin\,\varphi E_y \\
E_\varphi &= \cos\,\theta E_z
\end{aligned}
\tag{4.55}
$$

Hence, on applying geometrical relationships, we observe from Fig. 4.4a, that in general:

$$\begin{aligned} r_1 &= \left[r^2 + d^2 - 2rd\cos\theta_1\right]^{1/2} \\ r_2 &= \left[r^2 + d^2 - 2rd\cos(\pi - \theta_2)\right]^{1/2} \end{aligned} \tag{4.56}$$

For far field radiation simulations ($r > > d$, Fig. 4.4b), $\theta_1 = \theta = \theta_2$ and Eqs. (4.56) reduce, using binomial series, to:

$$\begin{aligned} r_1 &\approx r - d\cos\theta \\ r_2 &\approx r + d\cos\theta \end{aligned} \tag{4.57}$$

In the far field the $d\cos\theta$ term is negligible by comparison to r which implies that r_1 and r_2 are essentially parallel vectors as depicted in Fig. 4.4b. This, in turn, means that from an amplitude variations perspective we can, with little error, assume that:

$$r_1 \approx r_2 \approx r \tag{4.58}$$

Consequently, Eq. (4.54) can expressed more simply as:

$$E_\theta = \frac{j\omega\mu I_0 l}{4\pi}\frac{e^{-jkr}}{r}\sin\theta[2\cos(kd\cos\theta)] \tag{4.59}$$

for $z > 0$.

The term preceding the square brackets in Eq. (4.59) represents the E_θ field distribution for a single z-directed current element located at the origin of the coordinate system in Fig. 4.4, which is modified by an array factor (term inside square brackets—see Chap. 5) accounting for the fact that the source is at distance d from the perfectly conducting ground plane. By employing the modified free space Green's function which accommodates the presence of an infinite perfectly conducting ground plane—the half-space Green's function—the array factor in Eq. (4.59) results. Equation (4.59), normalised to a 0 dB maximum field strength, is plotted in Fig. 4.5 for four d values, namely $d = 0$, $\lambda/4$, $\lambda/2$, λ. The strong influence of the ground plane on the radiation pattern is clearly illustrated.

In summary, where a modified Green's function exists for a space containing a radiating source then a solution for the electromagnetic fields radiated into the designated volume can be constructed using Eq. (4.15). Unfortunately the range of antenna configurations which can be solved in this way is limited by the not insignificant difficulty of generating modified Green's functions for other than relatively symmetrical or regular spaces. For example Tai [4] has generated dyadic Green's functions for the following regular spaces and for volumes containing simple geometrical obstacles:

(1) parallel plate waveguide,
(2) rectangular waveguide,
(3) cylindrical waveguide,

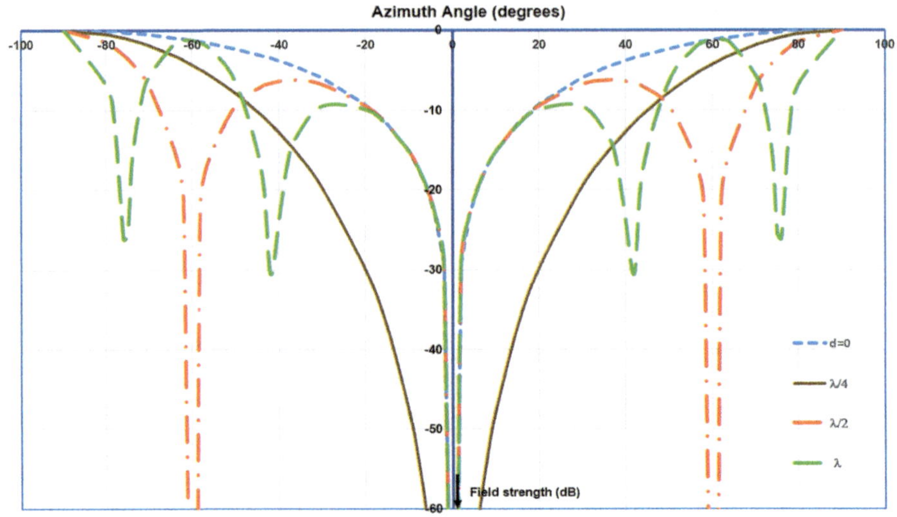

Fig. 4.5 Elevation plane field patterns in dB for a vertically orientated current element located at different heights above the perfectly conducting ground plane

(4) circular cylindrical obstacle in free space,
(5) conducting elliptical cylinder in free space,
(6) perfectly conducting wedge,
(7) the half-sheet,
(8) spheres and conducting cones.

While these examples can lead to radiation solutions which provide genuine insight into some fundamental antenna issues they are nevertheless limiting in scope. In reality, an electromagnetic antenna can comprise two or more linked regions, each of which may be represented by a Green's function, but no region separately represents a solution to the antenna problem. For example a compact slot radiator fed from waveguide can be represented by two interlinked regions namely the waveguide, which can be rectangular, cylindrical or elliptical, the exterior radiation space (usually a half-space). They are electromagnetically linked through a slot 'cavity' if the waveguide wall has finite thickness. If the coupled regions possess Green's function representations, the problem can be solved by a technique termed the moment method [1], which involves discretisation of the *slot* field and iterating toward a stable solution. It is summarised in Sect. 4.3.

We have chosen here to emphasise the moment method because generally the technique combines accuracy with computational speed which is very important in antenna array contexts where calculations for individual radiators may have to be repeated many times over as the simulation as a whole advances toward a stable solution. An alternate simulation technique termed the variational method [3] is arguably computationally swifter but it lacks the level of accuracy required by modern antenna design practice. On the other hand high modelling accuracy is provided by the finite element method (see bibliography [C]) but at the expense of computational speed arising from a very direct computational method which simply discretizes the space to be solved and then performs perhaps millions of field calculations over the finite element grid until stability is secured. This approach represents a distinct disadvantage in the context of array modelling.

4.3 Concise Formulation of the Moment Method

For linear differential equations, represented in general terms by Eq. (4.9), the method of moments (MoM) solution to this type of problem can be enunciated briefly as follows. For a linear operator **L** in an equation (e.g. equation 4.7) for which g is known, the determination of the unknown f can readily be computed as we have seen if the region supporting f possesses a Green's function (see Eq. 4.15). However if the region for which f is the solution is (say) perturbed away from a regular shape and does not exhibit a Green's function, then a solution is still possible by resorting to a discretisation of f [6]. That is we expand f in a series of functions f_1, f_2, f_3, ... defined in the domain of **L**. Therefore we express f as follows:

$$f = \sum_n \alpha_n f_n \tag{4.60}$$

where α_n are multipliers and are constants to be determined. The f_n terms are generally referred to as basis functions or expansion functions. For an exact solution to the relevant differential equation, the summation in Eq. (4.60) would normally be infinite and the f_n form a complete set (Appendix C) of basis functions. Otherwise the summation is truncated, within defined mathematical rules, to minimise computational effort.

For a linear operator, substitution of Eq. (4.45) into Eq. (4.9) yields:

$$\sum_n \alpha_n \mathsf{L}(f_n) = g \tag{4.61}$$

For a linear **L**, the relationship between f and g must also be linear in which case an inner product (denoted by $\langle f, g \rangle$) must exist. For example a common, but not unique, formulation is:

$$\langle f, g \rangle = \int_0^1 f(x) g(x) dx \tag{4.62}$$

The moment method is constructed by defining a set of testing, or weighting functions, w_1, w_2, w_3, … in the domain of **L**. So if we form inner products of each testing function w_m (m = 1, 2, 3…) with both sides of Eq. (4.55) we obtain:

$$\sum_n \alpha_n \langle w_m, \mathsf{L} f_n \rangle = \langle w_m, g \rangle \tag{4.63}$$

Equation (4.63) actually represents a set of equations which when expressed in matrix form encapsulates the essence of the moment method. That is:

$$[\mathsf{l}_{m,n}][\alpha_n] = [\mathsf{g}_m] \tag{4.64}$$

where

$$[\mathsf{l}_{m,n}] = \begin{bmatrix} \langle w_1, \mathsf{L} f_1 \rangle & \langle w_1, \mathsf{L} f_2 \rangle & ------ \\ \langle w_2, \mathsf{L} f_1 \rangle & \langle w_2, \mathsf{L} f_2 \rangle & ------ \\ -------- & -------- & ---- \\ -------- & -------- & ---- \end{bmatrix} \tag{4.65}$$

$$[\alpha_n] = \begin{bmatrix} \alpha_1 \\ \alpha_2 \\ \alpha_3 \\ \cdot \\ \cdot \\ \cdot \end{bmatrix} \tag{4.66}$$

$$[g_m] = \begin{bmatrix} \langle w_1, g \rangle \\ \langle w_2, g \rangle \\ \langle w_3, g \rangle \\ \cdot \\ \cdot \\ \cdot \end{bmatrix} \tag{4.67}$$

Generally the matrix (4.59) will be non-singular (i.e. have a non-zero determinant), and hence will be invertible. We can therefore set up an equation for the unknown α_n constants which takes the form:

$$[\alpha_n] = [\mathrm{I}_{nm}^{-1}][\mathsf{g}_m] \tag{4.68}$$

Once the α_n constants have been determined the solution for f is given by Eq. (4.45). The result may be exact, or approximate, depending on the original choice of f_n and w_n. It is relevant to note here that the common choice of equating f_n to w_n ($f_n = w_n$) is referred to as the Galerkin method [1]. Commercial software packages which have been constructed around the moment method are listed in references [7–11]. An antenna format which is particularly amenable to the MoM, and is used as a textbook example, is the dipole antenna. For this antenna, the method yields rapid and accurate solutions by employing well suited sinusoidal basis functions [1].

4.4 Chapter Summary

The moment method, which has been comprehensively evaluated in this Chapter, is shown to be an elegant mathematical technique and that it is particularly suited to the solution of a class of electromagnetic boundary value problem created by antennas and radiation. Importantly, in computational terms, the method provides a source of 'high speed' electromagnetic solvers, by offering a problem-specific approach which closely reflects the underlying physics. Furthermore, the rapid growth in the application of array antennas to new roles in mobile and satellite communications, has created design requirements for the electromagnetic modelling of arrays, and this calls for the embedding of boundary value solvers within array simulations. For these embedded MoM solvers computational speed is essential, since they may need to be accessed many thousands of times during the course of a design calculation.

Expressed in words, it can be averred that the MoM provides a solution to the problem of determining the electromagnetic fields generated by a known source located in a volume bounded by accurately defined surfaces. The mathematical processes required to achieve this outcome are summarised in this chapter. It is demonstrated that, in essence, the method relies on knowledge of a simpler electromagnetic wave solution for a comparable bounded space when excited by a unit intensity point source within it. It is shown that mathematically such a source is represented by a delta function. The solution produced is termed the Green's function for the specified region and Green's functions for a range of bounded spaces, which are relevant to antenna problems, are tabulated for easy referencing. The total field solution for a defined source distribution can then be formed by integrating the point source solution over all source locations and over all time intervals over which it acts. Further, it is shown that the linearity of the resultant integral equations permits discretisation of the relevant functions to form matrix representations which yield readily to computer processing. This procedure is embellished further in Chap. 5.

References

1. R.F. Harrington, *Field Computation by Moment Methods* (MacMillan, New York, 1968)
2. P.H. Morse, H. Feshbach, *Methods of Theoretical Physics* (McGraw-Hill Book Co., New York, 1953)

3. L. Cairo, T. Kahan, *Variational Techniques in Electromagnetism* (Blackie & Sons Ltd., London, 1965)
4. Chen-To Tai, *Dyadic Green Functions in Electromagnetic Theory* (IEEE Press, New Jersey, 1994)
5. R.E. Collin, F.J. Zucker, *Antenna Theory, Part 1* (McGraw-Hill Book Co., New York, 1969)
6. R.F. Harrington, Matrix methods for field problems. Proc. IEEE **55**(2), 136–149 (1967)
7. Wave 3D, CEM Works, Winnipeg, Manitoba, Canada
8. HyperLynx 3D EM, Mentor Graphics, Fremont, California, USA
9. Antenna VLab, EM Cos, Tbilisi, Georgia
10. MMANA-GAL, DL1PBD & DL2KQ Partnership, Sebaststrasse 8, 53115 Bonn, Germany
11. R.C. Hansen (ed.), *Moment Methods in Antennas and Scattering* (Artech House Inc., London, 1990)

Bibliography

1. R.E. Collin, *Field Theory of Guided Waves* (McGraw-Hill Book Co., New York, 1960)
2. R.F. Harrington, *Time-Harmonic Electromagnetic Fields* (McGraw-Hill Book Co., New York, 1961)
3. J.-M. Jin, *Finite Element Methods in Electromagnetics* (Wiley, New York, 2014)

Chapter 5
Moment Method Models of Compact Slot Antennas

5.1 Introduction

In Chap. 4 it has already been noted that for open, or closed, spaces which support time varying electromagnetic fields, excited by sources which can readily be expressed mathematically, then the electromagnetic boundary value problem presented can be solved without undue difficulty if the Green's function for the space in question is available. Unfortunately, this is seldom the reality, particularly in antenna developments. The typical problem which is encountered tends to comprise two or more interconnected regions each capable of supporting electromagnetic waves. A solution to such a problem requires that the fields in each region satisfy the local boundary conditions and simultaneously the boundary conditions at interfaces. An example of a problem of this type is illustrated in Fig. 5.1a. It comprises a narrow and inclined slot located in the upper broadwall of the rectangular waveguide (internal dimension; width a; height b and wall thickness T). The waveguide walls are generally assumed to be perfectly conducting. Figure 5.1b shows the slot detail (length L; width W) and the local slot coordinate system (ξ, η, ζ).

Electromagnetically the antenna structure represented by Fig. 5.1a comprises two regions; the free-space exterior to the waveguide (above the slot and usually modelled as a half-space), and the waveguide region (usually modelled as infinitely long in the z-direction). These regions are coupled electromagnetically through the slot. The computationally most efficient, and accurate, procedure for solving this problem, which is not untypical of a wide range of compact radiator geometries, is the moment method. Other such geometries include slots fed from transmission lines such as stripline, microstrip line, coplanar waveguide, but also slot fed patches, slot modified patches, cavity backed slots and combinations of all of these. Many will be addressed in later chapters.

Other powerful electromagnetic solvers based on the finite element and finite difference techniques can undoubtedly also be used to simulate compact slot radiators. But these techniques, while they are in principle unlimited in their

A. J. Sangster, *Compact Slot Array Antennas for Wireless Communications*, Signals and Communication Technology, https://doi.org/10.1007/978-3-030-01753-8_5

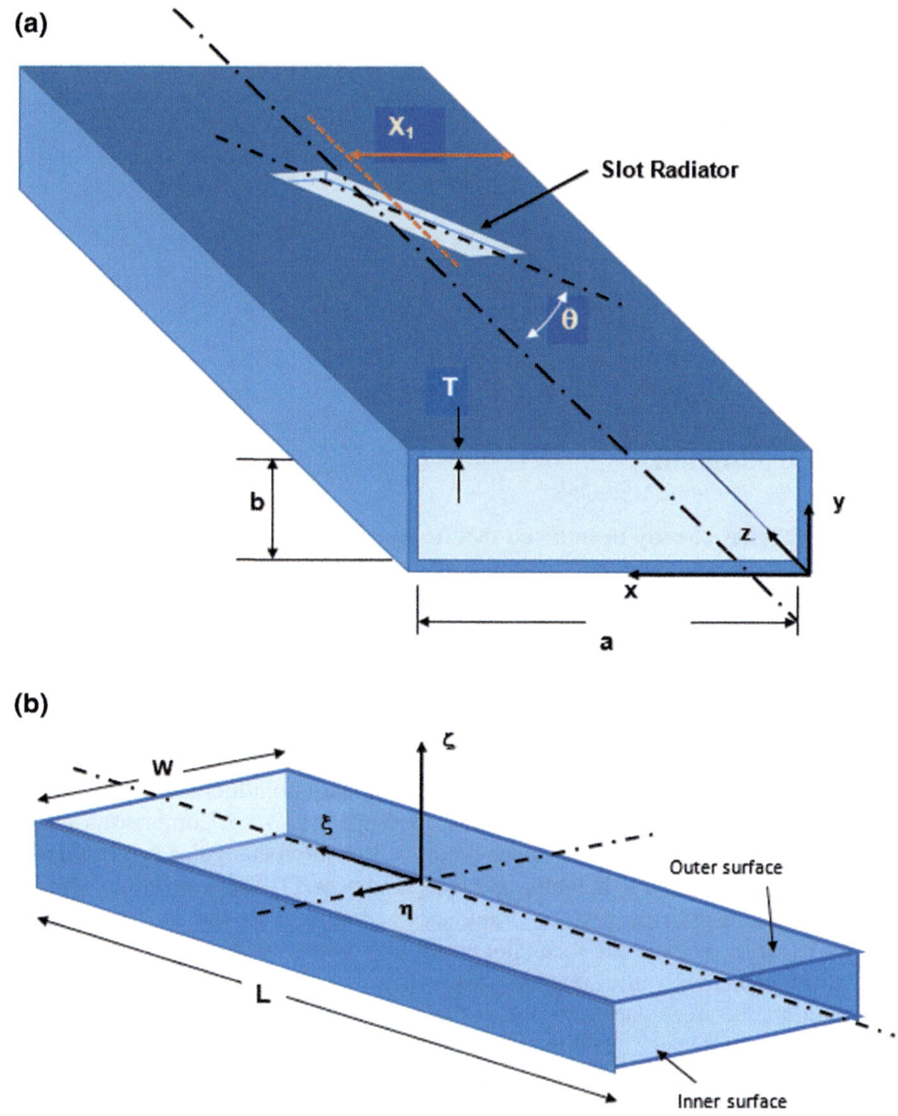

Fig. 5.1 **a** Waveguide fed compact slot radiator and **b** slot detail

applications, exhibit a consequential 'downside' of very large memory requirements and relatively 'slow' computational speeds. While very accurate simulations are possible, these methods, unlike the moment method, take no a priori account of model geometry and are as a result much more computationally demanding. This can represent a major disadvantage in the context of, for example, slot antenna array development [1]. In synthesising such an array to achieve target radiation pattern specifications each slot in the array, and there may be over one thousand, has to be accessed and adjusted many times. This means a slot simulation has to be entered repeatedly and clearly the computational speed offered by the moment method generates design advantages in this scenario. Recent developments in the use of artificial neural networks to facilitate the optimisation of compact slot antennas [2] involves 'training' and 'testing' of the neural system by repeated reference to the compact slot-antenna model. So again we encounter a design situation where we require efficient electromagnetic simulations to avoid excessive, or time consuming, computational effort.

5.2 Moment Method Solution

The essence of the moment method is summarised mathematically in Chap. 4, Sect. 4.3. Here we will apply the technique outlined there to the slot geometry presented in Fig. 5.1a. It is clear from a cursory examination of Fig. 5.1a that the primary coupled regions, namely the waveguide (region a) and the external half-space (region c) are regularly shaped spaces thus possessing Green's function solutions if excited from a point source. However, in the geometry shown, the behaviour of the fields in the slot (region b) has to be determined before the scattered fields in the waveguide and the radiated fields in the external space can be computed. Note that in the absence of the slot (i.e. if the slot were closed by a perfectly conduction insert) then the waveguide would support only an incident wave travelling from a remote matched source at one or other of its distant terminations. This wave can generally be presumed to be represented by the fundamental TE_{10} mode. On the other hand, if the slot is open it becomes a source of fields scattered into the waveguide and fields radiated into the half-space. In theory these scattered and radiated fields could be computed using a form of Eq. (4.15) if the instantaneous field distributions on the slot surfaces S and S′ were fully known. While approximate estimates of the slot field distribution for a very narrow slot in a 'zero thickness' waveguide wall have been reported in some early papers [3, 4], the field magnitude still remains to be computed for the wide range of possible slot locations, orientations, or shapes, which differ from those in the original papers. To secure a solution in the general case of a waveguide slot radiator, it is necessary to resort to discretizing the slot fields on both surfaces. Whereupon, on application of an iterative technique such as the moment method, a search for slot field values is progressed until a steady state solution to the boundary value problem is established. If the problem is properly posed, electromagnetically, then this iterative process should converge on the unique solution.

In formulating a moment method solution for any given slot radiator geometry, a number of conditions must be incorporated into the model, if a valid solution is to be generated. Firstly, and obviously the simulated fields must satisfy the Maxwell equations. Secondly, the external radiated fields must satisfy the radiation condition at infinite distance from the radiator. Thirdly, it is essential that at the inner and outer surfaces of the slot (S and S′) that field continuity conditions are applied. The first and second of these three conditions are immediately enforced by directing the analysis toward a solution of the relevant inhomogeneous Helmholtz equations for E and H (Eqs. 2.19 and 2.20) and by forming the relevant Green's functions for the coupled regions (a) and (c). Any solution constructed around the Green's functions ensures that the eventual field solutions emanating from the analysis satisfy both the electromagnetic boundary conditions on all conducting surfaces, and the radiation condition at infinity in region (c). Consequently, the key to formulating a set of equations which correctly represent the radiation problem depicted in Fig. 5.1a lies in proper establishment of the field boundary conditions at those slot surfaces which connect regions (a), (b) and (c).

The analysis begins by identifying the electromagnetic fields which must exist in each region in Fig. 5.1a. We have already noted that if the slot is 'closed' by a fictitious perfectly conducting insert a single incident wave (TE_{10} mode) continues to propagate through the waveguide. The fields of this wave are defined as $\mathbf{E}_i$, $\mathbf{H}_i$. When the slot is 'open', and capable of interfering with the wall currents of the TE_{10} mode, it becomes a discontinuity in the waveguide scattering fields back into it. These fields we will term $\mathbf{E}_a$, $\mathbf{H}_a$. So, in region (a) the total fields are $\mathbf{E}_i + \mathbf{E}_a$, $\mathbf{H}_i + \mathbf{H}_a$. The discontinuity fields on the surface S also scatters fields $\mathbf{E}_{b1}$, $\mathbf{H}_{b1}$ into the slot cavity thus generating further components $\mathbf{E}_{b2}$, $\mathbf{H}_{b2}$ on surface S′. These are the source of radiation into region (c) expressed through $\mathbf{E}_c$, $\mathbf{H}_c$. In order to complete the electromagnetic modelling process on the structure of Fig. 5.1a it is necessary to ensure that the above fields satisfy the boundary conditions on both surfaces S and S′. For aperture problems (see Chap. 3) the discontinuity fields on these surfaces are the electric field components tangential to the relevant surface. Consequently, the boundary conditions can be formulated as:

$$\hat{\mathbf{n}} \times \mathbf{E_a} = -\hat{\mathbf{n}} \times \mathbf{E_{b1}} \quad \text{on S} \tag{5.1}$$

$$\hat{\mathbf{n}} \times \mathbf{E_{b2}} = -\hat{\mathbf{n}} \times \mathbf{E_c} \quad \text{on S}' \tag{5.2}$$

Note that on the left side of Eq. (5.1) the unit vector $\hat{\mathbf{n}}$ points into the waveguide whereas on the right it is directed into the slot cavity. In Eq. (5.2) the opposite is true. Also $\mathbf{E}_i$ exhibits a zero tangential component on the wall containing the slot and so does not appear in the field continuity equations.

To simplify the process of representing this problem mathematically, it is advantageous here to employ the field equivalence principles outlined in Chap. 3 (see Figs. 3.2 and 3.3). This principle, firstly, allows us to replace the open slot surfaces S and S′ by an 'electric current wall' (i.e. a perfectly conducting membrane shown as chain-dotted line in Fig. 5.2), which means that the regions (a), (b) and (c) become respectively a classical closed rectangular waveguide, a perfect rectangular cavity and a classical half-space. All three regions now possess Green's function

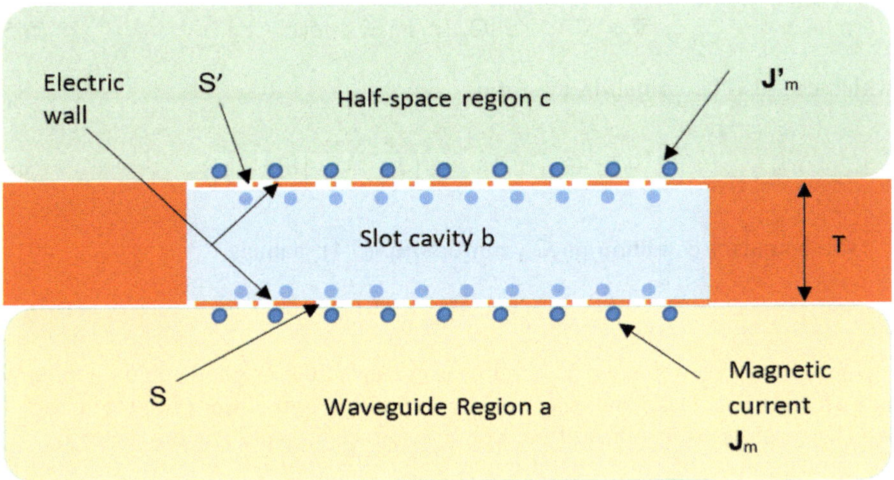

Fig. 5.2 Field equivalence representation of slot radiator in thick walled waveguide

representations as required by the moment method. Secondly, the representation is completed with each region being excited by magnetic current densities $\mathbf{J_m}$, $\mathbf{J_m}$' on S and on S′. These replace the aperture fields **E** and **E**' on S and S′ in the original problem. Thus we have:

$$\begin{aligned}
\mathbf{J_m} &= \hat{\mathbf{n}} \times \mathbf{E} \quad \text{on S in region a} \\
\mathbf{J_m} &= -\hat{\mathbf{n}} \times \mathbf{E} \quad \text{on S in region b} \\
\mathbf{J'_m} &= \hat{\mathbf{n}}' \times \mathbf{E}' \quad \text{on S}' \text{ in region b} \\
\mathbf{J'_m} &= -\hat{\mathbf{n}}' \times \mathbf{E}' \quad \text{on S}' \text{ in region c}
\end{aligned}$$

This theoretical model is illustrated in Fig. 5.2. Each region is now separately excited by a magnetic current and can be represented mathematically by the inhomogeneous Helmholtz equation rewritten below [see Eq. (2.19) and Table 4.1].

$$[\nabla \times \nabla \times \; -k^2]\mathbf{H}(\mathbf{r}) = \; +j\omega\varepsilon\mathbf{J}_m(\mathbf{r}') \tag{5.3}$$

with zero tangential electric field on the conducting surfaces, which can be expressed through the equation:

$$\hat{\mathbf{n}} \times \nabla \times \mathbf{H}(\mathbf{r}) = 0 \tag{5.4}$$

on the metal surfaces. The appropriate Green's equation for each region (see Table 4.1) is:

$$[\nabla \times \nabla \times -k^2]\bar{\mathsf{G}}_m(\mathbf{r}/\mathbf{r}') = -\bar{\mathsf{I}}\delta(\mathbf{r}-\mathbf{r}') \tag{5.5}$$

which can also be written in the form

$$[\nabla^2 + k^2]\bar{\mathsf{G}}_m(\mathbf{r}/\mathbf{r}') = -\left(\bar{\mathsf{I}} + \frac{\nabla\nabla}{k^2}\right)\delta(\mathbf{r}-\mathbf{r}') \tag{5.6}$$

The boundary condition on $\bar{\mathsf{G}}_m$ mirrors Eq. (5.4); namely

$$\hat{\mathbf{n}} \times \nabla \times \bar{\mathsf{G}}_m(\mathbf{r}/\mathbf{r}') = 0 \tag{5.7}$$

Once the as yet unknown $\mathbf{J}_m$ (or $\mathbf{J'}_m$) is evaluated the magnetic field distribution in each region is calculable using an equation of the form of Eq. (5.8) (see Chap. 4) which is expressed in terms of an arbitrary surface S_0 and location vector $\mathbf{r}_0$:

$$\mathbf{H}(\mathbf{r}) = j\omega\varepsilon \iint_{S_o} \bar{\mathsf{G}}_m(\mathbf{r}/\mathbf{r}_0) \cdot \mathbf{J}_m(\mathbf{r}_0) dS_0 \tag{5.8}$$

In order to determine the variables $\mathbf{J}_m$ and $\mathbf{J'}_m$ we need to formulate a pair of simultaneous equations for these two unknowns. These can be constructed by applying the boundary conditions for the magnetic fields $\mathbf{H}$ at the slot surfaces supporting $\mathbf{J}_m$ and $\mathbf{J'}_m$, namely S and S′.

The formulation leads to the following:

On surface S, which supports the magnetic current density $\mathbf{J}_m$, magnetic field continuity yields:

$$\begin{aligned} \hat{\mathbf{n}} \times \mathbf{H}_i(\mathbf{r}) &= j\omega\varepsilon_0 \hat{\mathbf{n}} \times \iint_{\mathbf{S}} \left[\bar{\mathsf{G}}_{ma}(\mathbf{r}/\mathbf{r}_0) + \bar{\mathsf{G}}_{mb}(\mathbf{r}/\mathbf{r}_0)\right] \cdot \mathbf{J}_\mathbf{m}(\mathbf{r_0}) dS_0 \\ &+ j\omega\varepsilon_0 \hat{\mathbf{n}} \times \iint_{\mathbf{S}} \left[\bar{\mathsf{G}}_{mb}(\mathbf{r}/\mathbf{r}_0)\right] \cdot \mathbf{J'}_\mathbf{m}(\mathbf{r_0}) dS_0 \quad \text{for } \mathbf{r} \text{ on } S \end{aligned} \tag{5.9}$$

While, on surface S′ which supports the magnetic current density $\mathbf{J'}_m$ the continuity condition produces:

$$\begin{aligned} 0 &= j\omega\varepsilon_0 \hat{\mathbf{n}} \times \iint_{\mathbf{S'}} \left[\bar{\mathsf{G}}_{mb}(\mathbf{r}/\mathbf{r}_0) + \bar{\mathsf{G}}_{mc}(\mathbf{r}/\mathbf{r}_0)\right] \cdot \mathbf{J'}_\mathbf{m}(\mathbf{r_0}) dS_0 \\ &+ j\omega\varepsilon_0 \hat{\mathbf{n}} \times \iint_{\mathbf{S'}} \left[\bar{\mathsf{G}}_{mb}(\mathbf{r}/\mathbf{r}_0)\right] \cdot \mathbf{J}_\mathbf{m}(\mathbf{r_0}) dS_0 \quad \text{for } \mathbf{r} \text{ on } S' \end{aligned} \tag{5.10}$$

In Eqs. (5.9) and (5.10) $\bar{\mathsf{G}}_{ma}$, $\bar{\mathsf{G}}_{mb}$, and $\bar{\mathsf{G}}_{mc}$ represent the dyadic Green's functions for regions (a), (b) and (c) respectively. Similar simultaneous equations are discernible in reference [5]. The relative complexity of Eqs. (5.9) and (5.10) make direct solution unfeasible. Consequently, a numerical method designed for the task has to be adopted; such a method is the moment method developed in the previous chapter [6, 7].

Electromagnetic radiation from a narrow rectangular slot fed from waveguide is, as we have already intimated, not a new problem. It has been studied widely for use as a compact radiating element in linear and planar array antennas. In today's world of burgeoning mobile communications, 'compactness' has become more critical than array applications, and the waveguide feed arrangement has inevitably been superseded by spatially constrained alternatives. These will be addressed in a later Chapter. Nevertheless, the waveguide fed slot provides a cogent example of how modelling by means of the moment method is developed in slot radiation problems.

Some of the original theoretic work on slot antennas, most notably by Oliner [3] and Das [4], has demonstrated that the electric field distribution along the slot length is well modelled by a single trigonometric function. Examples, generally evolved through the application of variational techniques, are shown in Fig. 5.3. The indicated differences are largely attributed to the focus on near resonant slots by Oliner, and on marginally non-resonant slots by Das. This precedent suggests that there could be potentially significant advantage, from a computational speed perspective, in adopting basis functions for the discretized slot field with trigonometric properties. Mathematically this leads to the following constructions (see Eq. 4.45):

$$\hat{\mathbf{n}} \times \mathbf{E} = \hat{\mathbf{n}} \times \sum_{s=1}^{N} \alpha_s f_s \hat{\mathbf{a}}_\eta \quad \text{on S} \tag{5.11}$$

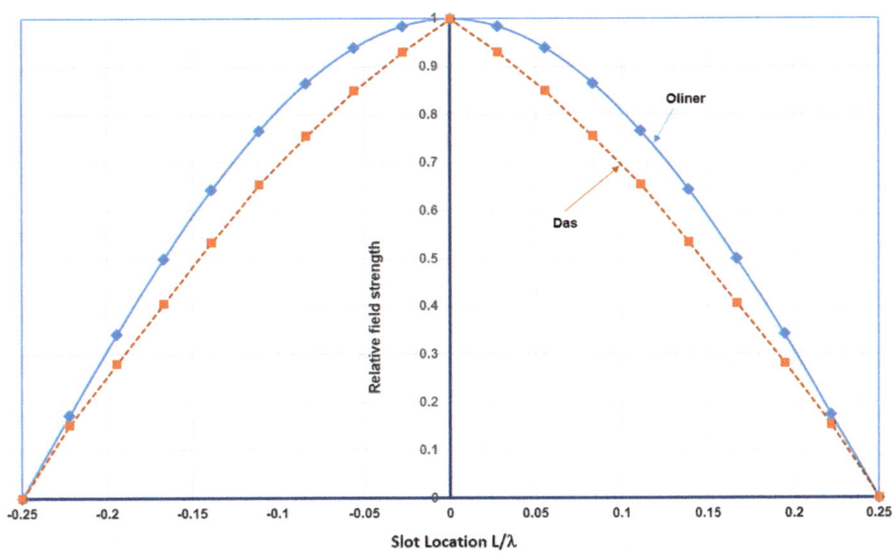

Fig. 5.3 Slot field approximations for a narrow (w/L < 0.1), near resonant slot (L/λ = 0.5) in the broadwall (T = 0) of rectangular waveguide, as adopted in early radiation computations [Solid line with diamond symbols = Oliner: Dashed line with square symbols = Das]

$$\hat{\mathbf{n}}' \times \mathbf{E}' = \hat{\mathbf{n}}' \times \sum_{s=1}^{N} \alpha'_s f_s \hat{\mathbf{a}}_\eta \quad \text{on S}' \tag{5.12}$$

where, referring to Fig. 5.1b:

$$f_s(\xi) = \sin \frac{s\pi(\xi + L/2)}{L} \qquad s = 1, 2, 3, \ldots, N \tag{5.13}$$

Following the procedures outlined in Chap. 4, Sect. 4.3, which are developed in more detail in [6], we choose testing functions [Eq. (5.14)] which mirror the basis functions, namely the Galerkin method

$$f_i(\xi)\hat{\mathbf{a}}_\eta \quad i = 1, 2, 3, \ldots \ldots, N \tag{5.14}$$

In order to generate a matrix representation of the simultaneous equation. This has the form:

$$\begin{bmatrix} [A] & [C] \\ [D] & [B] \end{bmatrix} \begin{bmatrix} [\alpha] \\ [\alpha'] \end{bmatrix} = \begin{bmatrix} [h] \\ [0] \end{bmatrix} \tag{5.15}$$

where:

$$A_{is} = -j\omega\varepsilon_0 \iint_S f_i \hat{\mathbf{a}}_\eta \cdot \{\hat{\mathbf{n}} \times \iint_{\mathbf{S}} \left[\bar{\mathsf{G}}_{ma}(\mathbf{r}/\mathbf{r}_0) + \bar{\mathsf{G}}_{mb}(\mathbf{r}/\mathbf{r}_0)\right] \cdot f_s \hat{\mathbf{a}}_\xi dS_0\} dS \tag{5.16a}$$

$$B_{is} = -j\omega\varepsilon_0 \iint_{S'} f'_i \hat{\mathbf{a}}_\eta \cdot \{\hat{\mathbf{n}}' \times \iint_{\mathbf{S}'} \left[\bar{\mathsf{G}}_{mb}(\mathbf{r}/\mathbf{r}_0) + \bar{\mathsf{G}}_{mc}(\mathbf{r}/\mathbf{r}_0)\right] \cdot f'_s \hat{\mathbf{a}}_\xi dS_0\} dS \tag{5.16b}$$

$$C_{is} = j\omega\varepsilon_0 \iint_S f'_i \hat{\mathbf{a}}_\eta \cdot \{\hat{\mathbf{n}} \times \iint_{\mathbf{S}'} \left[\bar{\mathsf{G}}_{mb}(\mathbf{r}/\mathbf{r}_0)\right] \cdot f_s \hat{\mathbf{a}}_\xi dS_0\} dS \tag{5.16c}$$

$$D_{is} = j\omega\varepsilon_0 \iint_{S'} f_i \hat{\mathbf{a}}_\eta \cdot \{\hat{\mathbf{n}}' \times \iint_{\mathbf{S}} \left[\bar{\mathsf{G}}_{mb}(\mathbf{r}/\mathbf{r}_0)\right] \cdot f'_s \hat{\mathbf{a}}_\xi dS_0\} dS \tag{5.16d}$$

$$h_i = -\iint_S f_i \hat{\mathbf{a}}_\eta \cdot [\hat{\mathbf{n}} \times \mathbf{H}_i(\mathbf{r})] dS \tag{5.16e}$$

The matrix entries A_{is}, B_{is}, C_{is}, D_{is} given mathematical expression in Eqs. (5.16a, 5.16b, 5.16c, 5.16d, 5.16e) are essentially electromagnetic coupling terms representing to the mutual coupling which occurs between the assumed current distributions ($f_{i,s}$) on the interior and external surfaces of the radiating aperture. In the particular case of the conventional slot (in Fig. 5.1) they can be ascribed the following specific meanings:

A_{is} represents self-coupling between magnetic current elements on the inner slot surface S.

B_{is} represents self-coupling between magnetic current elements on the outer slot surface S′.

C_{is} represents mutual coupling to the inner slot surface (S) due to magnetic current elements on the outer surface (S′).

D_{is} represents mutual coupling to the outer slot surface (S′) due to magnetic current elements on the inner surface (S).

Once the integrals in Eqs. (5.16a, 5.16b, 5.16c, 5.16d, 5.16e) have been evaluated, which is mathematically routine but tedious [8, 9], the matrix Eq. (5.15) is amenable to solution by a range of matrix manipulation tools. These can be found in standard mathematical textbooks [10, 11] and in convenient to use computer software packages [12]. The outcome of the process is plausible values for s, f_s, α'_s and α_s. These can then be used to compute the scattering parameters for the compact radiator as outlined below.

5.3 Scattering for Longitudinal Slot in Waveguide

The general broadwall inclined slot radiator depicted in Fig. 5.1a can be resolved into a longitudinal shunt slot when $\theta = 0°$, and into a transverse series slot when $\theta = 90°$. In fact mathematically it is more sensible to do this than attempt to solve the inclined slot example directly [9]. In practice the inclined broadwall slot has very few applications and most studies have been directed at modelling the series slot and the shunt slot. Below we revisit some validating results for the shunt slot, which is by far the most commonly used in array applications.

A cursory examination of the matrix entries in Eq. (5.16a, 5.16b, 5.16c, 5.16d, 5.16e) quickly establishes that the controlling matrix in Eq. (5.15) grows in size as the square of the number of basis functions s, i. The more basis functions selected, therefore, the slower will be the progress toward a computed solution. For conventional slots, at least, the advantage of the use of trigonometric basis functions is that accurate solutions are available for no more than three basis functions. This is demonstrated in Fig. 5.4 where the slot field distribution for a longitudinal slot in rectangular waveguide as computed by the moment method employing three basis functions (solid curve—round symbols), is compared with the predictions of Oliner (diamond symbols) and Das (square symbols) for the same geometry and discussed earlier. The moment method solution converges on a slot field distribution, which is approximately cosinusoidal as predicted by Oliner, but with a rather broader, less peaky, shape than the pure cosine. This modified shape is also predicted by Vu Khac (solid line—triangular symbols) employing a moment method constructed around thirty ‘pulse’ type basis functions [13]. The Das distribution with its sharp discontinuous peak can no longer be justified.

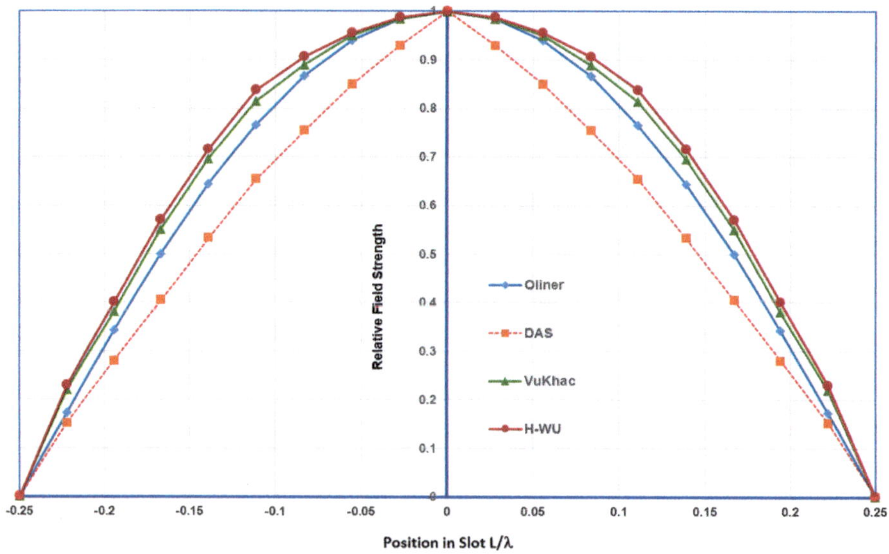

Fig. 5.4 E-field distributions for a shunt slot radiator in the broadwall of WG16 waveguide as predicted by the moment method (freq = 9.375 GHz, L = 0.361 λ, x_1 = 6.43 mm, w = 1.58 mm, T = 1.27 mm)

The key to determining slot scattering lies with Eq. (5.8) which provides the link to the fields generated by the slot discontinuity once $\mathbf{J}_m$ and $\mathbf{J}'_m$ are known. These become calculable [Eqs. (5.11) and (5.12)] once α_s and α_s' are evaluated through the moment method computation. Given that in employing the equivalence theorem, $\mathbf{J}_m$ and $\mathbf{J}'_m$ are positioned at the slot locations, but on 'unbroken' waveguide walls, the required waveguide Greens function for calculating internal scattering is the 'sum of modes' formulation [9] (see Appendix C). However, real power in the transmission and reflection signals within the waveguide is carried by the TE_{10} mode. Consequently we can write for the transmitted magnetic field component of the TE_{10} mode $\mathbf{H}_{10t}$:

$$\mathbf{H_{10t}}(\mathbf{r}) = \mathbf{H_{10i}} - \frac{1}{2}\sum_{s=1}^{N} \alpha_s \mathbf{H}_{10}^{+}(\mathbf{r}) \iint_S \mathbf{H}_{10}^{-}(\mathbf{r}) \cdot f_s \hat{\mathbf{a}}_\xi dS_0 \tag{5.17}$$

where $\mathbf{H}_{10}^{\pm}$ are the forward and backward propagating components of the magnetic field for the TE_{10} mode in the rectangular waveguide, while $\mathbf{H}_{10i}$ denotes the incident mode in the waveguide. The reflected field component $\mathbf{H}_{10r}$ can be calculated from:

$$\mathbf{H}_{10r}(\mathbf{r}) = -\frac{1}{2}\sum_{s=1}^{N} \alpha_s \mathbf{H}_{10}^{-}(\mathbf{r}) \iint_S \mathbf{H}_{10}^{+}(\mathbf{r}) \cdot f_s \hat{\mathbf{a}}_\xi dS_0 \tag{5.18}$$

The derivation of closed form, computer programming friendly, algorithms from Eqs. (5.17) and (5.18) is presented in Appendix C. It results in equations for the parameters of primary interest for compact antennas with 'known' radiation patterns, namely the reflection coefficient within the waveguide S_{11}, the transmission coefficient within the waveguide S_{12} and the relative power radiated into the half-space p_r/p_{in}. Typical results are graphed and discussed in the next section.

5.3.1 Computed and Measured Results

The moment method program formulated using trigonometric basis functions has been tested exhaustively [8, 9] to assess its sensitivity to the number of basis functions employed. For the computation of slot radiation, and slot scattering parameters, the sensitivity is demonstrated to be low, and it is safe to say that one basis function is generally satisfactory for modelling conventional waveguide slot radiators. This is demonstrated in Fig. 5.5 where the relative power radiated, as a function of slot length, for a conventional longitudinal slot in WG16 waveguide, as predicted by the moment method, is presented. Computations were performed for up to five basis functions. On the scale adopted in the graph there was no discernible difference between three or more basis functions, while the single basis

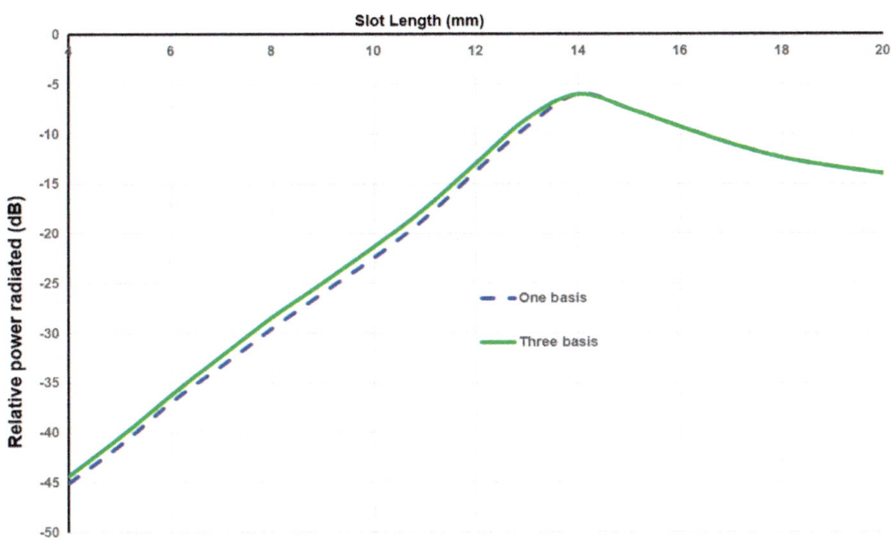

Fig. 5.5 Relative power radiated by a broadwall waveguide slot as a function of slot length L, with the number of basis functions as parameter (Freq = 10 GHz, a = 22.86 mm, b = 10.16 mm, x_1 = 2.00 mm, w = 1.5 mm, T = 0 mm)

function prediction is marginally lower than the three-basis case only for below resonance slot lengths. This is because close to resonance the slot internal field distribution is well represented by a simple cosine, and hence one basis function is all that is needed to generate accurate results. Needless to say this has had major significance for the speed and efficiency of array antenna simulation incorporating the method of moments. For example, in a multi-slot array simulation where the MoM is embedded into an iterative program designed to perform pattern synthesis [1], the MoM for each slot has to be accessed repeatedly, and computational speed becomes critical.

To emphasise the point further, the trigonometric formulation with three basis functions has been compared with the Vu Khac moment method [13] employing 30 pulse basis functions. The comparison was carried out on a longitudinal slot in WG16 waveguide at a frequency of 9.375 GHz. The WG16 had a wall thickness of 1.27 mm and the shunt slot, with w = 1.58 mm, was off-set from the sidewall of the waveguide by 1.43 mm. The results are presented in Figs. 5.6 and 5.7 which display reflection coefficient magnitude $|S_{11}|$ and transmission coefficient magnitude $|S_{12}|$ as a function of slot length L.

It is clear from Figs. 5.6 and 5.7 that the trigonometric moment method employing three basis functions computes values for $|S_{11}|$ and $|S_{12}|$ which are in very good agreement—largely within the computational errors—with the pulse basis moment method. Deviations outwith the computational error level only occur

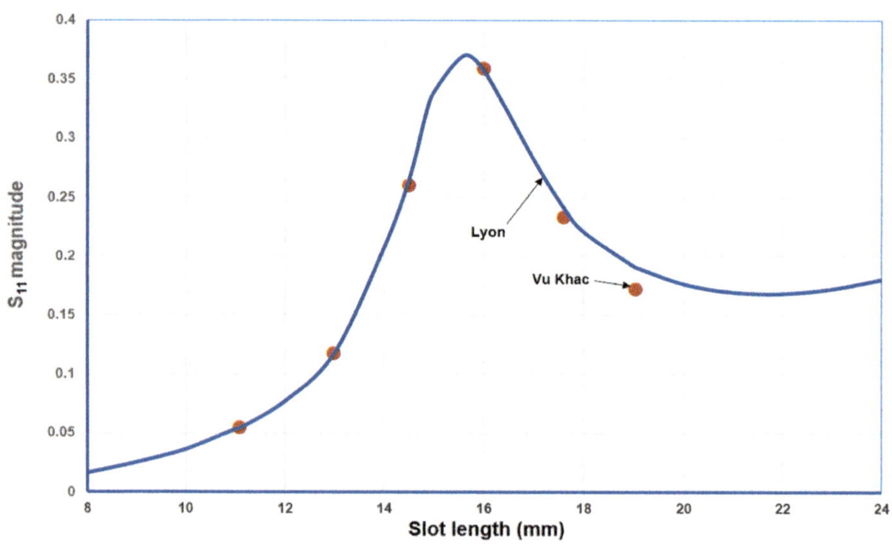

Fig. 5.6 Reflection coefficient magnitude ($|s_{11}|$) as a function of slot length as predicted by trigonometric basis (Lyon) and by pulse basis (Vu Khac) moment methods (WG16, Freq = 9.375 GHz, T = 1.27 mm, x_1 = 1.43 mm, w = 1.58 mm)

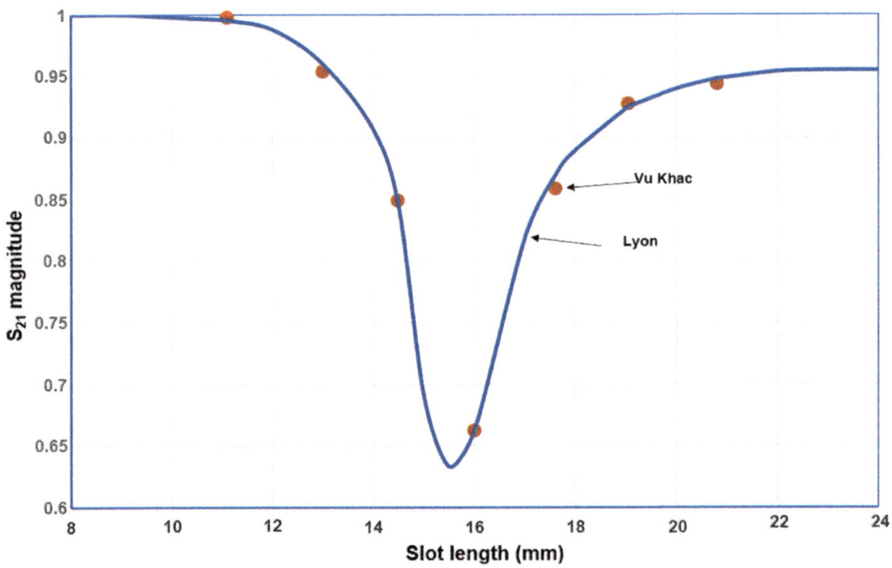

Fig. 5.7 Transmission coefficient magnitude ($|s_{12}|$) as a function of slot length as predicted by trigonometric basis (Lyon) and by pulse basis (Vu Khac) moment methods (WG16, Freq = 9.375 GHz, T = 1.27 mm, x_1 = 1.43 mm, w = 1.58 mm)

well above the resonant slot length (L ~ 15.5 mm). At such slot lengths, resonances of higher order than the half-cosine shape depicted in Fig. 5.4 begin to influence the calculations and it becomes necessary to include more than three basis functions used to generate Figs. 5.6 and 5.7. Phase computations, namely$/\underline{S}_{11}$ and $/\underline{S}_{12}$, have also been compared [8], and yield largely similar levels of agreement.

To complete the validation exercise on the trigonometric moment method, computed results have also been compared with measured characteristics for a typical longitudinal slot radiator in WG16 waveguide. Typical results for S_{12} magnitude and phase are presented in Fig. 5.8a, b. The measurements were procured by the application of a cavity technique [8]. This involves resonating the waveguide in a section containing the slot, which then becomes a detuning discontinuity, then measuring changes in resonant frequency and quality factor (Q) as a function of slot length. While the level of agreement is generally good, and certainly within the measurement errors, the displacement of the measured resonant frequency to a longer slot length than predicted by theory, was initially puzzling. But this was found to be almost wholly attributable to the fact that measurements were performed on round-ended slots for practical fabrication reasons. The theory assumed square-ended slots.

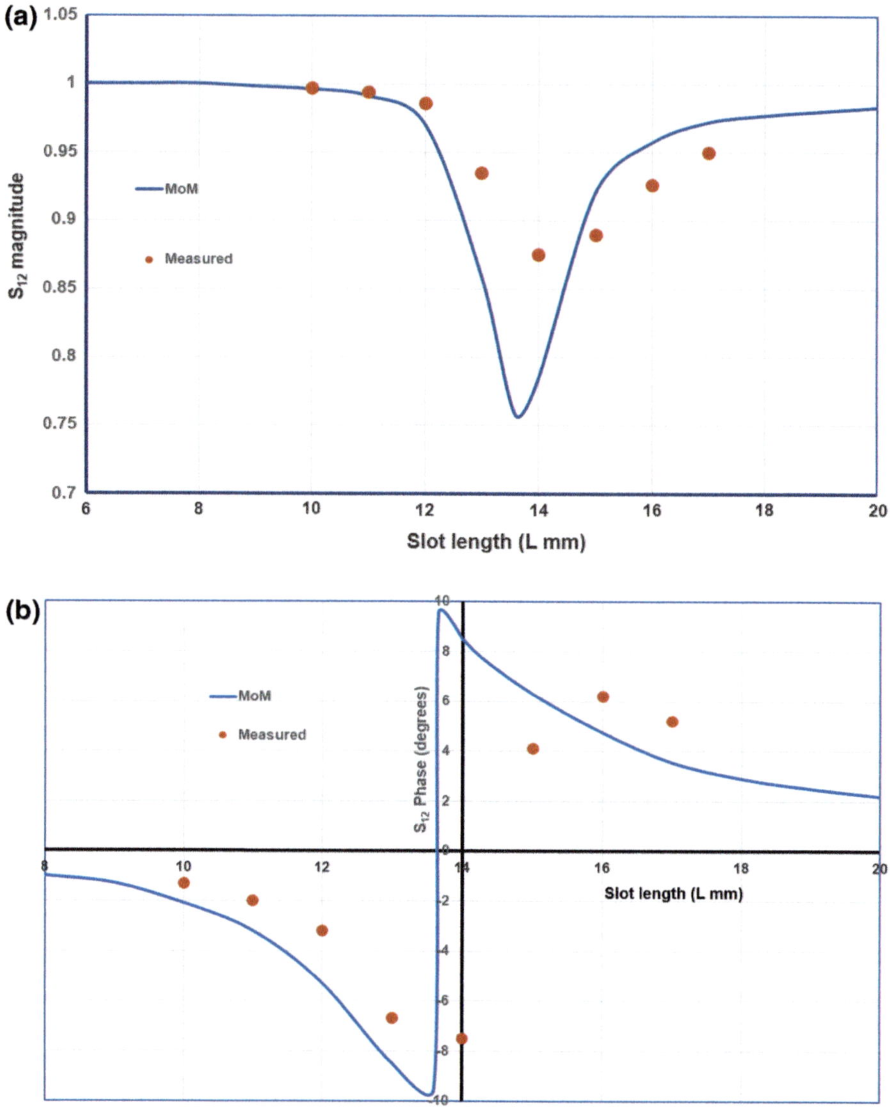

Fig. 5.8 **a** Transmission coefficient magnitude for a rectangular waveguide containing a longitudinal slot of changing length (Freq = 10.48 GHz, T = 1.32 mm, x_1 = 2.00 mm, w = 1.5 mm) **b** Transmission coefficient phase for a rectangular waveguide containing a longitudinal slot of changing length (Freq = 10.48 GHz, T = 1.32 mm, x_1 = 2.00 mm, w = 1.5 mm)

5.4 T-Shaped Slot in Bifurcated Waveguide

A more complex slot example, comprising a polarisation agile, T-shaped, slot fed from bifurcated waveguide has been included here to show that with insightful application of boundary conditions, and with particular attention to the slot field distribution and the basis function choice, the trigonometric moment method is of sufficient flexibility to establish accurate modelling tools for a wide range of compact slot geometries. The arrangement, which is depicted in Fig. 5.9, was devised originally to secure polarisation agility by independently exciting a pair of orthogonally aligned slots (see Chap. 1, Fig. 1.22) in dual-mode waveguide. Thus the slot pair is arranged to radiate axial or transverse linear polarisations if excited separately using mode switching, or if excited simultaneously and in phase quadrature, circularly polarised radiation is produced [14]. While the longitudinal, or shunt, slot portion of the orthogonal slot pair has been thoroughly examined in the literature, as outlined in Sect. 5.3 above, a transverse slot straddling the central

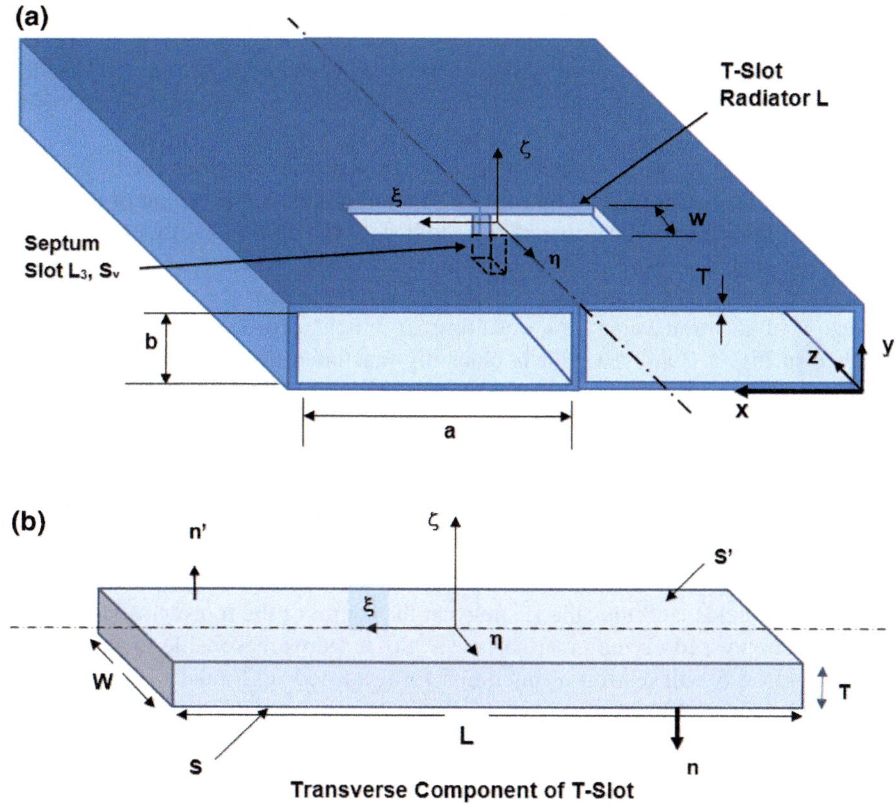

Fig. 5.9 **a** Transverse T-slot radiator in dual-mode rectangular waveguide. **b** Transverse element of T-slot showing coordinate system

wall of a bifurcated wall is relatively novel [15] and requires advanced electromagnetic modelling to establish its radiation characteristics. We revisit this case here to demonstrate the potential that exists in the moment method for addressing unconventional applications such as those encountered in evolving novel compact slot antenna geometries examined in ensuing chapters.

The matrix entries A_{is}, B_{is}, C_{is}, D_{is} defined in Eqs. (5.16a, 5.16b, 5.16c, 5.16d, 5.16e), have been evolved for a moment method formulation of the general case of electromagnetic power transfer from a transmission system (waveguide) through a slot cavity to a half-space where radiation occurs (Fig. 5.2). In the previous section (Sect. 5.3) it has been applied comprehensively to the well established example of a shunt slot in waveguide. We will now apply these Eqs. (5.16a, 5.16b, 5.16c, 5.16d, 5.16e) to the switchable T-slot radiator on dual mode waveguide, as depicted in Fig. 5.9.

From a mathematical perspective the fundamental difference between the symmetrically located T-shaped slot in bifurcated waveguide (Fig. 5.9a) and the conventional slot in regular waveguide shown in Fig. 5.1a, lies with the clearance slot which attracts additional mutual coupling between the inner surface S of the transverse slot and the septum slot surfaces S_ν. The computation of these additional coupling terms requires that a suitable trial magnetic field distribution can be formulated for the T-slot which encompasses all three surfaces S, S' and S_ν. Once this is accomplished the preliminary formulation of the moment method solution for the T-slot proceeds much as outlined in Sect. 5.3.

Consider, firstly, the tangential electric field distribution which the waveguide modal fields induce on the inner surface S of the transverse component of the T-slot (Fig. 5.9b). The functional form of this field will obviously be influenced by the septum slot and its depth L_3. For example if $L_3 \approx 0$ then the top of the waveguide bifurcation will suppress any induced electric field (E_η) near the centre of the inner surface S of the transverse slot, resulting in a field distribution not unlike that depicted in Fig. 5.10a. That is, it is plausibly presumed that when the septum slot approaches zero length the E_η field distribution on the inner surface S of the transverse slot is essentially the same as that of two slots, of length $L/2$, positioned end-to-end. Note that this is not necessarily true for the outer slot surface S' particularly if the waveguide wall thickness is not infinitesimally small. On the other hand, when the septum slot depth is finite, it becomes less easy to predict possible field distributions on the surfaces S and S_ν. However, from qualitative considerations it seems evident that the clearance slot will be parasitically excited by the radiating slot fields, and thus the E_η fields at the centre of the transverse slot will no longer be suppressed on the inner surface S. So, it seems reasonable to suggest that the E_η field on S will exhibit a functional form not too dissimilar to Fig. 5.10b.

With the above observations in mind we can proceed to the construction of suitable basis functions for the fields on S and on S_ν by noting the analogy between the E-field forms in the T-slot to the voltage profiles on an equivalent T-network of transmission lines [16]. Suppose that the transmission line depicted in Fig. 5.11a represents a conventional slot positioned wholly within the broadwall of a regular rectangular waveguide feed. The input impedance at the right hand extremity of the

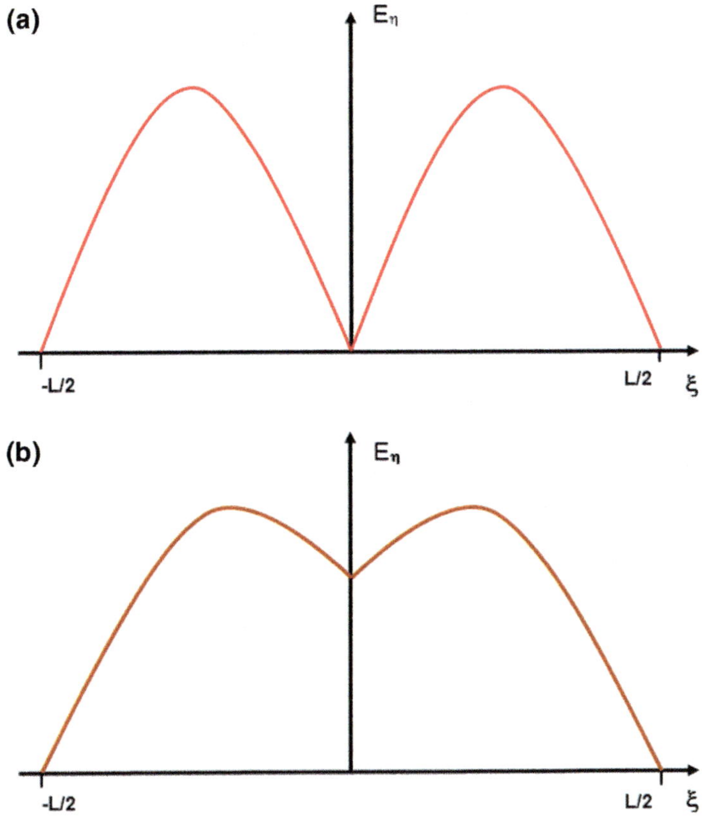

Fig. 5.10 Suggested electric field distributions in the transverse component of the T-slot **a** $L_3 = 0$, **b** $0 < L_3 < L/2$

normalised line, for which $Z_0 = 1$, can, from transmission line theory, be expressed as:

$$Z_{in} = j \tan \beta L \tag{5.17}$$

where $\beta = 2\pi/\lambda$ and λ is the line wavelength at the frequency of operation. However, with a short located at $x = L$ conditions must exist enforcing $Z_{in} = 0$. This in turn dictates that

$$\tan \beta L = 0 \tag{5.18}$$

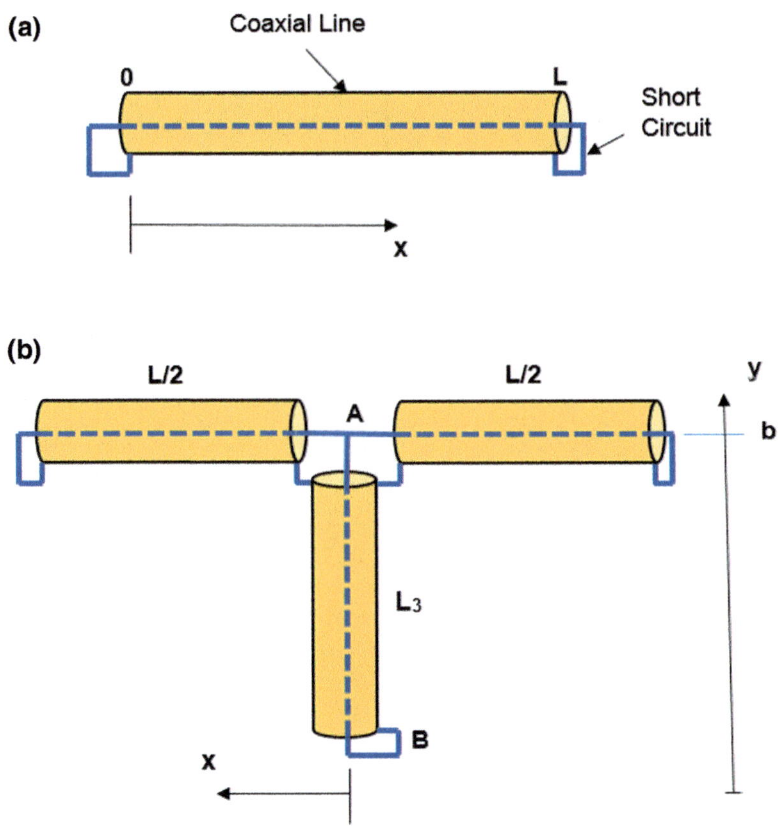

Fig. 5.11 Transmission line based equivalent circuits for waveguide slot radiators **a** conventional slot, **b** T-slot in bifurcated waveguide

leading to the result that

$$\beta = \frac{n\pi}{L} \quad (\mathrm{n} = 1, 2, 3\ldots) \tag{5.19}$$

The implication of this is that the line voltage (or current) forms a standing wave with the functionality

$$V = V_0 \sin\frac{n\pi x}{L} \tag{5.20}$$

By analogy the slot voltage $(V = E \times w)$ will have this same functionality, as will the E-field in the slot. This is precisely the trigonometric basis function form which was used successfully in Sect. 5.3.

In Fig. 5.11b we extend this idea to the T-slot of Fig. 5.9 with the aim of constructing a plausible basis function for this more complex slot. The transmission line model comprises two transmission lines of length *L*/2 representing the transverse slot portions on either side of the central bifurcation and a line, L_3 in length, representing the clearance notch. The three lines are connected in parallel at A, and all three are terminated in short circuits. From transmission line theory the input impedance at A, for the paralleled L/2 lines, is given by:

$$Z_A = \frac{j}{2}\tan\frac{\beta L}{2} \tag{5.21}$$

for normalised line impedances (Z_0 = 1). Hence, with Z_A as the load, the impedance at B can be expressed as:

$$Z_B = \frac{\frac{1}{2}\tan(\beta L/2) + j\tan(\beta L_3)}{1 - \frac{1}{2}\tan(\beta L/2)\tan(\beta L_3)} \tag{5.22}$$

But Z_B must be zero for the shorted line, which implies:

$$\tan(\beta L/2) = -2\tan(\beta L_3) \tag{5.23}$$

This transcendental equation can be solved to give n discrete values for β, in sympathy with the conventional case. However, given that we have already observed in Sect. 5.3 that the s = 1 basis function on its own can procure good results, up to the first resonance, for the conventional slot radiator, it seems sensible to follow the same route with the T-slot. Consequently, only the first root of Eq. (5.23) needs to be retained in the following moment method solution [15]. This root (β_1) can be determined quite straightforwardly by incorporating a *regula falsi* [10] procedure within the moment method analysis. It follows immediately that suitable basis functions on the inner surface *S*, which divides into two *S*/2 components, of the transverse slot are:

$$f_s = \sin\beta_1\left(\frac{L}{2} - |x|\right) = \sin\beta_1\left(\frac{L}{2} - |\xi|\right) \tag{5.24}$$

and preserving continuity with the conventional slot analysis we choose to employ testing functions (f_i) for the S/2 surfaces such that ($f_i = f_s$). On the outer surface of the transverse slot (*S*'), where the waveguide bifurcating wall is less influential with T > 0, the basis and testing ($f'_s = f'_i$) functions are unchanged from those used in Sect. 5.3. That is:

$$f'_s = \sin\frac{s\pi(\xi - \frac{L}{2})}{L} \tag{5.25}$$

with s = 1 for the fundamental term.

From the transmission line model in Fig. 5.11b it is quite clear that voltage across the two transverse arms will have the mathematical form:

$$V = \sin \beta_1 \left(\frac{L}{2} - |x| \right) \tag{5.26}$$

mirroring the trial field or basis function on S [Eq. (5.24)], while on the clearance or septum slot surfaces S_v Fig. 5.11b suggests that here the basis function form will be:

$$V = V' \sin \beta_1 [L_3 - (b - y)] \tag{5.27}$$

Since these voltages (or equivalent E-fields in slot terms) must equate at the junction A then we can write:

$$\sin \frac{\beta_1 L}{2} = V' \sin \beta_1 L_3 \tag{5.28}$$

Hence, by analogy, an appropriate basis function for the field forms on the S_v surfaces is:

$$f_{sv} = \frac{\sin(\frac{\beta_1 L}{2})}{\sin(\beta_1 L_3)} \sin \beta_1 (y - b + L_3) \tag{5.29}$$

With the trial field expressions now defined through Eqs. (5.24), (5.25) and (5.29) for the relevant surfaces of the T-slot, the analysis can proceed in a manner not too different from the development in Sect. 5.2. Commencing with Eqs. (5.9) and (5.10), which are generally applicable to slot radiators, a matrix equation for the T-slot can be formed with the following entries [see Eq. (5.15)]:

$$\begin{aligned} A_{is} = & -2j\omega\varepsilon_0 \iint\limits_{S/2} f_i \hat{\mathbf{a}}_\eta \cdot \{\hat{\mathbf{n}} \times \iint\limits_{S/2} [\bar{\mathbf{G}}_{ma}(\mathbf{r}/\mathbf{r}_0)] \cdot f_s \hat{\mathbf{a}}_\xi dS_0\} dS \\ & - j\omega\varepsilon_0 \iint\limits_{S} f_i \hat{\mathbf{a}}_\eta \cdot \{\hat{\mathbf{n}} \times \iint\limits_{\mathbf{S}} [\bar{\mathbf{G}}_{mb}(\mathbf{r}/\mathbf{r}_0)] \cdot f_s \hat{\mathbf{a}}_\xi dS_0\} dS \\ & - 2j\omega\varepsilon_0 \iint\limits_{S/2} f_i \hat{\mathbf{a}}_\eta \cdot \{\hat{\mathbf{n}} \times \iint\limits_{\mathbf{S}_v} [\bar{\mathbf{G}}_{ma}(\mathbf{r}/\mathbf{r}_0)] \cdot f_{sv} \hat{\mathbf{a}}_\xi dS_0\} dS \end{aligned} \tag{5.30a}$$

$$B_{is} = -j\omega\varepsilon_0 \iint\limits_{S'} f_i' \hat{\mathbf{a}}_\eta \cdot \{\hat{\mathbf{n}}' \times \iint\limits_{\mathbf{S}'} [\bar{\mathbf{G}}_{mb}(\mathbf{r}/\mathbf{r}_0) + \bar{\mathbf{G}}_{mc}(\mathbf{r}/\mathbf{r}_0)] \cdot f_i' \hat{\mathbf{a}}_\xi dS_0\} dS \tag{5.30b}$$

$$C_{is} = j\omega\varepsilon_0 \iint\limits_{S} f_i' \hat{\mathbf{a}}_\eta \cdot \{\hat{\mathbf{n}} \times \iint\limits_{\mathbf{S}'} [\bar{\mathbf{G}}_{mb}(\mathbf{r}/\mathbf{r}_0)] \cdot f_s \hat{\mathbf{a}}_\xi dS_0\} dS \tag{5.30c}$$

$$D_{is} = j\omega\varepsilon_0 \iint\limits_{S'} f_i \hat{\mathbf{a}}_\eta \cdot \{\hat{\mathbf{n}}' \times \iint\limits_{\mathbf{S}} [\bar{\mathbf{G}}_{mb}(\mathbf{r}/\mathbf{r}_0)] \cdot f_s' \hat{\mathbf{a}}_\xi dS_0\} dS \tag{5.30d}$$

$$h_i = -\iint_S f_i \hat{\mathbf{a}}_\eta \cdot [\hat{\mathbf{n}} \times \mathbf{H}_i(\mathbf{r})] dS \tag{5.30e}$$

Except for the A_{is} term, the above integral expressions are formally identical to the corresponding conventional slot versions. They differ only in the assumed functional form for the basis function f_i and the testing function f_s. On the other hand, A_{is} contains an additional term, relative to the conventional analysis, to accommodate the mutual coupling which can occur between the inner surface of the transverse slot and the septum slot surfaces. The Green's functions $\bar{\mathsf{G}}_{ma}, \bar{\mathsf{G}}_{mb}$, and $\bar{\mathsf{G}}_{mc}$ are essentially the same as for the conventional slot analysis, in Sect. 5.2, except that for $\bar{\mathsf{G}}_{ma}$ which must be treated in two parts with in-phase dominant modes on either side of the bifurcation. This is done by coordinate system adjustments. The previously defined rectangular slot Green's function $\bar{\mathsf{G}}_{mb}$ now applies to the transverse component of the T-slot. It is important to note that in the computation of B_{is}, C_{is}, and D_{is} higher order slot cavity modes will require to be included, more so than in the conventional slot solution, to accommodate basis functions such as f_s in Eq. (5.24) which will tend to generate such modes. The integral expressions delineated in Eqs. (5.16a, 5.16b, 5.16c, 5.16d, 5.16e) have been fully evaluated in references [8] and [9] and some typical results have been presented in Sect. 5.3. Needless to say the evaluation of Eqs. (5.30a, 5.30b, 5.30c, 5.30d, 5.30e) are not too different and are detailed in [8]. It should be noted here that the moment method computations have generally been performed for the $s = 1$ basis function on the inner slot surface S, and for the first basis function on the outer surface S' except where L_v is short, in which case the s = 3 basis function is introduced.

Below we present a selection of representative results to demonstrate the effectiveness of the moment method in modelling more complex and compact antenna designs. As with the longitudinal slot example in Sect. 5.2, once the matrix equation, with the entries detailed in Eq. (5.30a, 5.30b, 5.30c, 5.30d, 5.30e), has been solved by moment method interaction, plausible values for s, α_s, f_s, f_s', and f_v should become available. In a process little different to that adopted for calculating the longitudinal slot scattering values, those for the transverse T-slot can be similarly calculated. The scattering formulae are detailed in Appendix C.

5.4.1 Computed and Measured Results

The proof that a new engineering analysis, or that an analysis of a novel engineering advance, is mathematically and technically sound lies either by showing comparison with the results of other researchers, or by comparing the theory with measurements on carefully prepared models, as outlined in Sect. 5.3. Here we follow this measurement route, since at the time when the T-slot in bifurcated waveguide was proposed [15], no earlier results by other investigators were available.

Measurements were performed on a dual-mode waveguide fabricated from two lengths of standard WG16 (22.86 mm × 10.16 mm) brazed together along carefully machined sidewalls such that the common wall, or bifurcation, was 1.5 mm in thickness. The T-slot radiators were formed by machining dual-mode waveguide samples. Slots ranging in length from a minimum case with L = 10 mm and L_3 = 0.0 mm, and increased incrementally in approximately 1 mm length steps up to slots with L = 20 mm and L_3 = 5.5 mm.

Normally, slot discontinuities in waveguide can be measured and characterised using network analysis procedures. These can be very accurate if 'de-embedding' [17], of the unavoidable coaxial line to waveguide junctions, is undertaken carefully. However, for the dual-mode waveguide feed (Fig. 5.9) in addition to any coax-to-waveguide transitions which may be necessary, transitions from conventional waveguide to dual-mode waveguide are also required, resulting in an overall transition arrangement which is too difficult to de-embed. This was certainly true when the T-slot radiator was under development. A solution which was found to be accurate for below resonance slots involved resonating the device under test, including its conventional-to-dual waveguide transition, and performing measurements of the resonant frequency and the quality factor (Q) of the system as a whole [17, 18]. The arrangement is depicted schematically in Fig. 5.12.

In the figure the device under test (DUT)—namely the dual-mode waveguide feed supporting a T-slot radiator—was sandwiched between a pair of folded 'magic-tee' waveguide couplers. These components provide wideband even-mode (direct port) and odd-mode (orthogonal port) excitation of the bifurcated waveguide. Only the even-mode port was used for T-slot measurements. The odd-mode port can be, and has been, used for the characterisation of parallel shunt slot radiators, such as those illustrated in Fig. 1.22. At the even-mode port of each of the 'magic tees' a precision machined iris was attached thus forming a waveguide resonator. The irises were designed to be as identical as possible and each was accurately calibrated for its transmission coefficient as a function of frequency. The system was driven by a crystal controlled microwave source through a calibrated precision attenuator and a precision multi-hole waveguide sampling coupler (10 dB coupling). A similar coupler was also introduced into the output port of the system with the transmitted power subsequently dissipated in a well matched waveguide load. The signals emanating from the sampling ports of the two couplers were fed via diode detectors to a standard microwave display, which was controlled from the source. The measurement of cavity resonant frequencies and the cavity Q's for a broad range of T-slot geometries is relatively routine and has been implemented. Furthermore, the results have been converted into reflection coefficient (magnitude and phase) and transmission coefficient (magnitude and phase) by standard procedures [19]. The results are summarised in Figs. 5.13, 5.14, 5.15 and 5.16.

In Fig. 5.13 the power radiated by a 16 mm long T-slot with a 5.48 mm clearance slot is plotted as a function of frequency for both the moment method simulation and measured results. Equation (C.25) has been used to convert the reflection and transmission results into power radiated. The figure demonstrates,

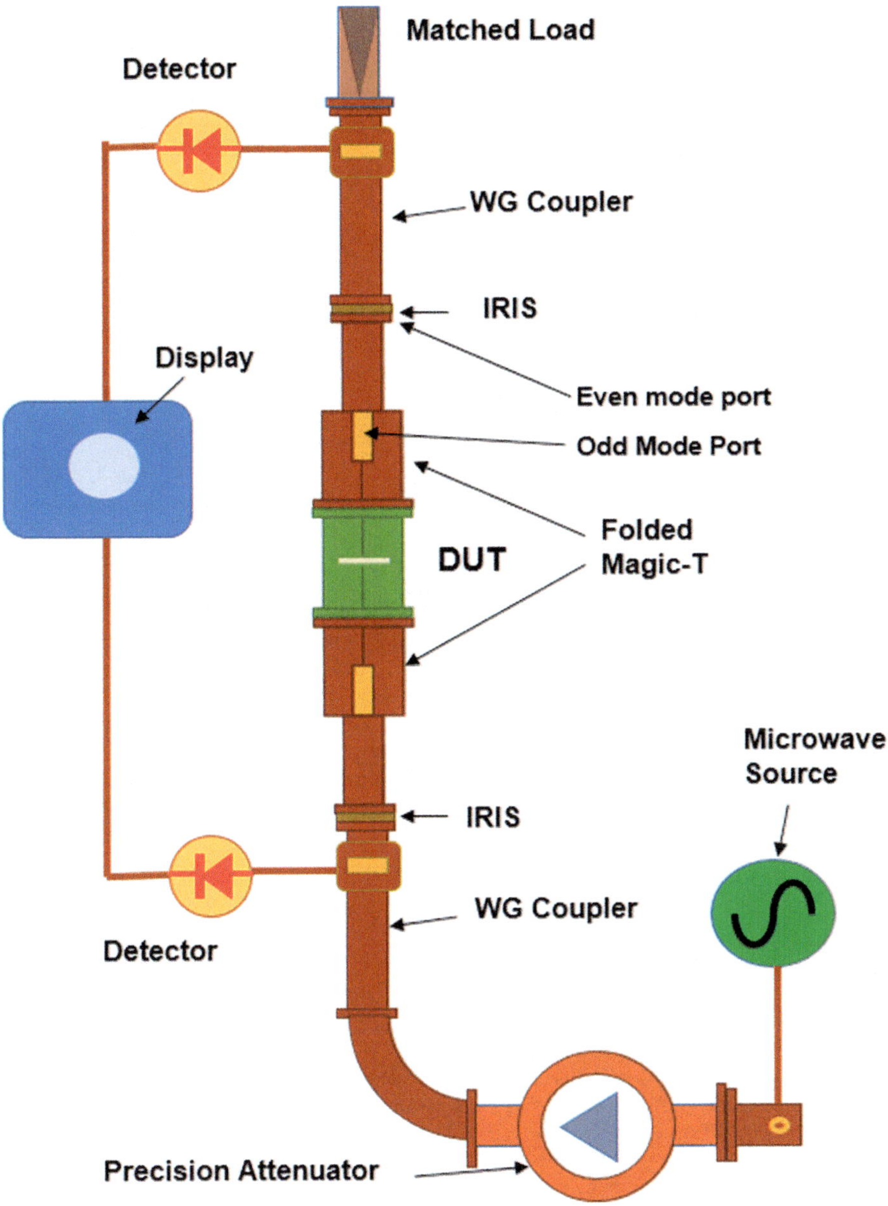

Fig. 5.12 Resonant cavity set-up for T-slot scattering in dual mode waveguide

rather as expected from perusal of the conventional slot case, that radiation is weak at low frequencies which are well below the T-slot resonance, and then grows steadily in strength as the source frequency approaches this resonance—predicted

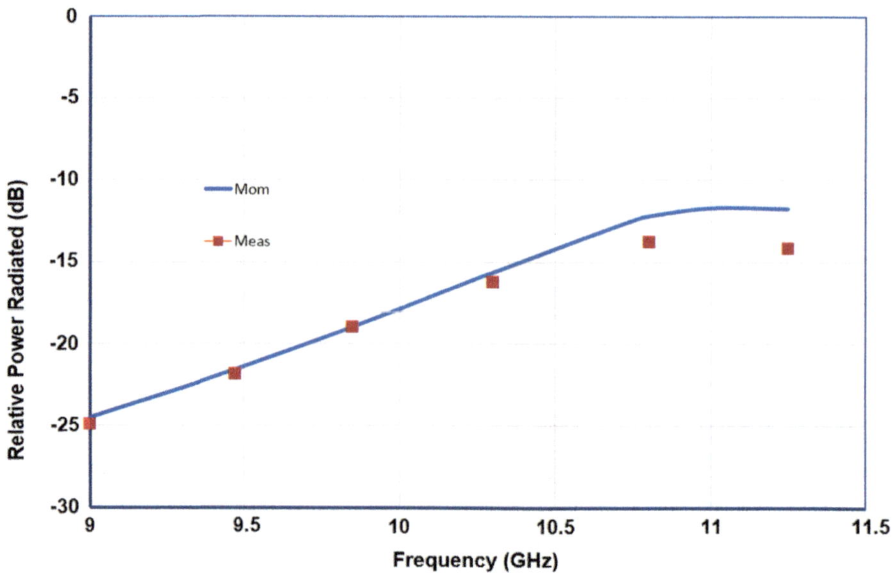

Fig. 5.13 Relative power radiated (P_{rad}/P_{inc}) versus frequency for a transverse T-slot in bifurcated waveguide in even-mode (WG16 wall thickness T = 1.32 mm; slot width w = 1.5 mm; slot length L = 16 mm; septum slot L_3 = 5.48 mm)

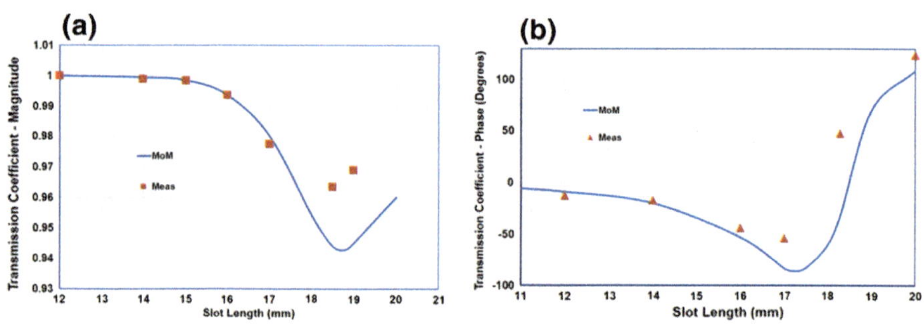

Fig. 5.14 T-slot transmission coefficient magnitude **a** and phase **b** as a function of slot length L (Even mode excitation; WG16; T = 1.32 mm; L_3 = 5.48 mm; bifurcation thickness = 1.5 mm; frequency = 9.89 GHz)

by the moment method to be at 11.1 GHz for the given dimensions. It is noticeable that agreement between theory and measurement is good, and well within the estimates of experimental error at frequencies below ~10.5 GHz where the T-slot is distinctly non-resonant. Close to the slot resonance, around 11 GHz, the measured results begin to deviate significantly from the moment method predictions. This can be attributed to a basic limitation in the cavity based measurement technique. For weakly radiating non-resonant slots the amount of power extracted from

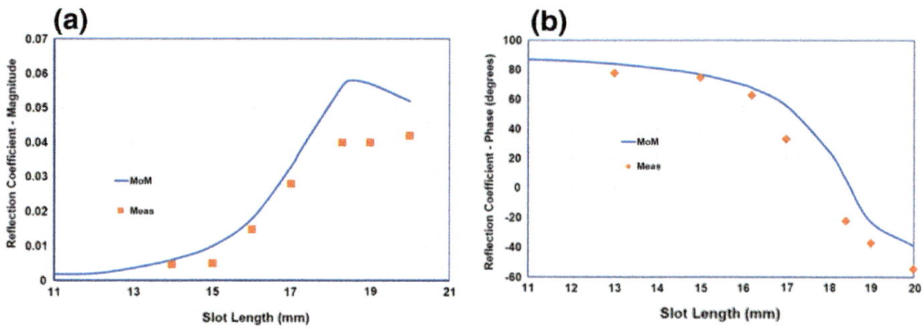

Fig. 5.15 T-slot reflection coefficient magnitude **a** and phase **b** as a function of slot length L (Even mode excitation; WG16; T = 1.32 mm; L_3 = 5.48 mm; bifurcation thickness = 1.5 mm; frequency = 9.89 GHz)

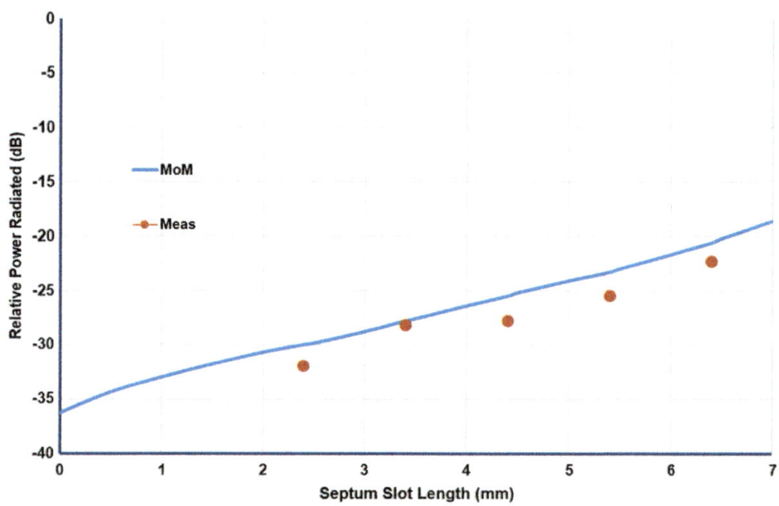

Fig. 5.16 Relative power radiated in dB as a function of clearance slot depth (L_3) for a symmetrically located T-slot—even mode excitation (WG16; T = 1.32 mm; L_3 = 5.48 mm; bifurcation thickness = 1.5 mm; frequency = 9.89 GHz)

the cavity is insufficient to 'dampen' the fundamental dual-mode waveguide resonance, which means that sharp, peaky, Q-curves with well defined, easily measured, characteristics continue to be observed. However, as the drive frequency approaches the T-slot resonant value, and the slot begins to radiate strongly, the dual-mode waveguide cavity becomes increasingly heavily loaded, thus 'flattening' the Q-curves. This then makes it very difficult to establish the precise cavity resonant frequency and to determine a value for the very low Q-level. The measurement error becomes large, as is reflected in the deviation between theory and experiment in the vicinity of the slot resonant frequency.

The above comments are reinforced in Figs. 5.14 and 5.15, which present, respectively, plots of transmission coefficient amplitude and phase and reflection coefficient amplitude and phase as functions of slot length. The plots are typical of a series resonant slot in waveguide. While the computed transmission coefficient magnitude falls from approximately unity to almost 0.94 at resonance, the reflection coefficient magnitude mirrors this, rising from almost zero for non-resonant lengths to almost 0.06 at resonance. The phase shift predictions establish that the slot, in transmission line terms, could be modelled by an RLC circuit in series with the line, with L $\ll$ C for short slots. At resonance L = C and the slot behaves as a series resistance in the line.

The comparisons between theory and measurement provided in Figs. 5.14 and 5.15 fully confirm the observation made with respect to Fig. 5.13 with 'good' levels of agreement for non-resonant slots (within experimental error). It is perhaps pertinent to note that while the measured points in Figs. 5.14a and 5.15a offer poor confirmation of the predicted resonant behaviour of the slot, this is less true of the phase measurements. These indicate an experimental resonant frequency of ~18, 0.5 GHz below the theoretical value, and well within the measurement error estimate.

Arguably, a feature of primary importance for potential users of the T-slot radiator is the degree to which its radiation level is influenced by the clearance slot length (L_3). A typical result, depicting relative power radiated as a function of septum slot length, is presented in Fig. 5.16. Not unexpectedly for L_3 approaching zero length the radiation is very weak at about −36 dB. It is then predicted to rise rapidly as L_3 is increased, to almost −20 dB when L_3 = 7 mm. The measured results confirm the theoretical prediction to within the experimental error. The experimental errors arise from several sources one of which is cavity damping for near resonant slots. Others are attributed to the square ended slots used in the theory. These were difficult to replicate experimentally, and the slot used were generally round-ended. Roughness of the machined surfaces can have a deleterious effect on measurements as can imprecise positioning of the slots within the dual-mode waveguide test piece.

Generally the degree of agreement between theory and experiment is sufficient to assert that the moment method is correctly simulating the T-slot radiator located in dual-mode rectangular waveguide. Needless to say, today, we can establish the operational integrity of any new computational analyses based on the moment method by performing a 'one-off' simulation of the geometry of interest using a modern commercially available finite element package such as Concerto [20], and comparing results. This has been done in Fig. 5.17 for an 'extreme' T-slot radiator with a very short radiating element (L = 2 mm) and a relatively long clearance notch (L_3 = 13, 14 and 15 mm) in a deep dual-mode waveguide structure. The plots show strong resonant behaviour with the resonant frequency shifting to lower values as the overall slot length increases—in accordance with Eq. (5.23). More importantly the degree of agreement between the moment method (solid lines) and the finite element technique (dashed lines) is excellent and well within modelling errors, particularly given the limited number of basis functions adopted in the moment method.

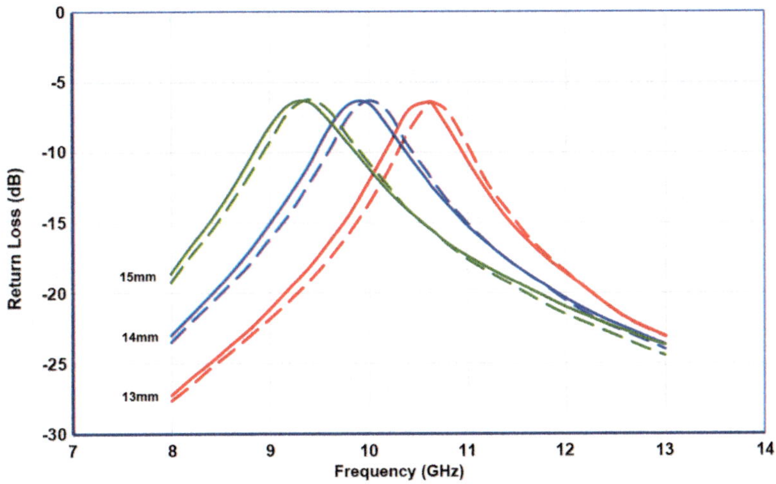

Fig. 5.17 Moment method solution (solid lines) for T-slot compared with finite element software (dashed lines) for deep dual-mode waveguide (L = 2 mm, L_3 = 13, 14, 15 mm, w = 1.5 mm, t = 1.27 mm)

5.5 Chapter Summary

Computer models for simulating electromagnetic scattering by elemental radiators have become key components for modern antenna design techniques, and in this chapter we have aimed at establishing the important role of the moment method in this trend. To do this two illustrative examples are explored in detail. While the examples focus on waveguide fed slots, it is important to recognise that the analytical procedures described are applicable more widely to other compact radiators such as dipoles and patches, driven from other transmission line forms, such as stripline and microstrip. Any limitations which may exist revolves around the availability of appropriate Green's functions.

Firstly, a conventional well established example of electromagnetic wave scattering by an inclined slot in the broadwall of waveguide is investigated. This entails the elaboration of a full solution for the conventional 'shunt' slot in rectangular waveguide. The relevant Green's functions for the coupled regions created by this slot geometry are formed. They comprise the Green's functions for rectangular waveguide, a half-space and a cavity. Once these are established it becomes possible to construct the integral equations describing the boundary value problem. Subsequently, the essence of the solution of these equations is the thoughtful choice of efficient trial functions or basis functions for the unknown aperture field. This is illustrated by describing and detailing the selection and formulation of a trigonometric option. It is shown that the trigonometric choice optimises calculation speed without sacrificing the accuracy of the scattering predictions for a range of shunt slot geometries. Secondly, the T-slot radiator in bifurcated waveguide is also subjected to the MoM to emphasise the flexibility of the technique. Without significant amendments to the computational steps pursued in modelling the conventional shunt slot in waveguide, the key to the evolution of a MoM for this quite different more complex scattering geometry is the careful construction of an appropriate basis function.

It is pertinent to note here, that the geometries of many compact radiators, which are amenable to solution by the method of moments, are undoubtedly solvable using finite element methods (FEM) such as HFSS and COMSOL. However, while these solvers permit model construction with minimal knowledge of electromagnetic theory, essentially by exchanging user modelling choices for computational intensity, they exhibit the significant downside of being relatively slow computationally in reaching a solution. As has already been intimated this rules out the use of FEM in embedded modules within array simulation packages.

Finally, it is arguably true to say that accurate computational design of aperture antennas, and in particular array antennas, made possible using the method of moments, has virtually obviated the traditional need for laborious back-up measurement procedures. In the not too distant past, 25–30 years ago, it was necessary to provide precise design data on waveguide slots, ranging over a multiplicity of sizes and orientations, in order to facilitate array synthesis calculations. At that time measurements were the only source of such data.

References

1. A.J. Sangster, A.H.I. McCormick, Theoretical design/synthesis of slotted waveguide arrays. Proc. IEE **136H**, 39–46 (1989)
2. T. Khan, A. De, M. Uddin, Prediction of slot size and inserted air gap for improving the performance of rectangular microstrip antennas using artificial neural networks. IEEE Antennas Wirel. Propag. Lett. **12**(1), 1367–1371 (2013)
3. A.A. Oliner, The impedance properties of narrow radiating slots in the broad face of rectangular waveguide. IRE Trans. **AP-5**, 4–20 (1957)
4. B.N. Das, G.S. Sanyal, Network parameters of a waveguide broad wall slot radiator. Proc. IEE **117**(1), 41–44 (1970)
5. T. Vu Khac, C.T. Carson, Coupling by slots in rectangular waveguide with arbitrary wall thickness. Electron. Lett. **8**, 456–458 (1972)
6. R.F. Harrington, *Field Computation by Moment Methods* (MacMillan, New York, 1968)
7. R.F. Harrington, Matrix methods for field problems. Proc. IEEE **55**(2), 136–149 (1967)
8. R.W. Lyon, An investigation of apertures in dual mode rectangular waveguide for variable polarisation arrays, Ph.D. thesis, Heriot-Watt University, 1979
9. R.W. Lyon, A.J. Sangster, Efficient moment method analysis of radiating slots in thick walled rectangular waveguide. IEE Proc. **128**(4), 197–205 (1981)
10. M.G. Salvadori, M.L. Baron, *Numerical Methods in Engineering* (Prentice-Hall International, London, 1961)
11. T.J. Akai, *Numerical Methods* (Wiley, Canada, 1993)
12. R.S. Esfandiari, *Numerical Methods for Engineers and Scientists using MATLAB* (C.R.C. Press, Boca Raton, USA, 2013)
13. T. Vu Khac, Solution of some waveguide discontinuities by the method of moments. IEEE Trans. **MTT-20**, 416–417 (1972)

14. J.S. Ajioka, D.M. Tang, N.W. Wong, Arbitrarily polarised slot radiators in bifurcated waveguide. IEEE Trans. **AP-22**, 196–200 (1974)
15. A.J. Sangster, R.W. Lyon, Moment method analysis of a T-shaped slot radiator in bifurcated waveguide. IEE Proc. **129**(6), 299–306 (1982)
16. R.J. Chignell, J. Roberts, Compact resonant slot for waveguide arrays. IEE Proc. **125**(11), 1213–1216 (1978)
17. A.E. Bailey, *Microwave Measurements* (Peter Peregrinus, 1989)
18. N.F. Wong, R. Tang, G.A. Okubo, *Investigation of Characteristics and Practical Implementation of Arbitrary Polarised Radiators* (Hughes Aircraft Co., Ground Systems Group, Fullerton, California, 1971)
19. D.M. Pozar, *Microwave Engineering*, 4th edn. (Wiley, 2011)
20. CONCERTO, *Vector Fields Ltd.* http://metamaterials-modelling.net/software-concerto.php

Chapter 6
Resonant and Non-resonant Linear Slot Arrays

6.1 Introduction

At microwave frequencies, array antennas generally comprise a linear or planar arrangement of periodically spaced aperture antennas. These may be of the slot type, open-ended waveguide type, or horn type, as described in Chap. 3. To simulate the electromagnetic wave nature of such periodic distributions of radiators it is usual, firstly, to model an equivalent periodic array of isotropic radiators, and secondly, to employ the principle of pattern multiplication to determine the radiation pattern of the original array. As the name implies, this principle suggests that if the far-field pattern of the isotropic array is multiplied with that of an individual, or typical, aperture antenna which forms an element of the actual array, the resultant pattern will be that of the original array. However, care must be exercised in following this procedure. Unless the element pattern has been obtained in the array environment this process ignores element-to-element interaction (mutual coupling) and the resultant pattern becomes an approximation to the required pattern. Nevertheless, the technique is a powerful one and is widely used to assess the properties of potential array designs.

Array antennas will also be examined using the now conventional z-transform method, since it gives access to the study of elementary array synthesis procedures. Although the z-transform method again neglects the effects of mutual coupling between closely spaced neighbouring radiating elements of the target array, for many compact microwave antennas of the array type, the method remains tolerably effective because mutual coupling is often too small to cause significant deviation from the modelled result.

A. J. Sangster, *Compact Slot Array Antennas for Wireless Communications*, Signals and Communication Technology, https://doi.org/10.1007/978-3-030-01753-8_6

6.2 Uniform Linear Array of Isotropic Elements

A linear array is one that is formed from a single line of multiple radiating sources. When these individual sources radiate equally in all directions the array is said to comprise isotropic elements. In addition the array is described as uniform when the elements are equally spaced and fed with equal signal levels. Such an array is depicted in Fig. 6.1. Although planar array (an array of linear arrays) antennas are more commonly used in mobile communications systems, it is advantageous to study the radiation patterns of linear arrays which are mathematically less complex. Nevertheless, the knowledge gained from linear array studies remains entirely relevant to, and representative of, planar arrays.

At a distant point from the antenna the total E-field, at radius r due to n elements can, by superposition, be expressed as:

$$E_T = E_0(1 + e^{j\psi} + e^{2j\psi} + \cdots + e^{j(n-1)\psi}) \tag{6.1}$$

where E_0 is the field magnitude at the distant point (r) due to a single element of the array, and very importantly:

$$\psi = k_0 d \sin\phi + \alpha \tag{6.2}$$

Here d is the element-to-element spacing, k_0 is the free-space phase coefficient, ϕ is the real angle from broadside, and α is the progressive element-to-element phase shift impressed on the array by the feed structure. Note that the factor $(\frac{\exp(-jk_0 r)}{r})$, which is common in all radiation field representations, has been suppressed for convenience. If we normalise Eq. (6.1) to the optimum field E_0 we obtain:

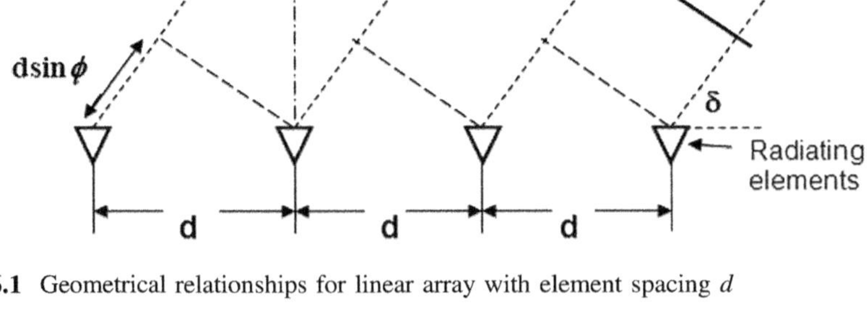

Fig. 6.1 Geometrical relationships for linear array with element spacing d

$$\frac{E_T}{E_0} = 1 + e^{j\psi} + e^{2j\psi} + \cdots + e^{j(n-1)\psi} \tag{6.3}$$

By replacing the exponential terms by the variable z (= $e^{j\psi}$) the far-field pattern of the linear array can be expressed mathematically as a polynomial, as follows:

$$\varsigma(z) = 1 + z + z^2 + z^3 + \cdots + z^{n-1} \tag{6.4}$$

This polynomial form has been shown [1] to be a powerful mathematical tool in the analysis and synthesis of array antennas. However, before advancing to the treatment of arrays in general, it is perhaps instructive to consider some special cases.

When $\zeta(z)$ is multiplied by z in Eq. (6.4) we obtain:

$$z\varsigma(z) = z + z^2 + z^3 + \cdots + z^n \tag{6.5}$$

Hence subtracting Eq. (6.4) from Eq. (6.5) gives:

$$(z-1)\varsigma(z) = z^n - 1$$

giving

$$\varsigma(z) = \frac{z^n - 1}{z - 1} \tag{6.6}$$

With a little manipulation, Eq. (6.6) can be written in the form:

$$\varsigma(z) = \frac{z^{n/2}(z^{n/2} - z^{-n/2})}{z^{1/2}(z^{1/2} - z^{-1/2})} \tag{6.7}$$

If this equation is recast in its exponential form, with z = $e^{j\psi}$, while recognising the identity:

$$\frac{e^{j\theta} - e^{-j\theta}}{2} = \sin\theta$$

then we discover that:

$$E_T = \frac{E_0 \sin\frac{n\psi}{2}}{\sin\frac{\psi}{2}} \tag{6.8}$$

Element phase is referred to the array mid-point to secure this simple form. Functionally, Eq. (6.8) is not too dissimilar to Eq. (3.23), the spectrum function for a uniformly illuminated linear aperture. The primary difference is the oscillatory nature of the denominator of Eq. (6.8) which strongly influences the pattern distribution. The result is that the linear array pattern and the linear aperture pattern are comparable only for a limited range of ψ values, which implies a limited range of element spacings d, as illustrated in Sect. 6.2.1.

6.2.1 Radiation Patterns

Equation (6.8) represents the functional form of the far-field radiation pattern, or polar diagram, of the uniformly excited linear array of isotropic elements. Since the radiation in the forward half-plane $-\pi/2 < \phi < \pi/2$ is clearly identical to the radiation into the reverse half-plane $\pi/2 < \phi < 3\pi/2$, we need consider only the former in the following discussion.

The ratio of sines in Eq. (6.8) implies that the pattern must be multi-lobed. The primary or main lobe occurs, where simultaneously $\psi \to 0$ and $\sin(n\psi/2) \to n\psi/2$. For this lobe $E_T/E_0 = n$; that is at the peak of the primary lobe the E-field at a given radius for the array is n times the E-field at the same radius for a single isotropic element. However, note that if the element spacing d is equal to or greater than λ_o there may be more than one principal maximum—these secondary maxima are termed grating lobes and appear where $\psi = \pi, 2\pi$ .... It is apparent from Eq. (6.2) that the principal maximum at $\psi = 0$ occurs where:

$$\sin\phi = -\frac{\alpha}{k_0 d} \tag{6.9}$$

When this principal maximum appears in a direction normal to the line of the array, i.e. at $\phi = 0°$, the array is termed a broadside array (see Fig. 6.2). Clearly, for this case $\alpha = 0°$, which means that the element-to-element phase shift is set to zero degrees. In the diagram, ϕ_1 denotes the angle to the first null measured from the array normal, while δ_1 provides essentially the same information, but measured from the array base line - $\delta_1 = 90° - \phi_1$.

When $\phi = 90°$ the principal maximum lies along the line of the array (Fig. 6.3) and the array, in this case, is termed end-fire. The angle from the lobe maximum to the first null is denoted by δ_2 in the figure. From Eq. (6.9) it is clear that for an end-fire array we require $\alpha = -k_o d$.

6.2.2 Broadside Array

For a broadside array of equally space and equally excited radiators (uniform array) the nulls of the radiation pattern occur where the sine ratio in Eq. (6.8) exhibits zero values, i.e. where:

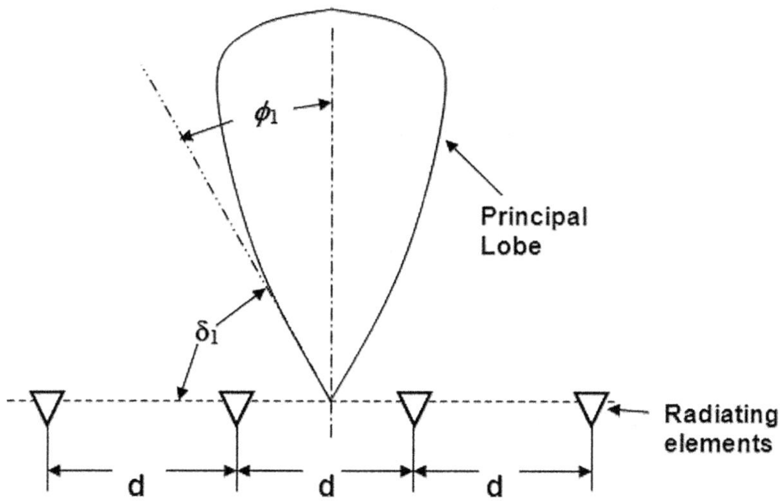

Fig. 6.2 Principal lobe—broadside radiation

Fig. 6.3 Principal lobe—end-fire array

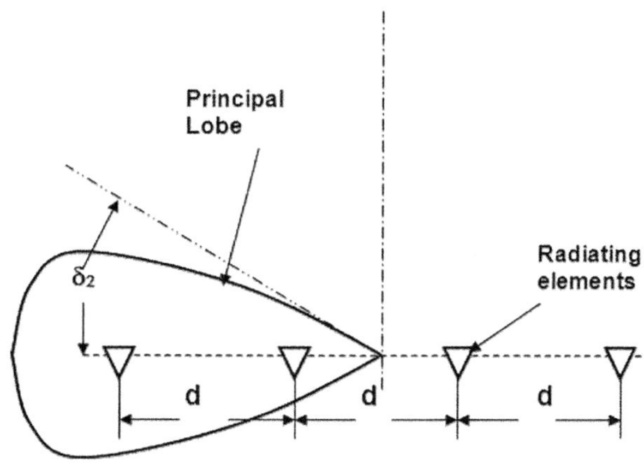

$$\frac{n\psi}{2} = \pm m\pi \tag{6.10}$$

Hence substituting for ψ using Eq. (6.2) and noting that for a broadside array $\alpha = 0$,

$$\frac{nk_0 d \sin\phi}{2} = \pm m\pi \tag{6.11}$$

which implies that:

$$\sin\phi = \pm\frac{2m\pi}{nk_0 d} \tag{6.12}$$

Therefore, the nulls in the pattern appear at angles given by:

$$\phi_m = \sin^{-1}\left(\frac{m\lambda_0}{nd}\right) \tag{6.13}$$

If the array is large, such that $nd \gg m\lambda_0$ then ϕ_m is small for the first few nulls nearest to the main lobe, and is given by:

$$\phi_m = \frac{m\lambda_0}{nd} \approx \frac{m}{L/\lambda_0} \tag{6.14}$$

In this relationship $L(\approx nd)$ is the total length of the array. Strictly, $L = (n - 1)$ d but the error is small for long arrays. The first nulls which define the principal lobe occur where m = 1, and for this broadside case the beamwidth between first nulls (BWFN) is given by:

$$BWFN = 2\phi_1 \approx \frac{2}{L/\lambda_0}\text{ radians} = \frac{114.6^\circ}{L/\lambda_0} \tag{6.15}$$

For reasons intimated earlier it is much more usual for antenna engineers to employ half-power beamwidth (HPBW) in defining the radiated main beam and for the broadside uniform array it is not difficult to show that

$$HPBW = 0.44 \times BWFN \tag{6.16}$$

So, for this array:

$$HPBW \approx \frac{2 \times 0.44}{L/\lambda_0}\text{radians} = \frac{50.4^\circ}{L/\lambda_0} \tag{6.17}$$

For this uniform broadside array of isotropic elements with $d < \lambda_o/2$ the radiated power density in the principal lobe can be approximated by:

$$p_{array} \approx \frac{P_T}{2\pi r^2 \theta_{3dB}}\ \text{W/m}^2 \tag{6.18}$$

where P_T denotes power transmitted and r is a representative far-field distance from the array. For a single isotropic radiator delivering the same power P_T, the power density at r is:

$$p_{iso} = \frac{P_T}{4\pi r^2}\ \mathrm{W/m^2} \tag{6.19}$$

Hence the directivity (D) of the array has the simple form:

$$D = \frac{p_{array}}{p_{iso}} \approx \frac{2}{\theta_{3dB}} \tag{6.20}$$

In the initial process of assessing array requirements in any given application, simple formulae such as Eqs. (6.17) and (6.20) are very useful for acquiring a quick appreciation of the trade off between the beamwidth desired and antenna size.

6.2.3 End-Fire Array

The only difference between a uniform end-fire array and a uniform broadside array is in the value of the progressive phase shift α, so the form of Eq. (6.8) is unaltered by the change. Consequently, the condition for pattern nulls remains as in Eq. (6.10). However, in the end-fire case, with $\alpha = -k_0 d$, the equation relating ψ and ϕ becomes:

$$\psi = k_0 d \sin\phi - k_0 d \tag{6.21}$$

This can be written in a more convenient form if we note that $\delta = 90° - \phi$, namely

$$\psi = k_0 d \cos\delta - k_0 d \tag{6.22}$$

Hence combining Eq. (6.22) with Eq. (6.10) we get:

$$\frac{n}{2} k_0 d(\cos\delta - 1) = \pm m\pi$$

or

$$\cos\delta - 1 = \pm\frac{2m\pi}{nk_0 d} = \pm\frac{m\lambda_0}{nd} \tag{6.23}$$

Thus in terms of δ the null angles are located at:

$$\frac{\delta_m}{2} = \sin^{-1}\left(\pm\sqrt{\frac{m\lambda_0}{2nd}}\right) \tag{6.24}$$

Once again, directing our attention to a long array with $nd \gg m\lambda_o$, the nearest pattern nulls occur at angles given by:

$$\frac{\delta_m}{2} \approx \sqrt{\frac{m\lambda_0}{2nd}} = \sqrt{\frac{m}{2L/\lambda_0}}$$

i.e.

$$\delta_m \approx \sqrt{\frac{2m}{L/\lambda_0}} \tag{6.23}$$

So, finally we obtain the useful relations:

$$BWFN = 2\sqrt{\frac{2}{L/\lambda_0}}\ \text{radians} = 114.6\sqrt{\frac{2}{L/\lambda_0}}\ \text{degrees} \tag{6.24}$$

For the end-fire array it is not difficult to show that HPBW = 0.66 × BWFN, and therefore we can write:

$$HPBW \approx 76.3\sqrt{\frac{2}{L/\lambda_0}}\ \text{degrees} \tag{6.25}$$

Note that the half-power beamwidth for the end-fire array is very much larger than for the corresponding broadside array. For example, when the ratio $L/\lambda_o = 50$ the end-fire array beamwidth is approximately 15 times greater than that of the equivalent broadside format. Some improvements in end-fire gain can be achieved without increasing the number of radiating elements by judicious adjustment of the element spacings d. The optimum spacing is termed the Hansen-Woodyard condition, which will be examined below.

6.2.4 Scanned Array

The above examination and discussion of broadside and end-fire arrays underlines a major property of array antennas, namely their ability to provide electronic beam scanning. The two cases represent the scanning extremes of zero degrees (broadside) and ninety degrees (end-fire) when the progressive phase shift α is set to, respectively, zero and $-k_o d$. At the $\phi = 0°$ ($\delta = 90°$) scan condition the principal beam exhibits minimum beamwidth and optimum directivity. As the beam is scanned away from broadside towards the end-fire condition both the beamwidth and the directivity deteriorate, slowly at first, and then increasingly rapidly as end-fire is approached. This behaviour is illustrated graphically in Fig. 6.4 which

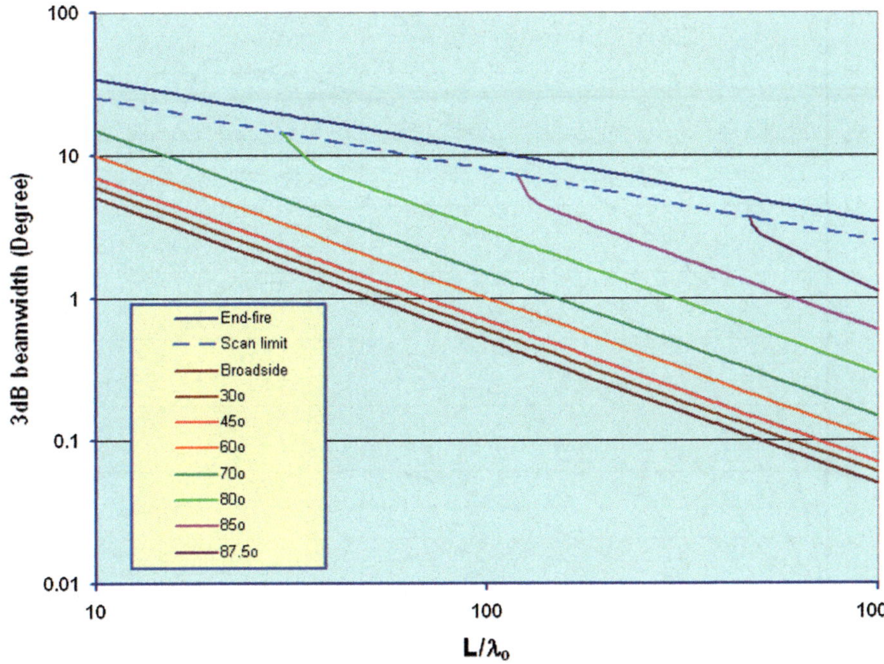

Fig. 6.4 Beamwidth variation with scanning (adapted from [2])

presents 3 dB beamwidth as a function of array length for a uniform linear array. These beamwidths were generated by forming the following relationship:

$$\frac{\sin\frac{n\psi}{2}}{\sin\frac{\psi}{2}} = 0.707 \tag{6.26}$$

for uniform linear arrays with $d = \lambda_o/2$. The curves in Fig. 6.4 represent varying scan angles (ϕ) and the scan limit defines the angle to which the principal beam can be scanned unmodified by the proximity of end-fire. For example, a very long array can be scanned much closer to end-fire than a short one. This figure makes it very clear that array length is the major determinant of the principal lobe beamwidth and hence of the array gain.

It is pertinent to ask here whether or not, for any given number of elements n, there is an element spacing d and an array length $L = (n - 1)d$ which procures optimum array gain and minimum beamwidth. For a broadside array calculation based on Eq. (6.8) it can be demonstrated that minimum beamwidth and maximum gain (allowing for power losses in sidelobes and grating lobes) occur where the element spacing is just less than one wavelength. This is shown clearly in Fig. 6.5 where gain is presented as a function of d/λ_o for a range of array options with values of n ranging from 2 to 24. The pronounced dip in the value of the gain at

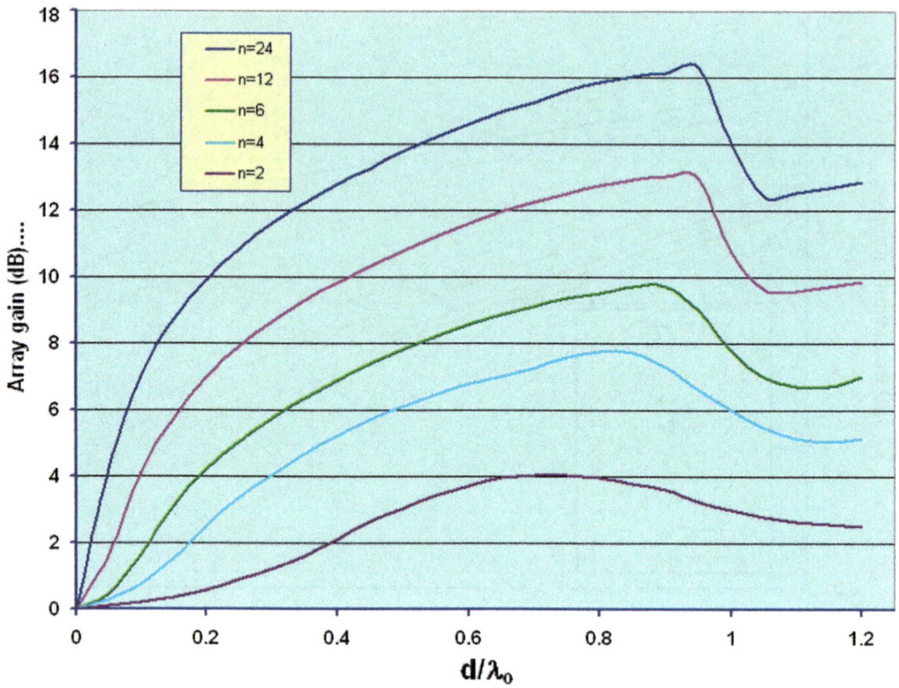

Fig. 6.5 Gain as a function of d/λ_0 for a uniform broadside array (adapted from [3])

$d/\lambda_0 \approx 0.95$ particularly where n is large, is produced by the appearance of the first pair of grating lobes in the end fire directions.

6.3 Array Design Using Theory of Polynomials

In this short section the intention is to introduce the reader to antenna synthesis, rather than analysis, where we have been directing our attention up to this juncture. The method which will be outlined applies strictly to uniform arrays. Nevertheless, the procedures developed here provide a good general introduction to antenna synthesis concepts without embarking on an overly complex mathematical journey. Actually, in practice, antenna design engineers generally use very sophisticated commercial software to perform the complex operations involved in synthesis, but to do this efficiently some awareness of the basic process is essential.

It has already been shown (Eq. 6.4) that the far-field pattern of an n-element uniform linear array can be expressed in the polynomial form:

$$\varsigma(z) = 1 + z + z^2 + z^3 + \cdots + z^{n-1}$$

where $z = e^{j\psi}$ and ψ is the difference in phase of waves radiated from any pair of adjacent elements. By a fundamental theory of algebra a polynomial of order (n − 1) has (n − 1) zeros (some of which may be multiple zeros) and can be factored into (n − 1) binomials [1]. Thus

$$\varsigma(z) = (z - r_1)(z - r_2)(z - r_3)\ldots(z - r_{n-1}) \tag{6.27}$$

The quantities r_1, r_2, …, r_{n-1} are termed the roots of the $(n - 1)$th order polynomial. Since $z = e^{j\psi}$ has magnitude unity, these roots can be represented graphically as located on a unit circle in the complex z-plane. The magnitude of ζ (essentially the far-field pattern) can then be determined by multiplication of the vector magnitudes (z − r) as illustrated in Fig. 6.6, which shows $z = e^{j\psi}$ with ψ measured clockwise from the real axis denoted by *Re*. The imaginary axis is denoted by *Im*. Complex vector $r_1 = e^{j\psi_1}$ is also represented on the diagram, and the vector $(z - r_1)$ is just the vector difference between z and r_1.

For arrays formed from isotropic elements, radiation will be symmetrically distributed around the axis of the array, and consequently only radiation angles ϕ in the range $-\pi/2$ to $\pi/2$ need be considered. From Eq. (6.2), this means that ψ, the argument of z, increases from $-k_o d + \alpha$ to $k_o d + \alpha$ as ϕ varies from $-\pi/2$ to $\pi/2$. Thus the range of ψ is $2k_o d$. For example, for a broadside array for which $\alpha = 0$, if the element spacing is set at half the free-space wavelength ($d = \lambda_0/2$) then ψ varies from $-\pi$ to $+\pi$, as shown in Fig. 6.7a. Note that z is presumed to travel in the

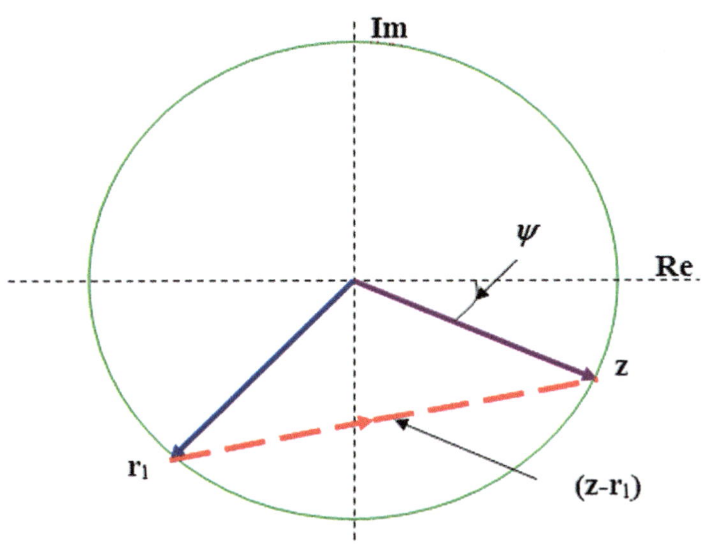

Fig. 6.6 Unit circle diagram

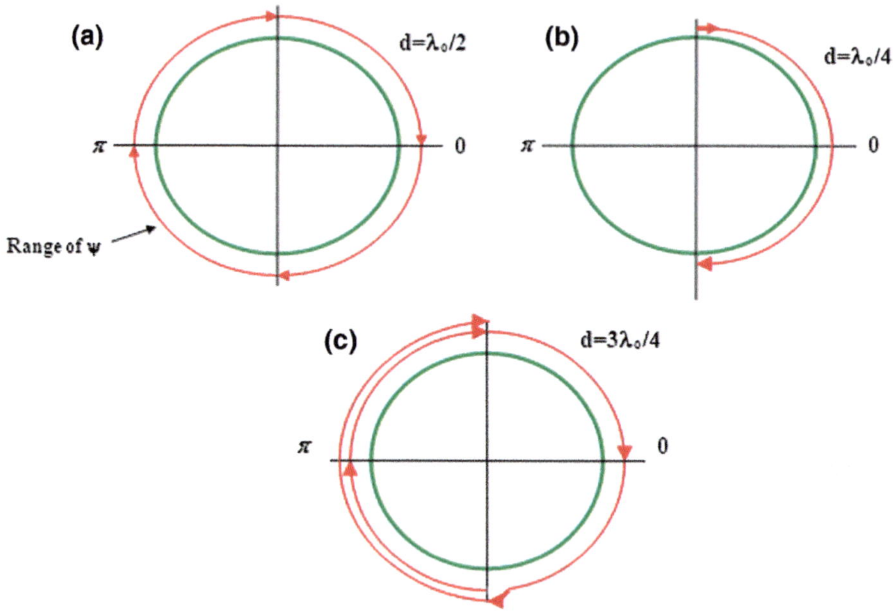

Fig. 6.7 Range of ψ for broadside array ($\alpha = 0$)

clockwise direction on the unit circle. For $d = \lambda_o/4$ (Fig. 6.7b) the range of ψ is from $-\pi/2$ to $+\pi/2$, while the range is from $+\pi/2$ to $-\pi/2$ when $d = 3\lambda_o/4$, with z traversing the unit circle 1.5 times (Fig. 6.7c). Given that mainbeams occur when $\psi = 0°, 360°, \ldots, 2\,m\pi$, arrays with large element spacings ($d \geq \lambda_0$) generate a grating lobe each time the locus of z passes zero on the circle diagram.

For a uniform linear array antenna the roots of its polynomial representation (Eq. 6.27) are not restricted as to where on the unit circle they lie, and embedded in this observation resides the basis of array synthesis. The roots r_n can be located at any angle ψ except $\psi = 0$ where the primary beam is always located. For the uniform array case the roots, plus the pole at the $\psi = 0$ point (black dot), divide the circle into n equal arcs. This is illustrated in Fig. 6.8a for a four element array.

The far-field pattern magnitude is obtained by forming the product of the vectors $(z - r_i)$. This can be done graphically by multiplying the lengths of the blue vectors in Fig. 6.8b for all positions of z (red dot) as it moves around the circle. It is not difficult to observe that the principal maximum occurs at $\psi = 0$ where all three vectors exhibit their maximum length. Subsidiary, less pronounced maxima, also occur mid-way between the zeros, where the pattern nulls occur. These features can be located in 'real space' (denoted by ϕ) by replacing ψ with ϕ, using Eq. (6.2) For example, the nulls of the four element array occur at the ψ values $-\pi/2, -\pi, -3\pi/2$, or more generally at $\psi_m = -2\,m\pi/n$ with $m = 1, 2, 3$. Therefore the nulls in the 'real space' radiation pattern occur at angles ϕ_m given by:

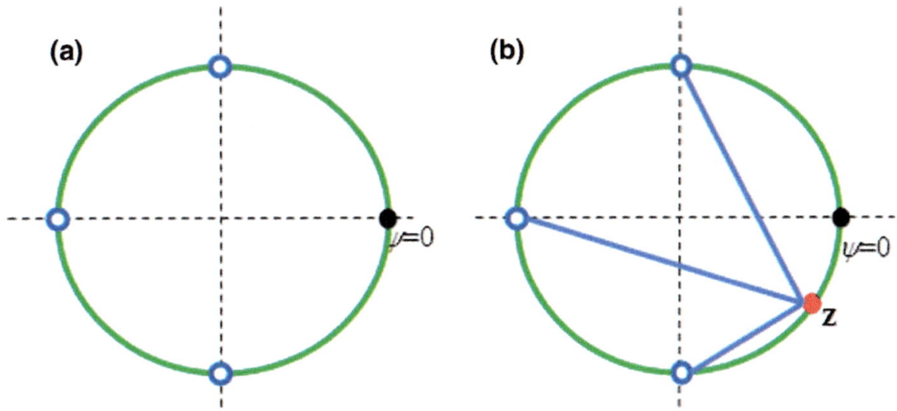

Fig. 6.8 Location of roots on circle diagram for four element uniform array

$$\sin\phi_m = -\frac{\alpha}{k_0 d} - \frac{2m\pi}{nk_0 d} \tag{6.28}$$

The 'real space' far-field pattern of the four element array can now be constructed either graphically, or by plotting the function ζ(z), which has the form:

$$\begin{aligned} \zeta(z) &= (z - e^{-j\pi/2})(z - e^{-j\pi})(z - e^{-j3\pi/2}) \\ &= z^3 + z^2 + z + 1 \end{aligned} \tag{6.29}$$

For $\alpha = 0$ and $d = \lambda_0/2$ an approximation to the resultant pattern as a function of ϕ is shown in Fig. 6.9. This technique for array pattern synthesis is generally attributed to Schelkunoff [1]. The range of ψ, in this case, is 2π.

For an end-fire array with four elements the root diagram remains unchanged, but with $\alpha = -k_o d$, this repositions the nulls in 'real space'. If in addition we reduce d to $\lambda_0/4$ the range of ψ ranges from $\psi = 0$ to $\psi = \pi$ around the lower half of the circle diagram as shown in Fig. 6.10. The root at $3\pi/2$ is now no longer contributing to the shape of the pattern in real space.

The resultant pattern for this case is shown in Fig. 6.12 (curve A). It has a very wide primary lobe and high side lobes, because of the non-contributing zero. However if we move all of the roots into the range of ψ as suggested in Fig. 6.11, a much improved pattern emerges for the same number of elements. This arrangement of roots on the circle diagram is generated by a polynomial of the form:

$$\begin{aligned} \varsigma(z) &= (z - e^{-j\pi/3})(z - e^{-j2\pi/3})(z - e^{-j\pi}) \\ &= 1 + 2ze^{-j\pi/3} + 2z^2 e^{-j2\pi/3} + z^3 e^{-j\pi} \end{aligned} \tag{6.30}$$

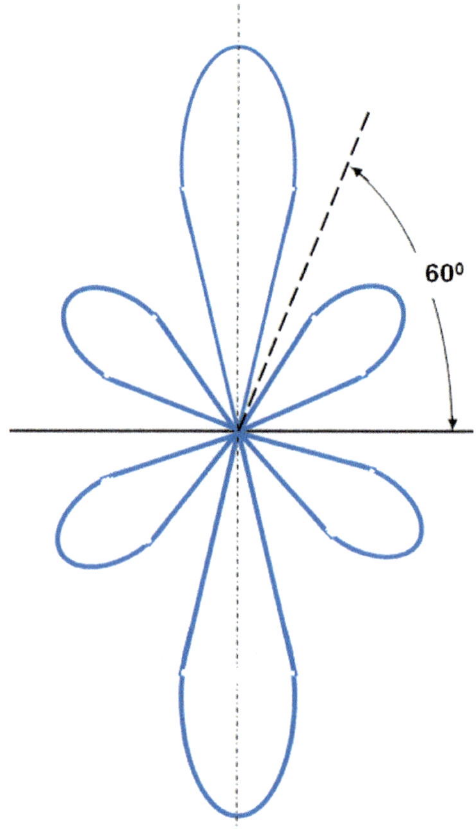

Fig. 6.9 Far-field pattern for a four element uniform broadside array with d = $\lambda_0/2$

By making the relation $z_1 = ze^{-j\pi/3}$ Eq. (6.30) can be written in the more convenient form:

$$\varsigma(z_1) = 1 + 2z_1 + 2z_1^2 + z_1^3 \tag{6.31}$$

This represents an array having radiating elements distributed in amplitude as 1: 2: 2: 1 and with a progressive phase shift of

$$\alpha = -\frac{\pi}{2} - \frac{\pi}{3} = -\frac{5\pi}{6} \tag{6.32}$$

This modified end-fire array pattern is sketched in Fig. 6.12 (curve B). Note the improvement in directivity.

If the overall length of the array is fixed but the number of elements it contains is allowed to increase, it becomes possible to improve the directivity still further if the nulls are optimally spaced within the range of ψ on the circle diagram. Curve C in Fig. 6.12 shows the pattern that results when the number of elements is increased to

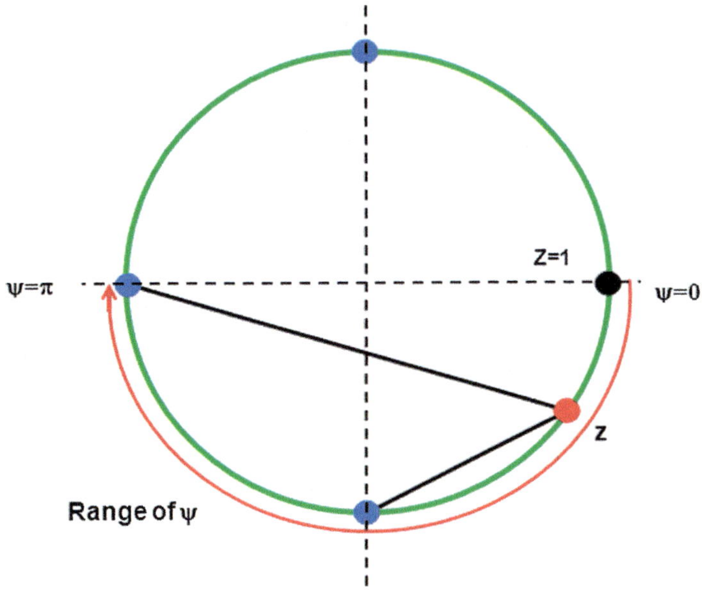

Fig. 6.10 Root location for four element end-fire array

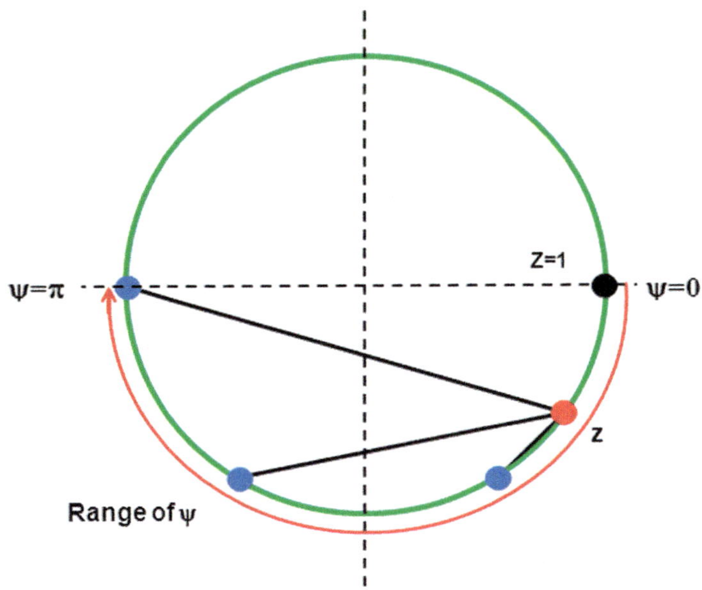

Fig. 6.11 Root locations for four element end-fire array with improved far field pattern

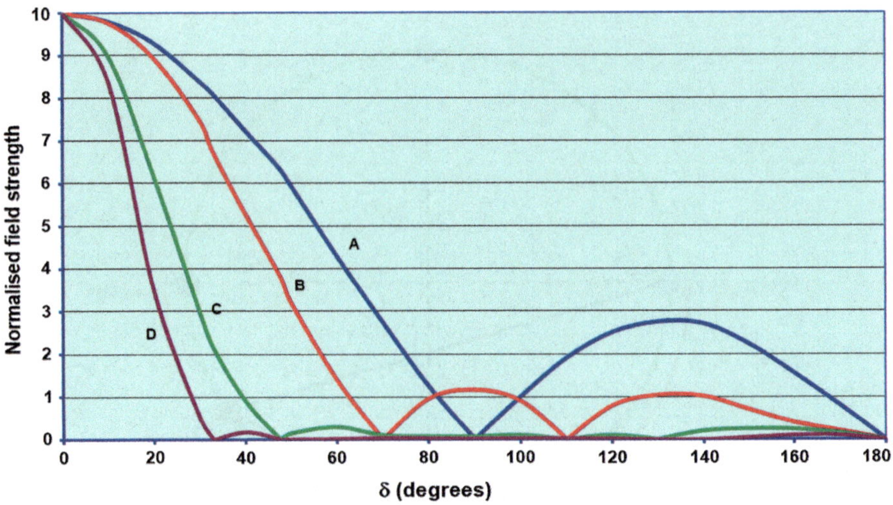

Fig. 6.12 End-fire array pattern improvements by root adjustments

seven by reducing the inter-element spacing to one eighth of a wavelength, so that the overall array length is still $3\lambda_o/4$. To obtain this result the nulls are equispaced in the range $\psi = 2k_o d = \pi/2$. Curve D demonstrates the pattern which results when the number of elements is increased even further (to 13) by reducing the inter-element spacing to $\lambda_o/16$. Again the nulls are equispaced in the range of ψ. However, these improvements in directivity as will be shown later tends to be achieved at the expense of other important factors of antenna performance—a typical engineering 'trade-off'.

For the uniform array, as we have seen, the maximum directivity and gain are directly related to the array length. In contrast, the above null adjustment procedures, which impinge on the element excitement levels and phase relations, appear to offer the possibility of arbitrarily improving directivity with an array of fixed length by simply using a sufficiently large number of elements. However, as always in technology, there is a balancing consequence. Closely spaced elements, phased and level controlled to achieve high directivity, present very low impedances to the input feed lines. This means very high currents in the feed lines—or in microstrip terms the feed lines become unfeasibly wide. In other words, with 'real' antennas the practical requirements of efficient delivery of power to each element imposes a limit to the directivity improvement which can be extracted from a fixed length array.

6.3.1 Optimum Element Spacing

Having discovered that the location of the roots, on the circle diagram for the array polynomial, has a major influence on the radiated pattern, it becomes possible to use this graphical technique to determine the optimum element spacing for a uniform array. For the broadside case the optimum pattern results from including the maximum number of nulls in the range of ψ. Therefore, for a large number of elements n, the maximum number of nulls are active when the range of ψ is 4π, as shown in Fig. 6.13a. Since the range of ψ for a broadside array is $2k_o d$ this implies that $4\pi d/\lambda_o = 4\pi$ and hence the optimum element spacing is $d = \lambda_o$.

In the case of the end-fire array the maximum number of nulls become active when the range of ψ approaches 2π, as suggested in Fig. 6.13b, which means that in this case the optimum element spacing d is almost $\lambda_o/2$. Note that in the limit where $d = \lambda_o/2$ there are two opposed end-fire mainbeams, and the array is bidirectional. Also, for the end fire array, if it is uniform, the roots are dispersed in an equally spaced format around the unit circle, and so their spacing is $2\pi/n$, except on either side of the pole at $z = 1$. Actually to further improve directivity the width of the principal lobe can be reduced further, than suggested in Fig. 6.13b, by simply adjusting the range of ψ so that for $\delta = 0$ the pole lies midway between $\psi = 0$ and the first zero of the polynomial. This is termed the Hansen-Woodyard condition [4]. To achieve this adjustment an extra inter-element phase shift $-\pi/n$ has to be introduced, or equivalently a phase shift of almost $-\pi$ distributed along the length of the array.

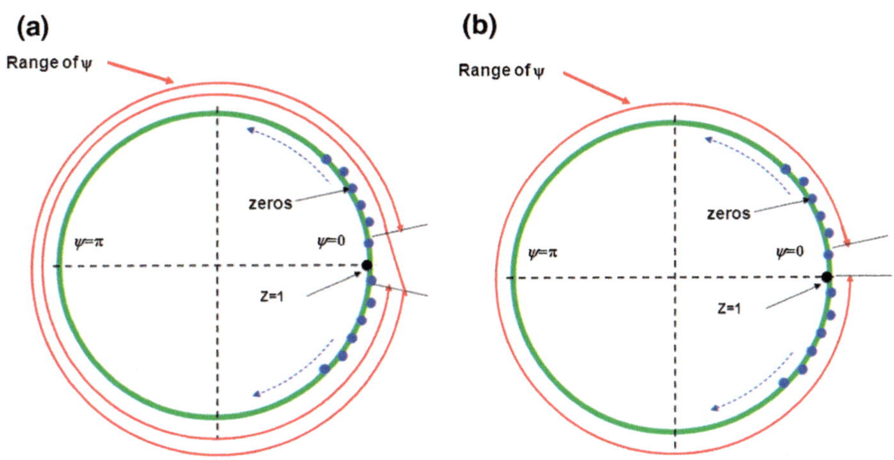

Fig. 6.13 Diagrams showing the range of ψ for optimum element spacing. **a** Uniform broadside array, **b** uniform endfire array. (The dotted arrows imply that the zeros occupy the entire circle)

6.3.2 Binomial Array

With fully adjustable arrays, in both element phase and in element power level, detailed radiation pattern control, encompassing even far-out sidelobes, becomes feasible. To emphasise this capability, here we will consider the formation of radiation patterns with no sidelobes.

If the roots of the polynomial $\zeta(z)$ are all equal and can be co-located at the π position on the unit circle, as suggest in Fig. 6.14, then the resultant pattern has only one null or zero at $\psi = \pi$ and consequently should possess no sidelobes. So does this root location scheme lead to a sensible polynomial which can be transformed into a realisable array? For the endfire array with the unit circle representation shown in Fig. 6.14, the polynomial must have the form:

$$\varsigma(z) = (z + 1)^{n-1} \tag{6.33}$$

The levels to which the individual elements of the array must be driven to secure this solution can be deduced by expanding Eq. (6.33) as a binomial series. For example for a four element array the relative magnitudes are in the ratios 1: 2: 2: 1. This element excitation scheme will result in no sidelobes for the four element array provided the element spacing d and the progressive phase shift α are also properly imposed.

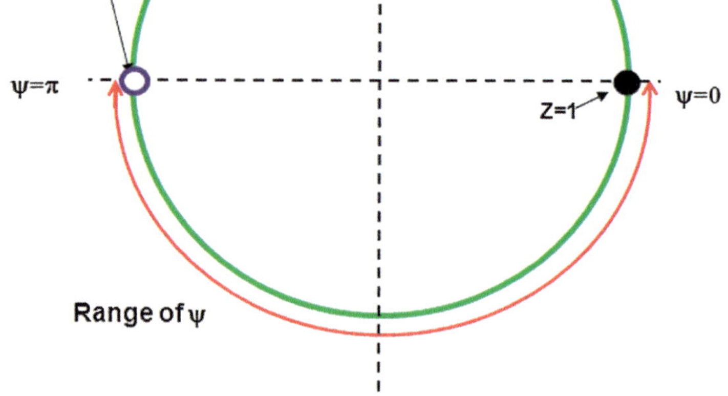

Fig. 6.14 Roots locations on unit circle for binomial array

6.4 Chapter Summary

Compact antennas for modern mobile communication systems generally require some degree of beam formation and steering in order to optimise signal strength in the direction of the satellite transmitter or relay antenna. This is particularly true in communication systems based on interaction with low earth orbit (LEO) satellites which traverse the sky in fifteen or so minutes. This can be done mechanically on vehicles, such as ships, which can support gyro-stabilised antennas. For more compact systems on much smaller platforms the ideal solution is electronic beam control using arrays, a topic which attracts significant research interest.

The chapter aims to demonstrate that the essence of array pattern control, associated with both uniform linear, and planar, array forms can be neatly summarised using polynomial representations of their basic periodic forms. This now classical analysis technique is employed throughout the chapter to examine the relationship between array structure, in particular element numbers, element spacing and element weightings, and how it impinges fundamentally on radiation pattern attributes such as directivity, gain, beamwidth, sidelobes and grating lobes. It is also emphasised that this powerful polynomial formulation presents an elegant route, not just to array analysis, but also to array synthesis. Several cases are examined to illustrate the technique.

References

1. S.A. Schelkunoff, W.T. Friis, *Antennas, Theory and Practice* (Wiley, New York, 1952)
2. R.S. Elliott, Beamwidth and directivity of large scanning arrays. Microwave J. (1963)
3. The Microwave Engineers Handbook, Page T-136 (1963)
4. M.M. Dawoud et al., Realization of superdirectivity from active and passive array antennas, in IEE Conference Publication No. 169 (1978)

Bibliography

1. R.E. Collin, F.J. Zucker, *Antenna Theory* (McGraw-Hill Book Co., New York, 1969)
2. R.F. Harrington, *Time Harmonic Electromagnetic Fields* (McGraw-Hill Book Co., New York, 1961)
3. C.A. Balanis, *Antenna Theory* (Wiley, New York, 1997)
4. R.S. Elliott, *Antenna Theory and Design* (Prentice-Hall Ltd., Englewood Cliffs, N.J., 1981)
5. R.C. Hansen, *Microwave Scanning Antennas* (Academic Press Ltd., New York, 1964)
6. E.C. Jordan, *Electromagnetic Wave and Radiating Systems* (Prentice-Hall Ltd., Englewood Cliffs, N.J., 1968)
7. A.W. Rudge, K. Milne, A.D. Olver, P. Knight (eds.), *The Handbook of Antenna Design: Volumes I and II* (Peter Peregrinus Ltd., London, 1986)
8. B.A. Munk, *Metamaterials: Critique and Alternatives* (Wiley, New York, 2009)

Chapter 7
Conventional Waveguide-Fed Travelling-Wave Slot Arrays

7.1 Introduction

In preceding chapters we have examined in detail the nature of radiation (Chap. 2) and how radiation is generated by individual aperture type radiators, including slots (Chap. 3). Antennas of this type can be modelled very accurately and efficiently using the moment method as we have demonstrated in Chaps. 4 and 5. Once accurate models of individual radiating elements have been established we have demonstrated in Chap. 6 how radiation can be focused and manipulated to procure high gain and radiation pattern control by embedding the elements within a periodic array. Unfortunately, the array techniques outline in Chap. 6 assume that rather complex methods of feeding electromagnetic power to the individual elements are available—namely by placing a stable source and phase shifter behind each elemental radiator, or by adopting a complex transmission line distribution network to channel power from a single source to the phase shifter controlled elements. Neither scheme is particularly appropriate for realising compact antenna structures which we are concerned with here.

Compact, efficient and inexpensive arrays are, arguably, most commonly secured by applying travelling-wave principles to a linear array of radiating elements. The technique has been explored in the literature for over sixty years [1–4], particularly in relation to slot arrays implemented in waveguide. Travelling-wave arrays employing radiators other than slots, and transmission systems other than waveguide [5] are also well established. Furthermore, planar array forms can be realised by combining together similar linear arrays in appropriately aligned parallel geometries. Examples of both linear and planar travelling-wave array antennas are depicted in Chap. 1, Figs. 1.13 and 1.14. It is well known that the rectangular waveguide provides an efficient low-loss method of transmitting electromagnetic waves over finite distance. But, in addition, if the waveguide contains more than one slot radiator, such as depicted in Fig. 1.12 all slots are excited by the same travelling-wave, potentially forming a simple and efficient antenna array. As we

A. J. Sangster, *Compact Slot Array Antennas for Wireless Communications*, Signals and Communication Technology, https://doi.org/10.1007/978-3-030-01753-8_7

have seen in Chap. 6, desirably focused and directed radiation patterns depend on the phase and radiation strength of the exciting electromagnetic wave at the position of each slot. In the case of excitement by a travelling wave, this means slots equi-spaced in the axial direction of the waveguide, generating ~360° progressive phase increases to secure an approximately broadside main radiation lobe. Lobe location is a key issue with travelling wave arrays as we shall see.

Radiation from an aperture or slot machined or etched into a conducting wall of a rectangular waveguide, or any other shape of waveguide for that matter, is maximised by orientating the slot into alignment with the TE_{10} mode magnetic field ($\mathbf{H}_{10}$) adjacent to the inner surface, or equivalently normal to the accompanying surface current density ($\mathbf{J}_{s10}$). This behaviour is illustrated in Chap. 1 and demonstrated theoretically in Chap. 5. For consistency, given that the longitudinal shunt slot is examined in detail in Chap. 5, we will direct our attention here toward the longitudinal slot array. Generally, for such an array resonant slots are employed to maximise array bandwidth. In Chap. 5, Figs. 5.6, 5.7 and 5.8 inform us that at resonance a longitudinal broadwall slot presents a real impedance to the waveguide —so in transmission line terms it can be modelled with a shunt resistance or conductance as shown in Fig. 7.1. A shunt, rather than a series, connection is dictated by the phase behaviour of S_{11} and S_{12} for the longitudinal slot.

The travelling wave array antenna formed from resonant longitudinal slots inserted periodically into a broadwall of waveguide has been examined widely in the literature. This interest has been encouraged by the early observation [2] that the off-set resonant slot, as depicted in Fig. 7.1a, can be modelled in transmission line terms by a simple shunt conductance (Fig. 7.1b). In fact Stevenson [2] has generated an elegant closed form equation for the normalised shunt conductance g of a narrow slot radiating into a half-space, namely

$$g = \frac{Y_s}{Y_0} = 2.09\frac{\lambda_g}{\lambda_0}\frac{a}{b}\cos^2\left(\frac{\pi\lambda_0}{2\lambda_g}\right)\sin^2\left(\frac{\pi\delta}{a}\right) \tag{7.1}$$

where $\delta = a/2 - X_1$, Y_s is the slot admittance in the waveguide (=G when real), Y_0 is the waveguide characteristic admittance, while $\lambda_g = \lambda_{g10}$ is the TE_{10} mode wavelength in the waveguide at the operating frequency. The validity of Eq. (7.1) as a 'trial' value in any travelling wave array synthesis procedure is confirmed by the moment method computations presented in Chap. 5. If the power entering the waveguide in the TE_{10} mode is known then the guide voltage and hence the equivalent transmission line value V (Volts) is available. Hence the power radiated by the slot, normalised to the input power, for the scenario in Fig. 7.1b is simply:

$$p_r = \frac{V^2 g}{2} \tag{7.2}$$

In words, the slot excitation level is proportional to g, which in turn [Eq. (7.1)] is proportional to $\sin^2\left(\frac{\pi\delta}{a}\right)$. The primary mechanism for controlling the radiation pattern for a shunt slot travelling wave array thus suggests itself. Namely, from

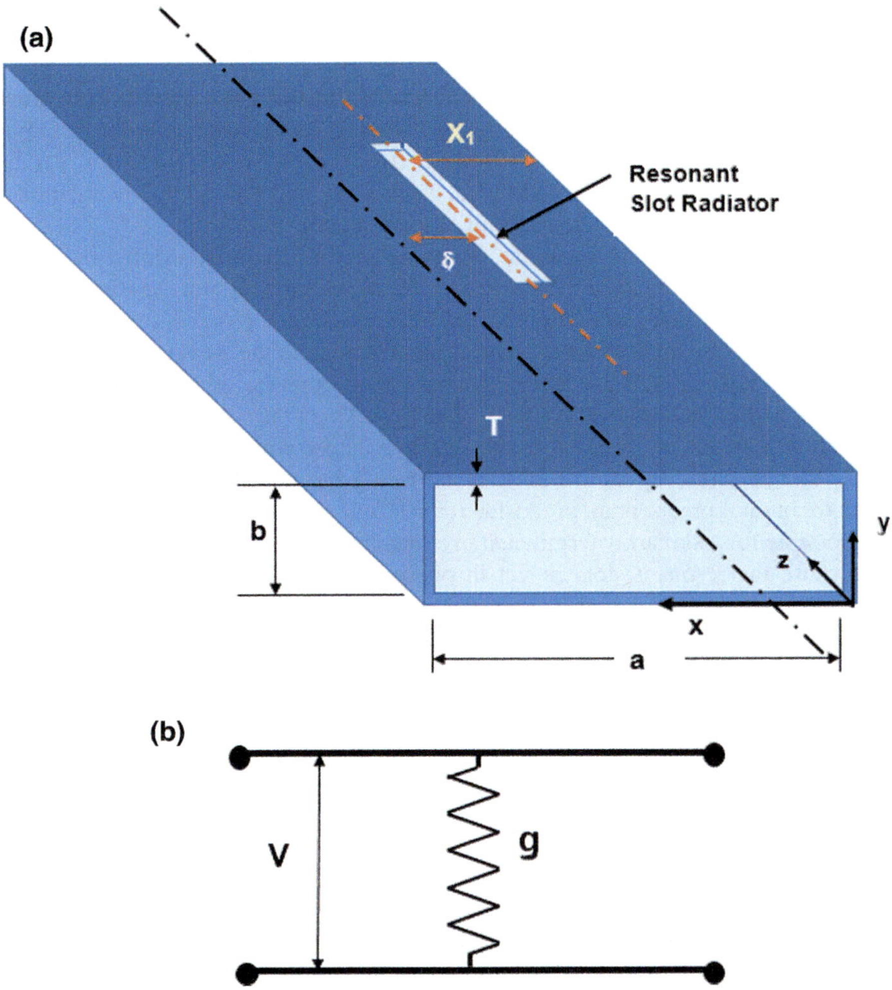

Fig. 7.1 Longitudinal (shunt) slot in TE_{10} mode rectangular waveguide: **a** waveguide geometry, **b** equivalent circuit for resonant slot with normalised $g = Y_s/Y_o$

estimated elemental power requirements, consistent with establishing a specified radiation pattern, we determine the conductances of the slots forming the array, and hence we can evaluate the slot off-sets δ. The procedure is elaborated upon in the next section.

7.2 Travelling-Wave Array—Design Principles

The essence of establishing a slotted waveguide travelling-wave array design can probably be explained most elegantly by focusing on the resonant slot example. This approach is competently developed in Johnson [6], borrowing from earlier, fuller treatments, in Silver [7], Collin and Zucker [8], and Elliott [9]. The method is summarised here for completeness and ease of reference.

The basic geometry of the shunt slot array antenna, as implemented in rectangular waveguide, is illustrated in Fig. 7.2. The longitudinal slots (z-directed) are off-set from the waveguide side wall by distances x_n for the n slots, and each resonant slot is assumed to present a real conductance to the waveguide g_1, g_2, ….. g_N, as indicated in Fig. 7.2b. Adjacent slots are on opposite sides of the centre line to avoid grating lobes (see Chap. 6, Sect. 6.3.1) in the y-z plane. If the axial separation distance between slots d is approximately equal to $\lambda_g/2$ then adjacent slots differ in phase by almost 360° thus placing the main beam in the y-z plane close to, but not precisely at, broadside (ϕ = 0°). It should be noted that $d = \lambda_g/2$ is not possible for a slot array terminated in a matched load, as shown in Fig. 7.2. This is because the resonant slots reflect in phase, resulting in unacceptable levels of

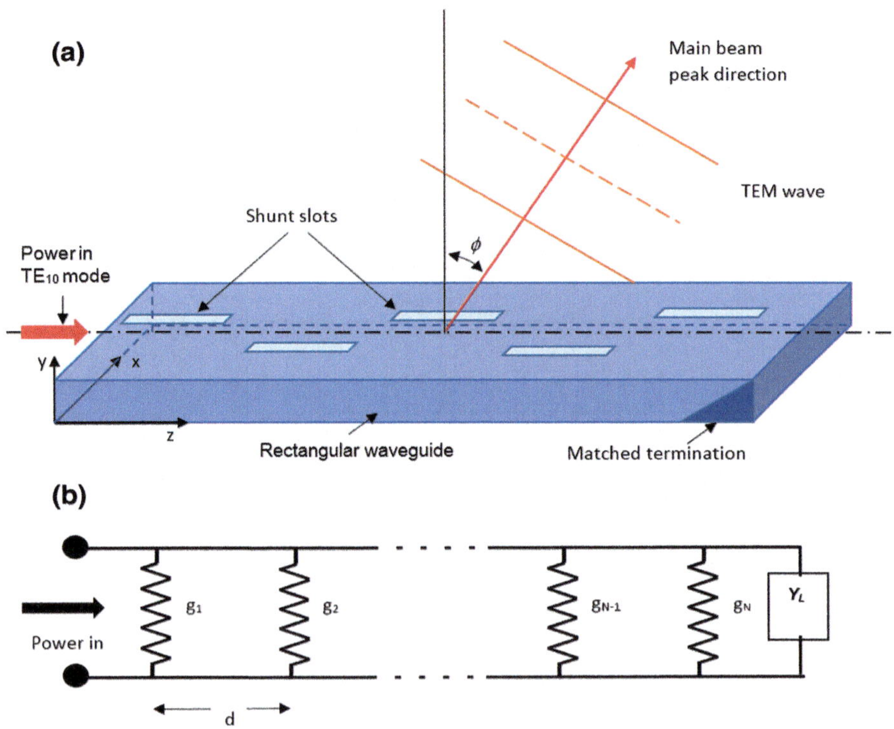

Fig. 7.2 Resonant slot travelling-wave array antenna (**a**) and equivalent transmission line model (**b**)

combined power being returned to the array input, and hence the microwave source. Operation of a travelling wave array exactly at broadside will be examined as a special case at the end of this section.

The main lobe of the travelling wave array radiation pattern is located in the angular range $-\pi/2 \leq \phi \leq \pi/2$ in the y-z plane. In the transverse x-y plane the pattern is essentially omni-directional in the forward half-space (+y-direction). Furthermore, the antenna is generally assumed to be terminated in a perfectly matched waveguide load represented by admittance Y_L in the equivalent transmission line.

For radiation pattern calculations in the ensuing section, we note here, referring to Fig. 7.2, that geometrically the centre of any given slot (n) can be defined as located at $\mathbf{r}_n$ where

$$\mathbf{r}_n = nd\hat{\mathbf{a}}_\mathbf{z} + x_n\hat{\mathbf{a}}_\mathbf{x} + \hat{\mathbf{b}}_\mathbf{y} \tag{7.3}$$

For a fixed height waveguide, b is common to all slots in the array, and hence it has no influence on pattern calculations and can therefore be omitted. Thus

$$\mathbf{r}_n = nd\hat{\mathbf{a}}_\mathbf{z} + x_n\hat{\mathbf{a}}_\mathbf{x} \tag{7.4}$$

7.2.1 Staggered Shunt Slot Array

In Chap. 6 it has been noted that the radiation pattern for an array antenna can, in the absence of mutual coupling, be elegantly formed from the product of the array factor and the element factor. The former applies to the radiation pattern of the equivalent array comprising isotropic radiators, while the latter alludes the radiation pattern of the isolated element. For a uniform array of isotropic elements it is shown that the radiation pattern can be simply represented mathematically [see Chap. 6, Eq. (6.3)] as:

$$\frac{E}{E_0} = 1 + e^{j\psi} + e^{2j\psi} + \cdots + e^{j(n-1)\psi} \tag{7.5}$$

where E_o is the field at a given radius for a single element, while E is the radiated field strength for the array at the same radius in the plane containing the array (y-z in Fig. 7.2). Also, for unstaggered elements, ψ the propagation phase difference between adjacent elements is given by (see Eq. 6.2) namely:

$$\psi = k_0 d \sin\phi + \alpha \tag{7.6}$$

However, for the travelling wave array, the slots are seldom equally excited and are seldom unstaggered, so Eqs. (7.5) and (7.6) have to be adjusted slightly to represent a non-uniform structure. Hence we obtain:

$$\frac{E}{E_0} = a_1 e^{jk_0\hat{\mathbf{r}}\cdot\mathbf{r}_1 + j\pi - j\beta_{10}d} + a_2 e^{jk_0\hat{\mathbf{r}}\cdot\mathbf{r}_2 + 2j\pi - 2j\beta_{10}d} + \cdots + a_n e^{jk_0\hat{\mathbf{r}}\cdot\mathbf{r}_n + jn\pi - jn\beta_{10}d} \tag{7.7}$$

where d is the element separation in the z-direction. It is also the distance from the origin to the first element. a_n is the field excitation amplitude at the nth slot while β_{10} is the propagation phase for the TE_{10} mode in the waveguide. The unit vector $\hat{\mathbf{r}}$ is given by:

$$\hat{\mathbf{r}} = \hat{\mathbf{a}}_\mathbf{x} \sin\Delta\cos\theta + \hat{\mathbf{a}}_\mathbf{y} \sin\Delta\sin\theta + \hat{\mathbf{a}}_\mathbf{z} \cos\Delta \tag{7.8}$$

with $\Delta = \frac{\pi}{2} - \phi$ in the y-z plane, and θ the angular rotation from the x-axis in the x-y plane. Consequently

$$\psi = k_0 d \sin\phi + \pi - \beta_{10}d + k_0 x_n \cos\phi\cos\theta \tag{7.9}$$

Generally we are interested in the radiation pattern in the y-z plane ($\theta = \pi/2$ plane) in which case the most relevant and useful form of Eq. (7.9) is:

$$\psi = k_0 d \sin\phi + \pi - \beta_{10}d \tag{7.10}$$

Evidently, and perhaps not surprisingly, Eq. (7.10) closely matches Eq. (7.6) except that the α phase term is replaced by the slot-to-slot phase delay ($-\beta_{10}d$) for the TE_{10} mode in the waveguide. The additional π phase shift term accounts for slot staggering. It disappears for an array with shunts slot located only on one side of the centre line.

Referring back to Chap. 6, we have observed there that the main radiating beam in the plane of the array (y-z plane) occurs where $\psi = 0$, while secondary beams or grating lobes occur at $\psi = 2\,m\pi$ ($m = 1, 2, \ldots M$). That is, for primary lobes we must have:

$$k_0 d \sin\phi + \pi - \beta_{10}d = 2m\pi \tag{7.11}$$

Hence primary lobes occur at values of the slot spacing d given by:

$$d = \frac{(2m-1)\lambda_0\lambda_{g10}}{2(\lambda_{g10}\sin\phi - \lambda_0)} \tag{7.12}$$

Equation (7.12) is plotted in Fig. 7.3, for low values m, for a resonant shunt slot travelling-wave array located in 22.86 mm × 10.16 mm (WG16) rectangular waveguide at an operating frequency of 10.0 GHz. Equation (7.12) reveals that the curves in Fig. 7.3 intercept the $\phi = -90°$ axis at

$$d = \frac{(1-2m)\lambda_0\lambda_{g10}}{2(\lambda_0 + \lambda_{g10})} \tag{7.13}$$

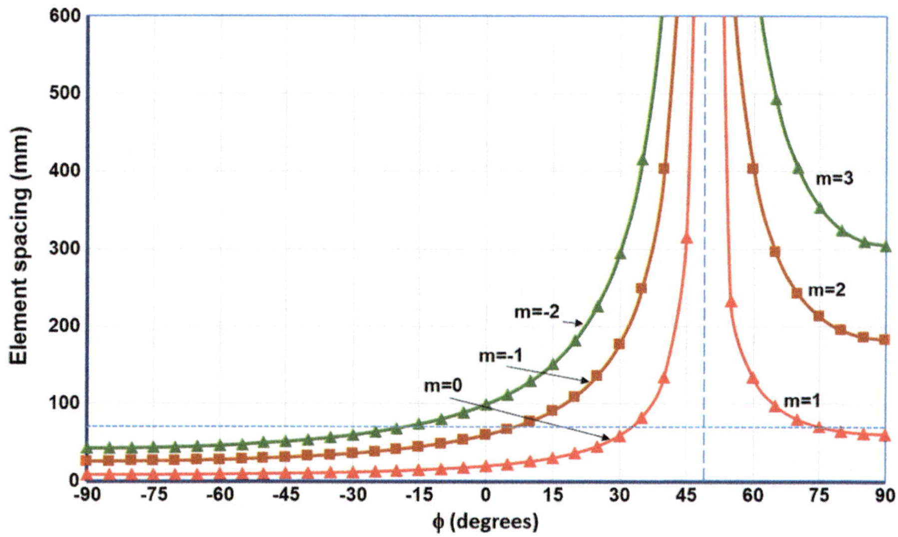

Fig. 7.3 Influence of slot spacing d on the principal beam locations (ϕ degrees) for a staggered shunt slot travelling wave array in WG16 waveguide at 10 GHz

and on the ϕ = +90° axis at

$$d = \frac{(2m-1)\lambda_0\lambda_{g10}}{2(\lambda_0+\lambda_{g10})} \tag{7.14}$$

while the infinity occurs where

$$\sin\phi = \frac{\lambda_0}{\lambda_{g10}} \tag{7.15}$$

The figure thus provides a clear and rapid indication of the number of principal beams generated by the array for a particular value of d. At a given value of d (60 mm say), if a horizontal line (blue dashed line) is inserted onto the figure at this value then the possible main beams, or grating lobes, in ϕ-space will be predicted to occur at approximately −20°, 7°, 33° and 75°. To secure a single main beam in the y-z plane the slot spacing d must be located between the m = 0 curve and the m = −1 curve at ϕ = −90°. That is, we must have:

$$\frac{\lambda_0\lambda_{g10}}{2(\lambda_0+\lambda_{g10})} \le d \le \frac{3\lambda_0\lambda_{g10}}{2(\lambda_0+\lambda_{g10})} \tag{7.16}$$

For WG16 operating at 10 GHz this implies that d should be in the range 8.55–25.65 mm. The m = 1 solution between ϕ = 75° and ϕ = 90° is not significant in the above range of d if $\lambda_g < 2\lambda_0$. This condition lies in the normal operating range for standard rectangular waveguides.

7.2.2 Unstaggered Shunt Slot Array

Although seldom adopted because of grating lobe issues a brief examination of the unstaggered shunt slot array provides further insight into the nature of the travelling-wave array antenna. As we have already noted, the fundamental phase relationship governing primary beam locations, namely Eq. (7.11), is modified by the absence of the π term when applied to the unstaggered case. Hence the equivalent d-equation [to Eq. (7.12)] locating main beams becomes

$$d = \frac{m\lambda_0\lambda_{g10}}{(\lambda_{g10}\sin\phi - \lambda_0)} \tag{7.17}$$

This equation is presented in graphical form in Fig. 7.4.

Figure 7.4 is superficially the same as Fig. 7.3, but careful examination reveals two main differences. The $m = 0$ solution in the staggered array case has become the $m = -1$ solution. However this does not mean that an $m = 0$ solution does not exist. From Eq. (7.17) it is clear that d is non-zero for $m = 0$ only where the denominator is also zero, namely at $\sin\phi = \frac{\lambda_0}{\lambda_{g10}}$. This condition is represented by

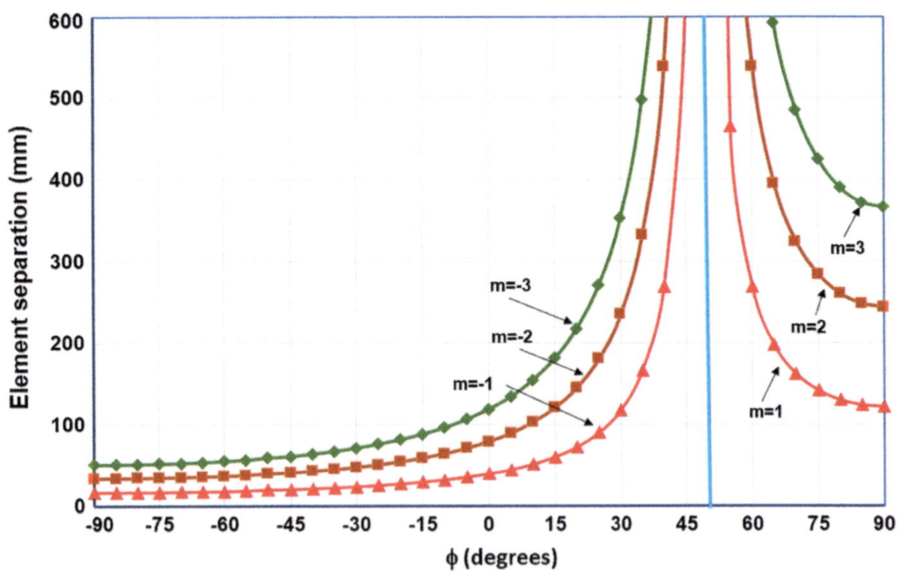

Fig. 7.4 Influence of slot spacing d on the principal beam locations (ϕ degrees) for a unstaggered shunt slot travelling wave array in WG16 waveguide at 10 GHz

the blue solid line in Fig. 7.4. It is solid because it now denotes a real solution, usually termed the d-independent solution. Physically this operating condition occurs when the phase velocity of the TE_{10} mode in the waveguide $v_{p10} = \omega/\beta_{g10}$ matches the z-component of the velocity of the radiated beam along the outer surface of the waveguide. For this solution to be a single beam operating mode, the element spacing d must be less than for the m = −1 solution at $\psi = -90°$, namely

$$d = \frac{\lambda_0 \lambda_{g10}}{(\lambda_{g10} + \lambda_0)} \tag{7.18}$$

or $d < 17.1$ mm in WG16 at 10 GHz.

This situation can be improved marginally by changing the frequency of operation to a lower value. This moves the TE_{10} mode toward its cut-off value (6.56 GHz) such that λ_{g10} increases more rapidly than λ_0. The result, as Fig. 7.5a shows, is that the blue vertical line, at the 'infinity', moves leftwards toward a lower value of ϕ, increasing the separation between the curves representing $m = -1, -2, -3$, and hence improving the d-range for single beam operation. The downside of this change is a more dispersive TE_{10} mode. The opposite trend is evident when the system frequency is raised as shown in Fig. 7.5b with the d-independent beam shifting to a higher value of ϕ.

7.2.3 *Travelling-Wave Array Antenna Design*

Long before the advent of electromagnetic wave simulation software, based on either moment methods or finite element methods, and before the appearance of desk top computers with both sufficient memory capacity, and computational speed, to accommodate complex antenna investigations, successful techniques were being developed to solve the specific problem presented by travelling-wave arrays of the shunt slot variety [9, 10]. For dependable and quick assessments of preliminary array designs, or for design engineers lacking access to modern simulation software and the means to run it, the above problem specific solutions continue to be helpful. Their relative simplicity mathematically, and their impressive accuracy, largely lies with the fact that shunt (end-to-end) slots in waveguide incur negligible mutual coupling, whereby the local fields from one slot, influence the fields in near neighbours (see Chap. 5).

The design method summarised here is attributable to Kaminow and Stegen [4], and is largely formulated by applying power conservation principles and transmission line theory, to a relatively complete two-wire transmission circuit model for the shunt slot array antenna, which is illustrated in Fig. 7.6. In the model the slots are represented by normalised admittances (y_n) rather than conductances (g_n) to accommodate the fact that longitudinal slots of finite length, even if resonant ($L = \lambda_0/2$), introduce phase shift into the transmission (τ_n) and reflection coefficients (ρ_n). Hence we define the complex admittance (normalised) of the nth slot as:

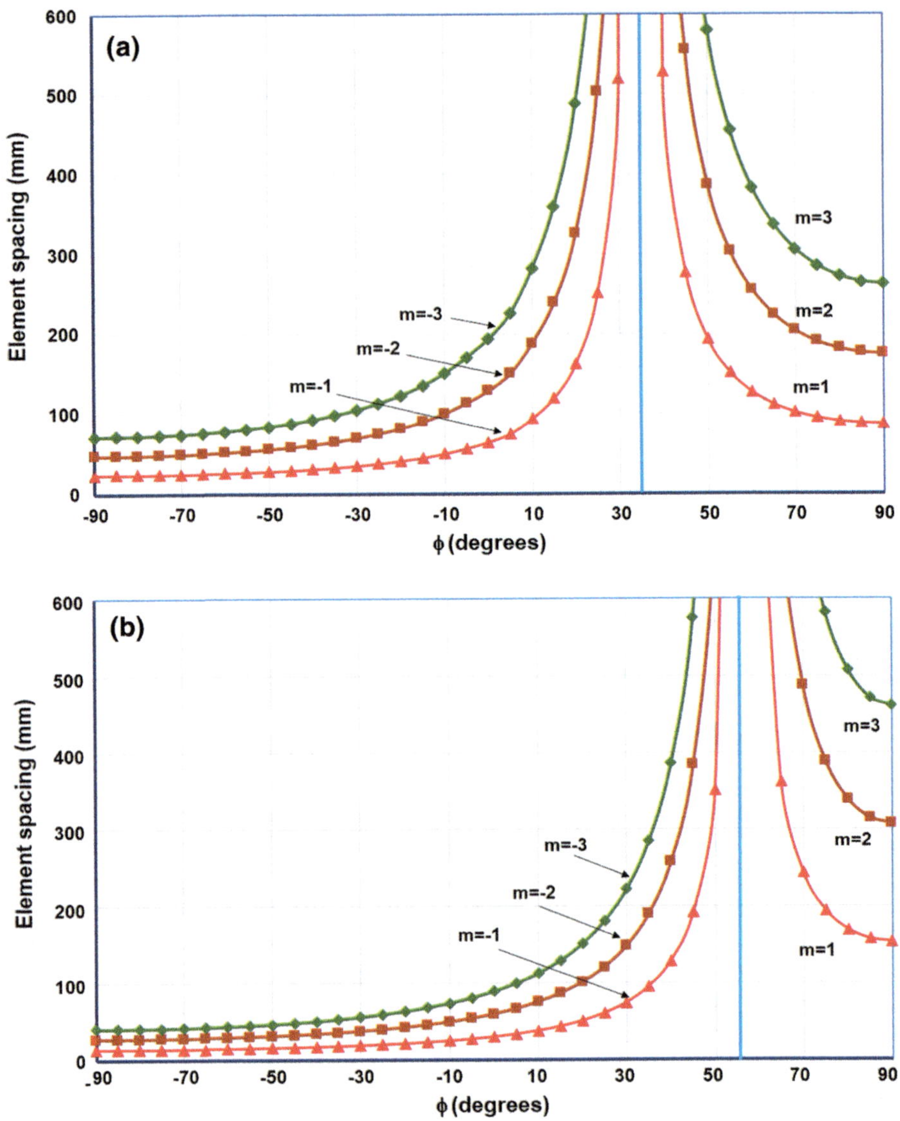

Fig. 7.5 Influence of slot spacing d on the principal beam locations (ϕ degrees) for a unstaggered shunt slot travelling wave array in WG16 waveguide at **a** 8 GHz and **b** 12 GHz

$$y_n = g_n + jb_n \tag{7.19}$$

The normalised admittance 'seen' by the nth slot 'looking' toward the termination is also defined as:

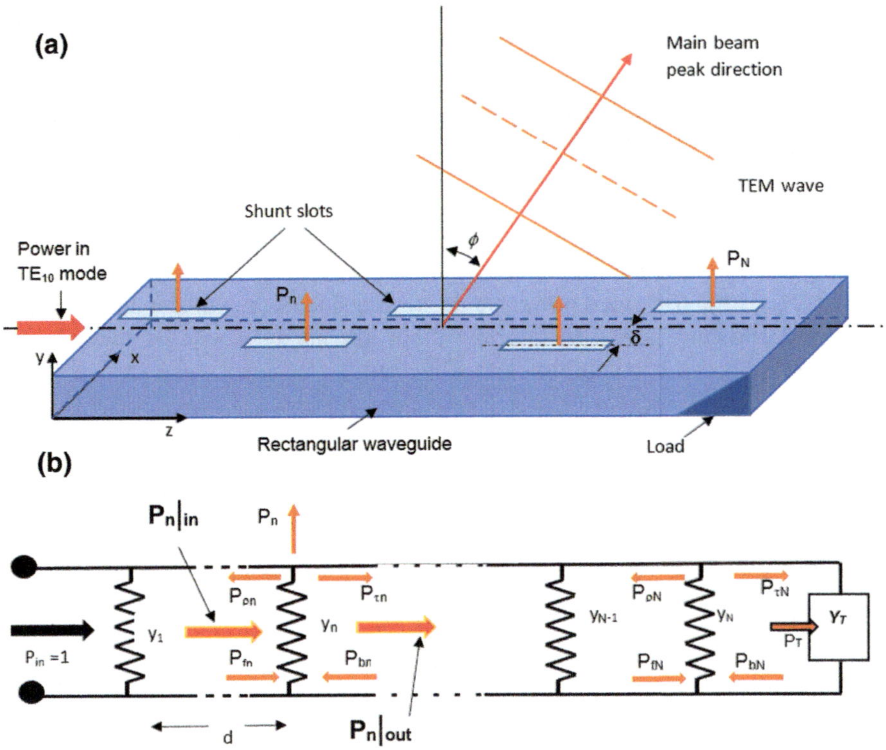

Fig. 7.6 Shunt slot travelling wave array antenna—design example **a** array configuration, **b** general transmission line model

$$y_{Tn} = g_{Tn} + jb_{Tn} \tag{7.20}$$

Consequently, 'looking' toward the nth slot from the $(n-1)$th slot, the combined admittance becomes:

$$y_{Tn-1} = y_n + y_{Tn} = g_{Tn} + jb_{Tn} + g_n + jb_n \tag{7.21}$$

Hence, using transmission line theory, the reflection coefficient seen by the nth slot can be expressed as:

$$\rho_n = \frac{1 - y_{Tn}}{1 + y_{Tn}} \tag{7.22}$$

For generality, the waveguide is assumed to be imperfect exhibiting a small level of wall loss. This is accommodated in the model by defining a complex propagation coefficient:

$$\gamma_{10} = \alpha_{10} + j\beta_{10} \tag{7.23}$$

Now the full range of power contributions at the nth slot is delineated in Fig. 7.6b, and hence applying power conservation there, we obtain:

$$P_n = P_{fn} - P_{\rho n} + P_{bn} - P_{\tau n} \tag{7.24}$$

where, P_n is the power radiated by slot n, P_{fn} is the power in the waveguide reaching slot n, $P_{\tau n}$ is the forward power in the waveguide on the load side of the nth slot, $P_{\rho n}$ is the backward power on the source side of the nth slot and P_{bn} is the 'backward' flowing power in the waveguide on the load side of the nth slot. Equation (7.24) can be simplified into (Fig. 7.6b):

$$P_n = P_n|_{in} - P_n|_{out} \tag{7.25}$$

where $P_n|_{in}$ and $P_n|_{out}$ represent the net powers in the waveguide flowing into and out of the nth slot. If the voltage on the transmission line representation of the slot array (Fig. 7.6b) at the nth slot is V_n then the following definitions apply:

$$P_n = \frac{1}{2}|V_n|^2 g_n \tag{7.26}$$

and

$$P_{fn} = \frac{1}{2}|V_n|^2 g_{Tn} \tag{7.27}$$

Also at the termination:

$$P_T = \frac{1}{2}|V_T|^2 g_T \tag{7.28}$$

The travelling-wave array design commences with the observation, as we have already noted, that the required radiation pattern for the array is dictated by the slot excitation levels a_n (see Chaps. 3 and 6), such that $P_n \propto a_n^2$. We also note that in a lossless waveguide $V_n = V_T = V$, for all n. Generally the travelling-wave array is created in low loss waveguide, in which case a trial starting point for a travelling-wave array design is to assume that the waveguide loss is sufficiently close to zero to let $V_n = V = constant$. In this case, from Eq. (7.26), we can state that:

$$P_n = C_g a_n^2 \tag{7.29}$$

The constant C_g can be estimated by introducing a trial value for that proportion of the source power reaching the load after passing through the array (r say). Thus

$$r = \frac{P_T}{P_{in}} = P_T \tag{7.30}$$

if power from the source P_{in} is normalised to unity. With these simplifications we can define [9] a continuous conductance per unit length g(z) such that:

$$g(z) = g(nd) = \frac{g_n}{d} \tag{7.31}$$

In this approximated 'array', now represented by a continuous conductance, we can also replace P_n by $P(z)$ and P_{fn} by $P_f(z)$. Consequently for a loss free waveguide and a perfectly matched load, if the input power is normalised to unity then at any position z:

$$g(z) = \frac{P(z)}{P_f(z)} \tag{7.32}$$

$$P_f(z) = 1 - \int_0^z P(z)dz = 1 - \int_0^z C_g(a(z))^2 dz \tag{7.33}$$

with $a(z) = a_n/d$ and using the continuous version of Eq. (7.29). Furthermore at the load end of the array ($z = L$) we must have, from Eqs. (7.30) and (7.33):

$$r = 1 - \int_0^L C_g(a(z))^2 dz \tag{7.34}$$

On combining Eqs. (7.32), (7.33) and (7.34) an elegant expression for the continuous conductance is generated—namely:

$$Lg(\varsigma) = \frac{(a(\varsigma))^2}{\frac{1}{1-r}\int_0^L (a(\varsigma))^2 d\varsigma - \int_0^L (a(\varsigma))^2 d\varsigma} \tag{7.35}$$

where $\zeta = z/L$. Equation (7.35) is plotted in Fig. 7.7 for $a(\zeta)$ distributions with Taylor functionality in order to produce Chebyschev radiation patterns [10]—one which produces −20 dB flat sidelobes in the array radiation pattern, while the second case develops −25 dB sidelobes.

Four curves are presented in Fig. 7.7, two associated with −20 dB sidelobe designs (blue traces with circle markers) and two associated with −25 dB sidelobe designs (red traces with square markers). In both cases the curves are differentiated by the power (r), as a percentage of the source power, reaching the load. It is noticeable that for the low power cases (r = 5%) $g(\zeta)$ as a function ζ is much more sensitive to increasing ζ than for the cases where r = 35%. The reason for this

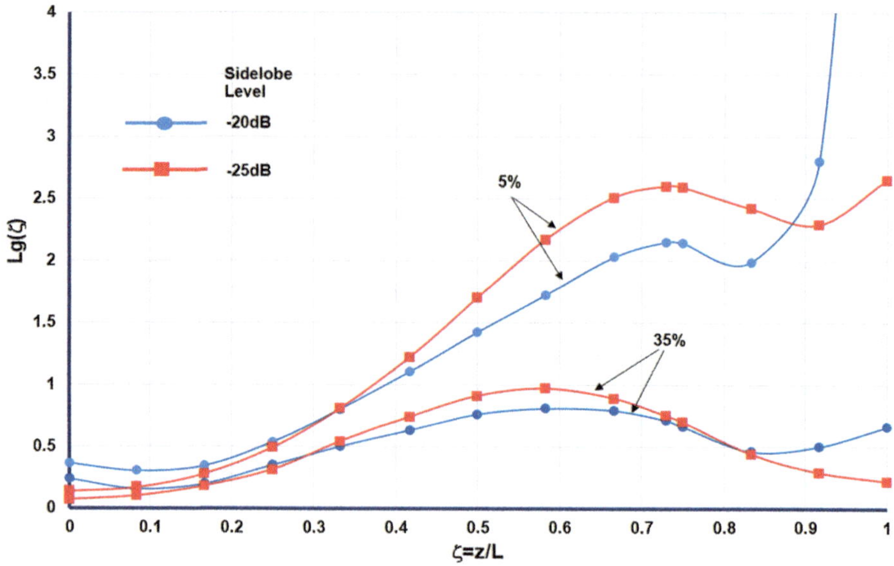

Fig. 7.7 Continuous conductance profile for lossless, perfectly matched, shunt-slot travelling-wave array antennas (with acknowledgement to [9])

difference in behaviour lies with the fact that with only 5% of the input power reaching the load the last few slots in the array are exposed to a very weak level of TE_{10} mode excitation. However, given that $a(\varsigma) \propto \sqrt{P(\varsigma)}$, it becomes clear from Eq. (7.31) that a low power level ($P_f(\zeta)$) requires a large $g(\zeta)$ value to secure the desired slot excitations $a(\zeta)$. This difficulty is exacerbated for the −20 dB sidelobe array which requires higher $a(\zeta)$ levels at the slots near the array extremities to procure the higher sidelobes. Note that this option equates to higher gain. In fact the way in which the magnitude of $g(\zeta)$ escalates as ζ approaches unity highlights a significant design limit for the shunt slot travelling-wave array antenna. Eventually, as r is reduced, the level of $g(\zeta)$ cannot be realised by a shunt slot in waveguide where the maximum $g(\zeta)$ is restricted by the optimum off-set ($\delta = a/2$) of a slot from the centre line of the waveguide [see Fig. 7.1 and Eq. (7.1)].

A compromise between the magnitude of r, array efficiency and the desired gain becomes inevitable. High efficiency (low r) is easier to procure with low sidelobe levels which equate to less than optimum gain. In general this balancing process becomes less pronounced for long arrays with large N. With r at the 35% level in Fig. 7.7 it is interesting to note that the $g(\zeta)$ distribution along the array follows closely the Taylor distribution applied to $a(\zeta)$. This is simply because $P_f(\zeta)$ remains relatively constant along the array in these cases. Finally, low sidelobe designs require care in implementing a terminating load which must be well matched to the feed waveguide forming the array. An imperfect match results in a reflected TE_{10} mode which becomes a source of a secondary radiation pattern with a 'main' beam pointing in a direction equal and opposite, relative to the array normal, to the designed beam. This 'reflection' beam together with the original sidelobes in the same direction should ideally combine to form a lobe which is higher than the designed sidelobe level. Calculations performed by Dion [10] suggest that at the −30 dB sidelobe level in an array designed on the basis of 'dumping' 5% of the source power in the termination, a load with a generous VSWR of ~ 1.34 would be satisfactory. On the other hand, at r = 35% a high quality load with a VSWR ≈ 1.1 is required to maintain the sidelobe specification. At the −20 dB sidelobe level the load requirements are, not unexpectedly, considerably more relaxed with a load with a VSWR ≈ 1.4 being acceptable, even at r = 35%.

For a slot array of predetermined length L, the Dion continuous conductance technique permits the allocation of trial values for the discrete conductances g_n in an ensuing practical design. For such an N element array design with $L = Nd$ we choose g_n as defined in Eq. (7.31) with $z = nd$, on the assumption that the first slot is located a distance d from the array input. An estimate for the parameter r is also furnished by the Dion analysis and hence the power P_T transmitted into the load is calculable.

For an array more accurately modelled by admittances (Fig. 7.6b) embedded in a waveguide, which is not loss free, Kaminow and Stegen [4] provide an iterative procedure to procure a design. This commences with the trial conductances from the above procedure and the known conditions at the load, then progresses slot by slot back toward the array input, by employing power conservation principles and transmission line theory. The power conservation requirement is defined in Eq. (7.24), with power definitions in Eqs. (7.26), (7.27) and (7.28). Since it is unlikely that the termination will be other than well matched we can reasonably set $g_T = 1$, in which case $V_T = \sqrt{P_T}$.

At slot N, which we assume is distance d from the load, the power conservation requires:

$$P_N|_{out} = \exp(\alpha_{10} d) P_T \tag{7.36}$$

and for a matched termination Eq. (7.32) applies, giving:

$$g_N = \frac{P_N}{P_{fN}} \approx \frac{P_N}{P_N|_{in}} \tag{7.37}$$

Now, for a lossy line, transmission line theory gives the voltage at slot N as:

$$V_N = V_L(\cosh(\gamma_{10} d) + y_T \sinh(\gamma_{10} d)) \tag{7.38}$$

while at slot $N - 1$ we obtain:

$$V_{N-1} = V_N(\cosh(\gamma_{10} d) + y_{TN} \sinh(\gamma_{10} d)) \tag{7.39}$$

using Eq. (7.21). At the *N*th slot transmission line theory also yields:

$$y_{\tau N} = \frac{y_T + \tanh(\gamma_{10}d)}{1 + y_T \tanh(\gamma_{10}d)} \tag{7.40}$$

Furthermore from Eq. (7.21)

$$y_{TN} = y_N + y_{\tau N} = g_N + jb_N + g_T \tag{7.41}$$

with b_N computed from:

$$\angle \frac{jV_n y_n}{jV_{n+1} y_{n+1}} = \angle \exp(-j\beta_{10}d) \tag{7.42}$$

This equation ensures that the slot to slot excitation phase shifts in accordance with array beam formation in the specified design direction. It leads to the following iterative relationship for phase, namely

$$\tan^{-1}\frac{b_n}{g_n} = \tan^{-1}\frac{b_{n+1}}{g_{n+1}} + \beta_{10}d + \angle\frac{V_{n+1}}{V_n} \tag{7.43}$$

For the *N*th slot, 'looking' into a matched termination, this reduces to:

$$\tan^{-1}\frac{b_N}{g_N} = \beta_{10}d \tag{7.43b}$$

With P_N, g_N and b_N at the final slot now established the iteration becomes focused on the ($N - 1$)th slot. Firstly the power levels at slot *N* can be transferred, in accordance with transmission line theory and power conservation, back to slot $N - 1$ by applying power conservation at a reference plane on the load side of slot N − 1 (see Fig. 7.6). This leads to

$$\begin{aligned} P_{N-1}|_{out} &= P_{\tau,N-1} - P_{b,N-1} = \exp(2\alpha_{10}d)P_{f,N} - |\rho_N|^2\exp(-2\alpha_{10}d)P_{f,N} \\ &= \left[1-|\rho_{\mathrm{N}}|^2\right]\exp(2\alpha_{10}d)P_{f,N} + |\rho_N|^2\exp(2\alpha_{10}d)P_{f,N} - |\rho_N|^2\exp(-2\alpha_{10}d)P_{f,N} \\ &= \left[1-|\rho_{\mathrm{N}}|^2\right]\exp(2\alpha_{10}d)P_{f,N} + 2|\rho_N|^2\sinh(2\alpha_{10}d)P_{f,N} \end{aligned} \tag{7.44}$$

And

$$P_N|_{in} = \left[1 - |\rho_N|^2\right]P_{f,N} \tag{7.45}$$

The following useful equation [10] relating the $N - 1$th slot to the *N*th slot then results:

$$P_{N-1}\big|_{out} = P_N\big|_{in}\left\{\exp(2\alpha_{10}d) + \frac{2|\rho_N|^2}{1-|\rho_N|^2}\sinh(2\alpha_{10}d)\right\} \quad (7.46)$$

where ρ_n is defined in Eq. (7.22). Hence with P_N known at the Nth slot and $P_N|_{in}$ available from Eq. (7.37) the power leaving the $N-1$th slot can be calculated using Eq. (7.46). The power on the source side of slot N − 1 is computed using Eq. (7.25) with g_{n-1} derived from the transmission line Eq. (7.40). V_{N-1} is determined from transmission line theory [Eq. (7.39)] which then permits the evaluation of $b_{N-1.}$ The process is then repeated for slot $N-2$, $N-3$, etc. back to slot *1*. Eventually, with all of the g_n's and b_n's evaluated the slot offsets δ_n, the slot spacings d_n and the slot lengths L_n can be deduced from Eqs. (7.1) and (7.12) with the slot resonant lengths 'tweeked' to accommodate the susceptance values. Actually, the sophisticated antenna simulation software packages which are available to today's designers, allows the electrical data generated above for the travelling-wave array antenna to be inserted into a waveguide slot analysis program (see Chap. 5) to ascertain accurate dimensional design parameters including slot widths and waveguide wall thicknesses.

7.2.4 Resonant Array Design

A disadvantage of the travelling-wave array antenna described in the preceding sections is that the operating beam, or main beam, is not located at broadside ($\phi = 0$), and furthermore it scans with changes in frequency, as is made clear in Figs. 7.3, 7.4 and 7.5. However, a broadside shunt slot array can be realised by 'resonating' the waveguide if bandwidth is not an issue. A typical application where a very narrow bandwidth is acceptable is in low cost antennas for search and rescue radar beacons. The concept is illustrated in Fig. 7.8. The array is 'shorted' exactly $\lambda_{g10}/4$ (or $3\lambda_{g10}/4$) beyond the last slot, while the staggered shunt slots forming the array are spaced exactly $\lambda_{g10}/2$ apart. The result is a standing-wave excited slot array rather than a travelling-wave excited array. Note that while the slots are also operated at their resonant length ($\lambda_0/2$), the array format is termed 'resonant' because of the standing-wave excitation mode. The essence of this mode is depicted in Fig. 7.9. The TE_{10} mode which is presumed to enter the waveguide from the left (Fig. 7.8a) is totally reflected at the shorting plate terminating the waveguide on the right. In the absence of slots, and in a lossless waveguide, a standing-wave with perfect nulls at distances $p\lambda_{g10}/2$ ($p = 1, 2, 3 \ldots$) in front of the shorting wall will be formed.

For resonant length slots separated by precisely $\lambda_{g10}/2$ transmission line theory dictates that the effective conductance at the input to the array (g_e), as depicted in Fig. 7.9, is given by:

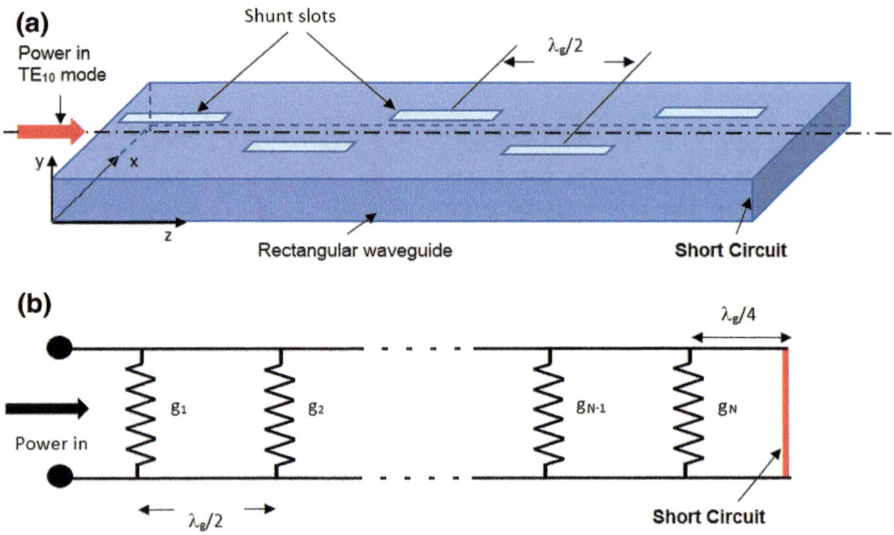

Fig. 7.8 Resonant form of the waveguide shunt slot array **a** schematic representation, **b** circuit representation

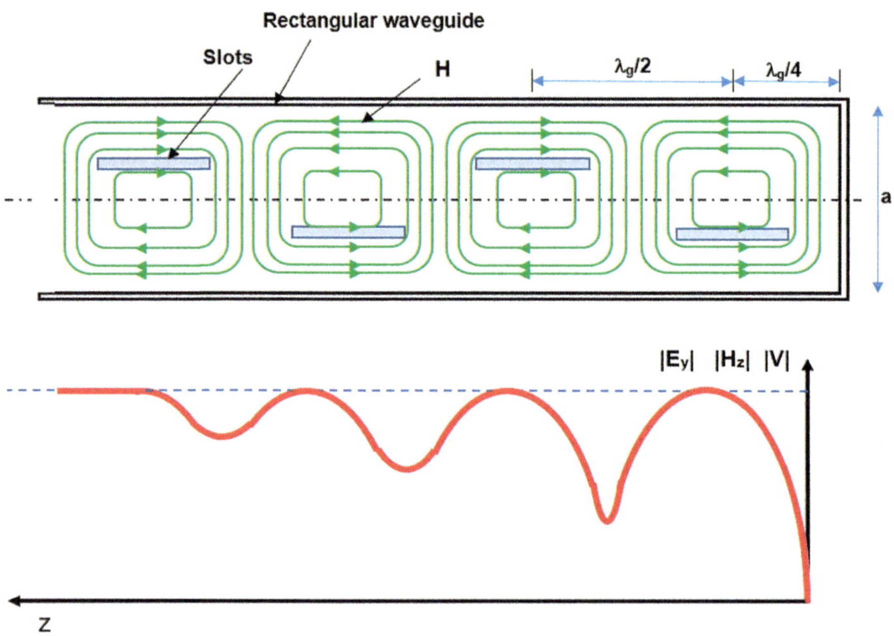

Fig. 7.9 Pictorial representation of the slot excitation mechanism in the resonant array

$$g_e = \sum_{n=1}^{N} g_n \tag{7.47}$$

where N is the number of slots and for a good array matched to the source, normally we would choose $g_e = 1$. Note that the $\lambda_{g10}/4$ short circuited line presents an open circuit across the final slot.

The matching condition alone is insufficient to determine all of the g_n values, but we have seen from Chap. 3, and from Chap. 6, that the field distribution across the aperture of the antenna, or the element excitation level in the case of an array, can be shaped to procure high directivity or low side lobe radiation patterns (see Fig. 3.7). Consequently, for the resonant array let the excitation level, or coefficient, for the nth slot be a_n. For a uniform array (see Chap. 6) the a_n's have a fixed level, but may be set to a cosine pattern (for example) if lower sidelobes are desired. Since the slot conductance is proportional to the radiated power [see Eq. (7.2)] we can therefore write:

$$g_n = C_g a_n^2 \tag{7.48}$$

where C_g is a constant yet to be defined. Hence combining Eqs. (7.47) and (7.48) yields:

$$C_g \sum_{n=1}^{N} a_n^2 = 1 \tag{7.49}$$

for a matched array. If the a_n's are established to meet pattern requirements then the constant C_g becomes calculable. Once C_g is known the g_n's can be determined and hence the remaining unknowns required to complete the array design, namely the slot off-sets d, can be computed by resorting to Eq. (7.1).

7.3 Some Slot Array Applications

With their relative simplicity, from a geometrical perspective, their robustness, their reliability and efficiency, slotted waveguide array antennas have found particular favour in airborne radar, radiometry and satellite communications roles. Furthermore, in both resonant and travelling-wave forms, control of slot excitation levels and phase relationships can be sufficiently precise, as we have seen, to permit array designers to procure antenna radiation patterns with tightly specified gains, and sidelobe levels. This leads to high operating efficiencies. It is fair to say that modern precision manufacturing techniques now permit relatively faithful replication of computer simulations, and this has resulted in array designs displaying sidelobe levels well below −20 dB. A typical 'flat plate' waveguide slot array is shown in Fig. 1.13. The flat plate diameter, in the case of a circular embodiment, can range from well over a metre at S and C band frequencies to little more than a few centimetres at mm-wave frequencies (>60 GHz). Applications for such antennas range widely from airborne tracking and seeking roles to ground based mobile phone systems which predominantly employ linear array versions of the slotted waveguide antenna (see Johnson [6] and Rudge [11]).

Following pioneering effort in the 1940s and 1950s [1–3], slot array antennas in radar, and in communications applications flourished, arguably most fruitfully and determinedly during the 'cold war'. However by the 1990s these developments had seemed to stall. The advent of ever more precise manufacturing methods, such as laser machining, electron beam technology, photographic etching and 3D printing, have recently begun to push applications into the mm-wave frequency range (circa 60 GHz). The low-loss advantage of waveguide, over alternative feed mechanisms, remains a major attraction for antenna designers, and unquestioningly helps to maintain interest in slotted waveguide arrays. At mm-wave frequencies relatively small arrays can procure the tightly controlled radiation patterns, which are instrumental in the realisation of atmospheric sensors, automotive radar sensors and gigabit/second wireless interfaces [12]. In the 'wireless' category, a number of applications are being planned such as high definition multimedia interfaces (HDMI) enabling (1) uncompressed high definition (HD) video, streaming wirelessly rather than by inconvenient cable technology, (2) computer mobility in distributed environments, (3) wireless docking stations, (4) wireless gigabit Ethernet operation, (5) fast transfers of bulky files, (6) wireless gaming [13]. Other applications will undoubtedly emerge in the near future.

Wireless systems at 60 GHz have materialised as one of the most promising candidates for multi-gigabit indoor communication links. Evidently, a deciding factor that encouraged this recent interest, lies with the huge unlicensed bandwidth (up to 7 GHz) available worldwide. While this is comparable to the unlicensed bandwidth allocated for ultra-wideband (UWB) systems, at 60 GHz the available bandwidth is continuous and power limits are much less restrictive. This difference is governed by the fact that an UWB system must operate in concert with other systems, and is thus subject to very much more demanding regulations [5].

In communication terms, the bandwidth available at 60 GHz band is one of the largest unlicensed bandwidths which has ever been allocated. The huge 7 GHz bandwidth has the potential to enable systems demanding very high levels of channel capacity and flexibility thus making 60 GHz technology particularly attractive for gigabit wireless applications. Furthermore, 60 GHz regulation allows much higher transmit power compared to other existing wireless local area networks (WLANs) and wireless personal area networks (WPANs) operating at microwave frequencies, so accommodating and counteracting the higher transmit power levels, which are necessary to overcome the inevitably higher path losses encountered at 60 GHz.

At the time of writing, it is true to say that a large number, and a wide range, of planar antennas have been studied for millimeter-wave radio and radar applications. The most obvious alternative to the waveguide slot array is the substrate mounted patch array [6]. While high gain operations have been demonstrated with the latter, microstrip substrates suffer from serious loss mechanisms at mm-wave frequencies. Consequently, the efficiency diminishes as the gain and/or frequency is pushed upwards. Despite the maturity of patch array technology these limitations are unavoidable. It has been estimated that the efficiency of microstrip arrays in the 60 GHz band, with gains of the order of 35 dBi, could be 20% lower than an equivalent waveguide slot array [16]. On-chip antennas have also been mooted for mm-wave wireless roles but these exhibit their own particular drawbacks. On conductive high-permittivity silicon their radiation efficiency is inevitably low. As a result, in a WPAN environment, where wireless communication distances of the order of 10 m need to be accomplished, multi-element arrays are necessary to achieve the 15–20 dBi gain levels required for reliable transmission. If implemented in MMIC chip technology such systems attract high cost [16].

It is, therefore, difficult not to conclude that, for wireless applications, waveguide slot antenna arrays represent the most attractive candidates where high-gain planar antennas, with high efficiency, are sought. Nevertheless waveguide fed slot arrays present their own difficulties where low cost is important. Primarily, the complicated three dimensional feed arrangements in early designs tended to limit their use to cost-insensitive military or professional applications. An array configuration with manufacturing costs potentially comparable to that of microstrip counterparts was required. The solution lay with single-layer waveguide slot array designs which lend themselves to mass production [13–18].

The impetus for this development was to note that, in principle, planar waveguide fed slot arrays can be reduced to two parts. These are the slotted top plate and the grooved bottom plate containing the waveguide channels (Fig. 7.10). Several versions of these single-layer waveguide arrays, intended for high efficiency over 12 and 20 GHz bands, were initially developed [13–17]. Subsequently, manufacturing methods improved sufficiently to permit the extension of single layer waveguide arrays to higher frequencies up to and including 60 GHz [18–20].

Slotted waveguide, single-layer, planar array antennas, aimed at satisfying mm-wave developments in communications, evolved rapidly after 1995, under the promptings of research scientists at the Tokyo Institute of Technology in particular. Initially these revolved around the conceptual arrangements depicted in Fig. 7.10 and in particular the alternate phase arrangement in Fig. 7.10b [13]. This anti-phase relationship between adjacent waveguides is counteracted by the 'mirrored' slot staggering, which is clearly seen by comparing Fig. 7.10a, b. So while the phase reversals have no influence on the beam forming, it has significant implications for the wave propagation in the waveguide grooves. With no net current flow in the internal walls the degree of electrical contact between the slot plate and the channelled plate could be greatly relaxed. This meant that more cost effective array manufacturing techniques could be explored [15, 16].

In Ref. [15] it was shown that while anti-phase operation of a waveguide fed planar slot array eliminated the need for 'good' electrical contact between the common internal waveguide walls, and the slotted top plate, the same did not apply to the peripheral walls. The solution was to introduce traditional 'choke' technology

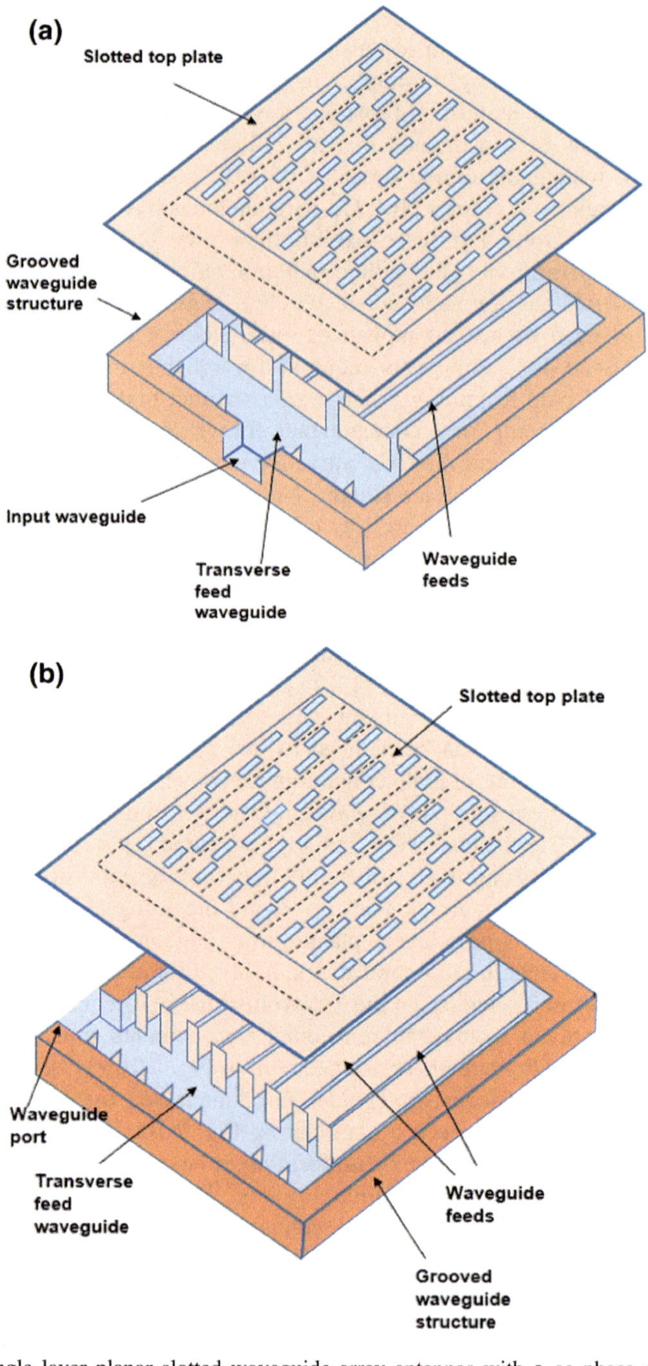

Fig. 7.10 Single layer planar slotted waveguide array antennas with **a** co-phase corporate feed **b** anti-phase transverse linear waveguide feed

(see Fig. 1.13), long used to suppress leakage between minimal contact waveguide flanges. With the addition of choke suppression, Ref. [16] demonstrated that mass production of the waveguide antenna at low cost was possible. Die-casting was adopted for the corrugated base plate of the antenna to circumvent high precision metal cutting which was considered to be too time-consuming and expensive. The dimensional tolerance associated with die-casting is typically only $\pm 50 \mu m$ and normally some further machining is necessary in the 26 GHz band. To avoid this, a high precision casting technique was developed (see Ref. [16]).

To confirm the fabrication process, a travelling-wave array antenna with a specified gain of over 30 dBi was designed, for operation at approximately 25 GHz. The aperture area occupied by slots was set to be 16 cm × 15 cm. It comprised 16 parallel waveguides, with a = 8 mm and b = 3 mm, each feeding 19 staggered shunt slots. For the die-cast antenna the gain exceeded 31 dBi over the frequency range and the corresponding efficiency was more than 65%. The return loss was a creditable −20 dB. At that juncture in the evolution of arrays for mm-wave systems (circa 2005), the planar slotted waveguide array antenna could justifiably be claimed to surpass competitive offerings in any given frequency and gain range.

More recent developments [17] acknowledged the ‘beam-squint’ problem encountered with travelling wave arrays (see Figs. 7.4, 7.5 and 7.6) which led to the evolution of matched shunt slots incorporated into a planar array of centre fed slotted waveguides. By feeding each linear sub-array, not from their ends as in Fig. 7.10, but from their mid-points, the long line effect was halved with concomitant bandwidth enhancement. But more significantly, this symmetrical feed scheme enforced the main beam of the array as a whole to remain on the boresight. Furthermore, the necessity for the sub-arrays to be designed for non-boresight operation to minimise return loss was greatly mitigated by employing matched slots. Reflection cancellation can be procured in a number of ways, usually by introducing an insert in the vicinity of the slot, which exhibits a reflection phase counter to that of the slot. The inductive post was found to be the most effective mechanism for suppressing the cumulative slot reflection at the input port, thus allowing the beam tilting technique to be discarded [17]. The resultant antenna displayed wide bandwidth and a main beam directed steadily at boresight. However sidelobes were higher than equivalent end-fed examples of the same antenna due to the blockage incurred by the centred feed geometry.

Later developments in the slotted waveguide array scenario were very much driven by poor transmission compatibility with microwave integrated circuits. This disadvantage does not arise to the same extent with patch arrays fed from microstripline or coplanar waveguide. Needless to say the realisation of single-layer waveguide arrays offering well designed well matched transition structures for efficient integration with other planar circuits became a high priority [18]. This so called substrate integrated waveguide (SIW), also called post-wall waveguide or laminated waveguide, was realised in a metal-clad dielectric substrate by introducing two rows of metallised via-holes by a routine printed circuit-board fabrication technique (see Fig. 7.11). Importantly this could be accomplished at low

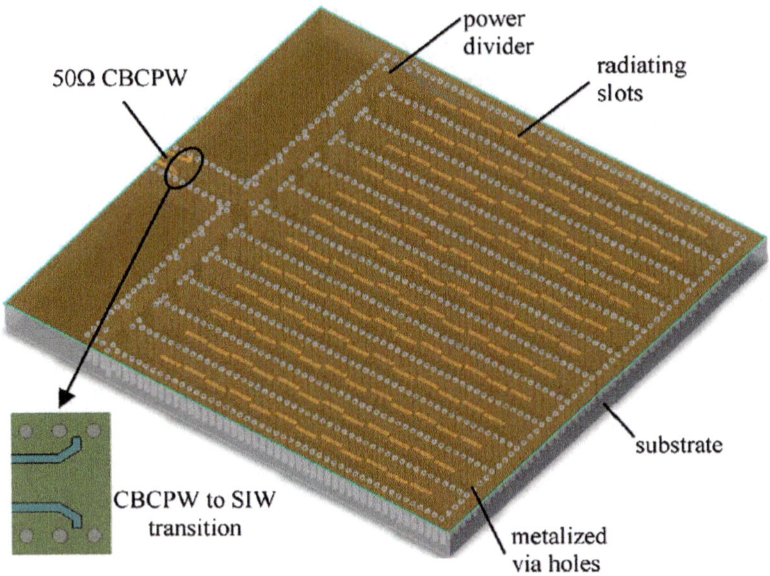

Fig. 7.11 Millimetre-wave slotted waveguide array antenna in substrate based post-wall or SIW waveguide [18]

cost. It has been demonstrated [18] that the SIW slot array antenna could easily be integrated with other circuits, resulting in cost-effective subsystems. A range SIW slot antenna arrays and beam forming networks have been developed [19]. In Ref. [19] in addition to the extension of the SIW concept from 60 to 79 GHz the designed antenna was formed on sufficiently thin substrate sheets (flexible PC foil) to limit waveguide leakage problems in the dielectric and at the post-wall, which can arise at mm-wave frequencies. In the polyimide foil the post wall is replaced by a closely packed 'forest' of nickel wires formed by a process termed ion track technology. Physical flexibility was a useful bonus.

The loss of gain and efficiency which is unavoidable with substrate based slot array solutions, led some researchers to persist with hollow waveguide options in order to retain its intrinsic low loss advantages [20–23]. To procure such a hollow waveguide antenna design, at a cost level compatible with substrate based alternatives, development gravitated toward a structure described as a double-layer, corporate-feed, waveguide slot array [20]. The typical antenna consisted of a corporate waveguide feed (see Fig. 7.10a) arranged in the lower layer, and 2×2 small subarray units in the upper layer. The 4 elements in the subarray are almost equally spaced from the coupling aperture between layers at the end of each branch of the corporate feed so that the array is substantially fed by a full corporate feed (Fig. 7.12). With constant element spacing of less than the free-space wavelength, grating lobes were effectively controlled.

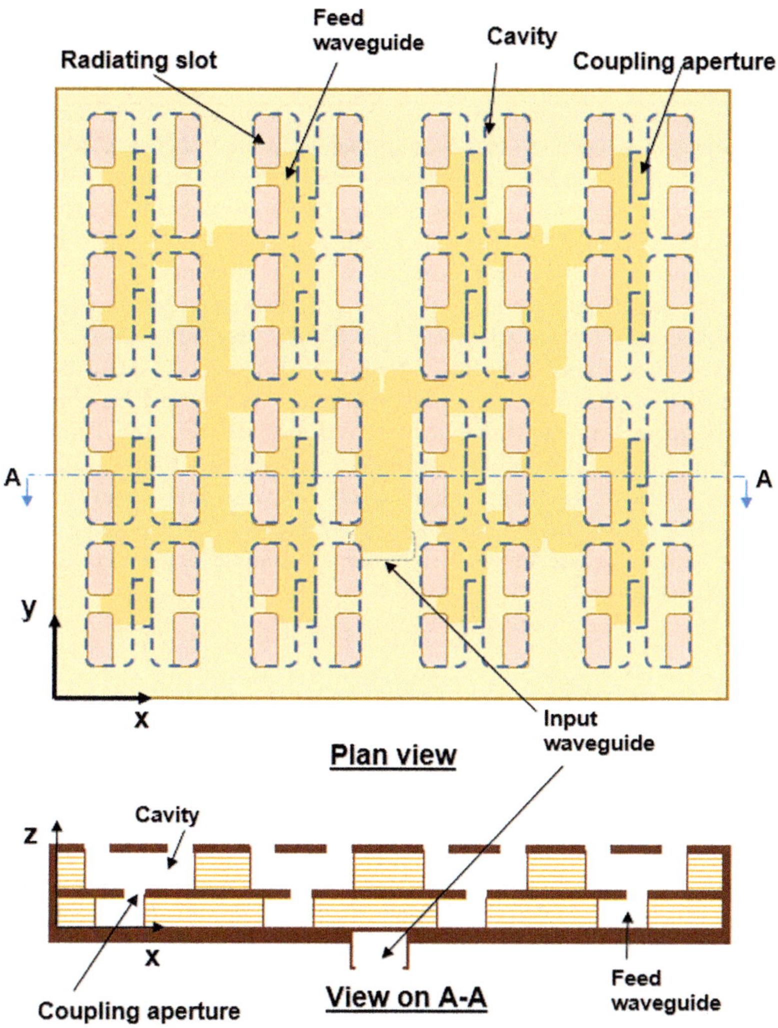

Fig. 7.12 Corporate feed and laminated waveguide slot array antenna

However, fabricating the low-loss hollow waveguide feed structure embedded within a solid metal plate was found to be verging on impossible by the available manufacturing techniques, such as machining or die-casting. The answer was to construct the structure from very thin metallic plates etched with carefully positioned apertures forming elements of the feed waveguide. When laminated, registered and secured by diffusion bonding, the layered structure incorporated the required hollow waveguides, coupling slots and radiating slots (Fig. 7.12) of the designed waveguide slot array antenna. This method was demonstrated to provide the necessary high precision at potentially low cost.

While in the 12 GHz band, a cavity-backed planar slot array antenna with 16 × 16-elements had demonstrated a gain of 33.7 dBi with 85% efficiency and 12% bandwidth (VSWR < 2) using conventional fabrication techniques, it was difficult to repeat these characteristics in the millimetre wave band. Greater fabrication accuracy was needed. This was further illustrated [24] in a 40 GHz band, two-layer, slotted-waveguide antenna array with 24 × 24-elements. While it displayed a creditable gain of 33.8 dBi, efficiency was appreciably lower at 51%, and a narrower bandwidth of 5.5% was measured, even when the array was carefully assembled and secured with screws. In contrast, later attempts [20, 21] which have employed the laminated fabrication method, achieved superior efficiency and bandwidth outcomes from a corporate feed hollow waveguide array. The figures compared very favourably with lower frequency examples, even in the high frequency band of 60 GHz. An illustrative photograph of the 60 GHz corporate feed hollow waveguide array antenna (16 × 16 slots), showing the laminated construction is presented in Fig. 7.13. The robust yet compact form for of the manufactured array is clearly seen particularly when compared with a coin similar in size to a UK sterling 10p piece. The inclination of the slots by 45° relative to the main axis of the almost square antenna, represents a simple yet effective scheme to minimise grating lobes [21].

7.4 Chapter Summary

The slotted-waveguide travelling wave array, which forms a primary source of robust, efficient, high gain and cost effective antennas, represents an ideal candidate for many communications roles. It has been examined in detail in this chapter. In particular, it is shown how the distinctive and basic dimensional choices presented by the slot array exert a strong influence the radiation pattern trade-offs of gain, sidelobe level and absence of grating lobes. Furthermore, they are demonstrated to have significant implications for frequency sensitive beam scanning, which inevitably dictates available bandwidth.

Iterative procedures, relying both on established transmission line models of the slotted waveguide and on power conservation requirements at every element of the array, are revisited to illustrate the fundamental design method. Typical results are evaluated.

The chapter is completed with a review of recent developments aimed at achieving the importation of slotted waveguide array techniques into antenna applications which have been propelled by the opening up of millimetre-wave frequency bands for exploitation by digital communication systems, atmospheric sensing and wireless networks.

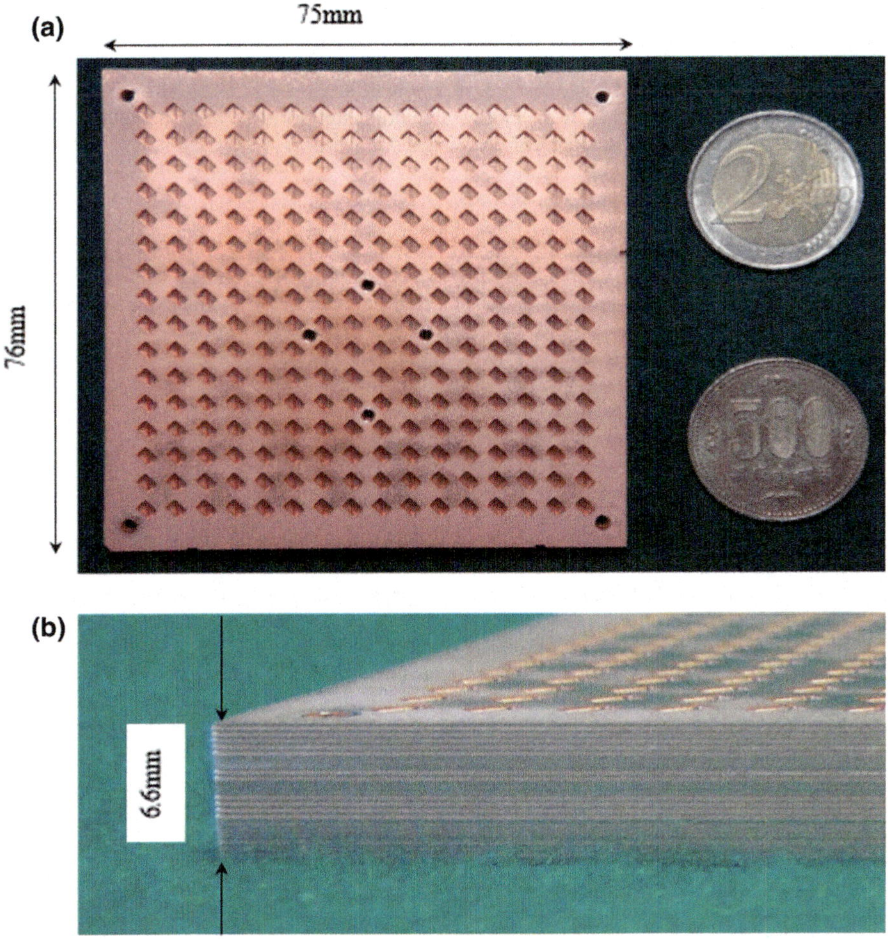

Fig. 7.13 Photograph of laminated hollow waveguide slot array **a** top view, **b** side view [21]

References

1. H.A. Bethe, Theory of diffraction by small holes. Phys. Rev. **66**(7–8), 168–183 (1941)
2. A.F. Stevenson, Theory of slots in rectangular waveguide. J. Appl. Phys. **19**, 24–38 (1948)
3. A.A. Oliner, The impedance properties of narrow radiating slots in the broad face of rectangular waveguide. IRE Trans. **AP-5**, 4–20 (1957)
4. I.P. Kaminow, R.J. Stegen, *Waveguide Slot Array Design* (Hughes Aircraft Co. Technical Memorandum, 348, Culver City, California, 1954)
5. A. Derneryd, Linear polarised microstrip antennas. IEEE Trans. Antennas Propag. **AP-24**, 846–850 (1976)
6. R.C. Johnson, *Antenna Engineering Handbook* (McGraw-Hill Inc., New York, 1993)
7. S. Silver, *Microwave Antenna Theory and Design* (McGraw-Hill Book Co., New York, 1949)
8. R.E. Collin, F.J. Zucker, *Antenna Theory* (McGraw-Hill Book Co., New York, 1969)

9. R.S. Elliott, *Antenna Theory and Design* (Prentice-Hall Ltd., Englewood Cliffs, N.J., 1981)
10. A. Dion, Non-resonant slotted arrays. IRE Trans. Antennas Propag. **AP-6**, 360–365 (1958)
11. A.W. Rudge, K. Milne, A.D. Olver, P. Knight (eds.), *The Handbook of Antenna Design: Volumes I and II* (Peter Peregrinus Ltd., London, 1986)
12. S.K. Yong, C.-C. Chong, An overview of multigigabit wireless through millimeter wave technology: potentials and technical challenges. EURASIP J. Wirel. Commun. Netw. **2007**, 1–10 (2007)
13. K. Sakakibara, Y. Kimura, A. Akiyama, J. Hirokawa, M. Ando, N. Goto, Alternating phase-fed waveguide slot arrays with a singlelayer multiple-way power divider. Proc. Inst. Electr. Eng. Microwaves Antennas Propag. **144**, 425–430 (1997)
14. M. Ando, J. Hirokawa, in *High Gain and High Efficiency Single Layer Slotted Waveguide Array in 60 GHz Band*. Proceedings of the 11th International Conference on Antennas and Propagation, Publication No. 436, vol. 1 (Edinburgh, UK, 1997), pp 464–468
15. Y. Kimura, T. Hirano, J. Hirokawa, M. Ando, Alternating-phase fed single-layer slotted waveguide arrays with chokes dispensing with narrow wall contacts. Proc. Inst. Electr. Eng. Microwaves Antennas Propag. **148**, 295–301 (2001)
16. Y. Kimura, Y. Miura, T. Shirosaki, T. Taniguchi, Y. Kazama, J. Hirokawa, M. Ando, A low-cost and very compact wireless terminal integrated on the back of a waveguide planar array for 26 GHz band fixed wireless access (FWA) systems. IEEE Trans. Antennas Propag. **53**(8), 2456–2463 (2005)
17. S. Park, Y. Tsunemitsu, J. Hirokawa, M. Ando, Centre feed single layer slotted waveguide array. IEEE Trans. Antennas Propag. **54**(5), 1474–1480 (2006)
18. X.-P. Chen, K. Wu, L. Han, F. He, Low cost high gain planar array antennas for 60 GHz band applications. IEEE Trans. Antennas Propag. **58**(6), 2126–2129 (2010)
19. S. Cheng, H. Yousef, H. Kratz, 79 GHz slot antennas based on substrate integrated waveguides (SIW) in a flexible printed circuit board. IEEE Trans. Antennas Propag. **57**(1), 64–71 (2009)
20. Y. Miura, J. Hirokawa, M. Ando, Y. Shibuya, G. Yoshida, Double-layer full-corporate-feed hollow-waveguide slot array antenna in the 60 GHz-band. IEEE Trans. Antennas Propag. **59** (8), 2844–2851 (2011)
21. T. Tomura, Y. Miura, M. Zhang, J. Hirokawa, M. Ando, A 45° linearly polarised hollow waveguide corporate-feed slot array antenna in the 60 GHz band. IEEE Trans. Antennas Propag. **60**(8), 3640–3646 (2012)
22. M. Zhang, J. Hirokawa, M. Ando, Fabrication of a slotted waveguide array at 94 GHz by diffusion bonding of laminated thin plates. IEICE Trans. Commun. **E93-B**(10), 2538–2544 (Oct 2010)
23. D. Kim, M. Zhang, J. Hirokawa, M. Ando, in Design of a Dual-Polarisation Waveguide Slot Array Antenna Using Diffusion Bonding of Laminated Thin Plates for 60 GHz Band. IEEE Antennas and Propagation Society International Symposium, 2012. [https://doi.org/10.1109/aps.2012.6348506]
24. S.S. Oh, J.W. Lee, M.S. Song, Y.S. Kim, Two-layer slotted waveguide antenna array with broad reflection/gain bandwidth at millimetre-wave frequencies. IEE Proc. Microwave Antennas Propag. **51**(5), 393–398 (2004)

Chapter 8
Frequency Scanned and Leaky-Wave Linear Slot Arrays

8.1 Introduction

The aim of this chapter is to provide the reader with a thorough introduction to the nature of the transition from 'simple' slot arrays in conventional waveguides to often quite complex configurations, generated using advanced simulation techniques. Such antennas cannot readily be analysed using the methods outlined in Chaps. 5, 6, and 7. Additionally it is intended that the following sections will provide a 'bridge' to the exploration in Chap. 9 of open periodic structures in so far as they are employed to realise compact array forms, and thereby radiation pattern optimisation. As we shall see, these antenna forms can be categorized as frequency selective surfaces and are sometimes referred to as meta-structures.

In the previous Chapter (Sects. 7.2.1 and 7.2.2) we have already discovered that for linear slot arrays implemented in conventional rectangular waveguide a beam squinting propensity is exhibited with changing frequency. This is usually considered to be problematic, particularly in bandwidth terms, and certainly not helpful. Nevertheless, a linear slot array which offers main beam positioning or scanning that is highly sensitive to the antenna operating frequency can have very definite and positive applications.

8.2 Frequency Scanned Slotted Waveguide Array

The evolution of the frequency scanned antenna was arguably propelled by the desire to achieve range, height and azimuth information on a target from a single radar platform. Unlike the active phased array antenna which can achieve this role readily, the frequency scanned antenna option, by being wholly passive, is very reliable and cost effective [1]. It is well known that slotted waveguide linear arrays are troubled by beam squinting if operated over a significant bandwidth.

A. J. Sangster, *Compact Slot Array Antennas for Wireless Communications*, Signals and Communication Technology, https://doi.org/10.1007/978-3-030-01753-8_8

The essence of the problem is illustrated in Figs. 7.4 and 7.5. Normally radar engineers would wish to minimise this difficulty, but in the frequency scanned antenna the beam squinting is purposefully enhanced. This can be done in several ways, given that scanning of the primary beam for a slot array in air filled waveguide is governed by Eq. (7.11). That is:

$$\sin\phi = \frac{\beta_{10}d - 2m\pi}{k_0 d} \tag{8.1}$$

Note that staggering (see Chap. 7) becomes irrelevant in waveguides modified for scanning.

8.2.1 Dielectric Loaded Waveguide

Enhanced scanning is most readily and obviously procured by introducing a low loss dielectric of relative permittivity ε_r into the waveguide to stretch the electrical length of the modal path between slots. Equation (8.1) becomes:

$$\sin\phi \approx \frac{\sqrt{\varepsilon_r}\beta_{10}d - 2m\pi}{k_0 d} \tag{8.2}$$

for operation well above cut-off. Obviously the larger we make the relative permittivity of the waveguide filling, the greater is the influence of $\beta_{10}d$ in Eq. (8.2) thus augmenting the influence of frequency on ϕ. Unfortunately, at typical frequencies for long range radar operating in the L, S and C bands, this solution results in feed structures which are unacceptably heavy and inefficient, when conventional dielectric materials are employed.

8.2.2 Sinuous Waveguide Feed

The logical alternative to raising the relative permittivity of the guide filling to boost the phase shift term in Eq. (8.1), is to simply increase the interior distance between radiating apertures to l_g (>d) so that Eq. (9.1) becomes:

$$\sin\phi = \frac{\beta_{10}l_g - 2m\pi}{k_0 d} \tag{8.3}$$

To achieve this operating mode it becomes necessary to physically fold the waveguide into a sinuous or serpentine shape [1, 2]. The basic arrangement is shown schematically in Fig. 8.1.

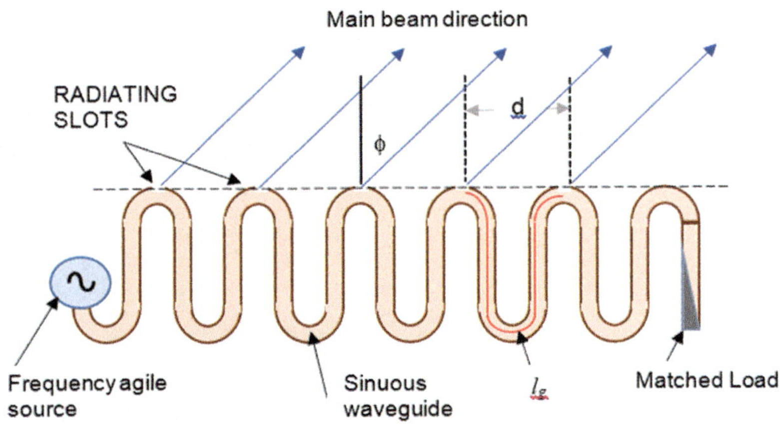

Fig. 8.1 Schematic demonstrating frequency scanning using a sinuous waveguide feed

The sinuous structure, sometimes referred to as a serpentine feed, is usually formed from conventional air filled rectangular waveguide. Fabrication can be made simpler if low height waveguide is employed but this means sacrificing power handling. The feed is excited at one end by a frequency agile source and is terminated in a wideband matched load. The radiating slots are located in the waveguide wall forming the outer curved surfaces, and are separated by distance d externally, and by distance l_g internally. The main beam direction is denoted as angle ϕ from broadside. For a broadside beam (ϕ = 0) the internal slot spacing l_g must be equal to $m\lambda_{g10}$ where m is an integer (m = 1, 2,..., M). Consequently $\beta_{10}l_g$ = 2 $m\pi$ and the right hand side of Eq. (8.3) becomes zero as required. At the frequencies where this condition holds we replace λ_{g10} by λ_{g10m}.

Consequently, for a primary beam (m = 1), the angle ϕ in Eq. (8.3) can be written:

$$\sin\phi = \frac{l_g}{d}\left(\frac{\lambda}{\lambda_{g10}} - \frac{\lambda}{\lambda_{g10m}}\right) \tag{8.4}$$

which is generally termed the frequency scan equation. To scan the primary beam over all of ‘real’ space from −90° to +90° then we must have $-1 < \sin\phi < 1$. Clearly this range can be achieved with low frequency deviation if $l_g \gg d$. Note that d and l_g can be adjusted independently thus permitting separate optimisation of scan sensitivity and of grating lobe suppression.

For grating lobes (see circle diagrams in Chap. 6) at angles ϕ_n and ϕ_{n+1} (say) Eq. (8.3) can be manipulated into the following grating lobe equation:

$$\sin\phi_n - \sin\phi_{n+1} = \frac{\lambda_0}{d} \tag{8.5}$$

Fig. 8.2 Double curvature parabolic reflector antenna illuminated by a sinuous feed. Courtesy of NRPL Group: www.radartutorial.com

Clearly this equation is independent of l_g which implies that the grating lobes of the linear sinuous array antenna simply obey the strictures for linear arrays as presented in Chap. 6.

In practice, frequency scanned array antennas based on the sinuous feed seldom exist as stand-alone linear arrays. It is much more common to see the sinuous feed combined with a parabolic reflector as shown in Fig. 8.2, or as an end feed to a planar array of parallel slotted waveguides (see Chap. 7) as presented in Fig. 8.3. These antennas provide 3-D scanning capability with azimuth information developed from the rotation of the structure as a whole while height information is generated from the frequency scanned feed. Early versions were by no means compact.

8.2.3 Periodically Loaded Waveguide Feed

As we have seen in the preceding sections the key to enhanced frequency scanning in a waveguide feed is the $\beta_{10}d$ term in Eq. (8.1) which needs to be much larger than the $k_o d$ term. Enhancement of β_{10} is possible by dielectric loading as demonstrated in Sect. 8.2.1, while by differentiating the internal distance $(d \to l_g)$ between apertures from the external spacing d (see Sect. 8.2.2) it is possible with the resultant sinuous structure to improve scanning sensitivity by making $l_g \gg d$. Unfortunately dielectric loading presents considerable antenna weight and poor electrical efficiency, while serpentine feeds are bulky, incur aperture blockage difficulties in reflector antennas (see Fig. 8.2), and the long sinuous waveguide structure attracts power loss issues.

Fig. 8.3 Planar array of slotted waveguide radiators end-excited by a sinuous feed (on the left of the planar array structure) (see [1])

A third possible solution is to enhance β_{10} relative to k_o by adopting periodically loaded waveguide, which has applications in many other areas of microwave technology. For example, iris loaded circular waveguide is widely used in linear accelerators [3–5], and in power travelling wave tubes [3, 4], while rectangular waveguide loaded with capacitive irises has traditionally formed the basis of high power, high frequency, bandpass filters [6–8]. More recently periodic loading of flared waveguide has become a common feature of low noise feeds for satellite systems [9]. The advantages of the 'slow-wave' feed waveguide, over the serpentine feed, for frequency scanning radar are mainly mechanical. The former is considerably smaller, and hence much less heavy for equivalent electrical performance. The role of frequency scanning feeds in 3-D radar are examined in detail in references [10, 11].

8.3 Transition from Uniform Waveguide to Slow-Waveguide Feed

The feed structure for a slotted waveguide linear array possessing enhanced scanning ability usually comprises a fin loaded rectangular waveguide which 'slows' the propagating mode relative to empty waveguide. Hence the term 'slow-wave' feed.

So, in order to progress, we need to be able to assess the influence of the periodically spaced metal fins or irises on the waveguide propagation characteristics. We will attempt to do this in a manner which is physically intuitive. The literature on the topic of slow-wave structures is very extensive and the early contributions, in the 1950s and 1960s, are highly mathematical. So the subject can be somewhat impenetrable without deep study.

8.3.1 Fast Waveguide Frequency Characteristics

For the original empty, perfectly conducting, rectangular waveguide of width a, and height b, we discovered in Chap. 1 that the waveguide supports a dominant TE_{10} mode, which cuts off at a frequency given by $k_o a = \pi$, that is $f_c = c/2a$ Hz. This is depicted in Fig. 7.14a (generally referred to as the ω–β diagram) where the modal frequency characteristics are presented diagrammatically, with the parabolic shape of the curve representing the characteristic equation for the TE_{10} mode (see Appendix B):

$$(\beta_{10}a)^2 = (k_0 a)^2 - \left(\frac{\pi}{a}\right)^2 \tag{8.6}$$

Actually, in general the ω–β diagram (Brillouin diagram) for empty rectangular waveguide can exhibit many higher order solutions, with cut-offs reaching increasingly higher values on the $k_o a$ axis, representing all of the possible TE_{mn} and TM_{mn} modes as encapsulated by Eq. (A2.13) in Appendix B.

Interestingly, when the rectangular waveguide is modified by the insertion of a periodic array of metal fins (period = p, height = h) as suggested in Fig. 8.4, the structure can still support an electromagnetic wave [6]. Not surprisingly its propagation characteristics change significantly to accommodate the influence of the fins. In short any valid modal solution for the periodic waveguide must satisfy the electromagnetic boundary conditions on **E** and **H** over the fin surfaces as well as over the waveguide inner surfaces, as encountered in the empty waveguide case. Note, that if the TE_{10} mode waveguide contained only one metal iris or fin, reflection and transmission of this mode at the iris would occur. However, while these modal constituents would continue to satisfy the electromagnetic boundary conditions on the waveguide walls they can only partially satisfy the boundary conditions in the vicinity of the fin. Complete satisfaction of the boundary conditions for the structure as a whole requires a suitable combination of higher order TE and TM modes [6, 7]. These modes will be evanescent (see Chap. 1) and contribute to capacitive stored energy fields in the immediate vicinity of the iris [4]. Now, when a multiplicity of fins in a periodic array are present in the waveguide, as hinted at in Chap. 7, the boundary requirements at each fin does not change and multiple TE and TM waveguide modes will continue to be necessary to meet them. However, the fact that the fins are actually closely spaced, in a periodic manner, sets a concomitant requirement of periodicity on those modal fields which are in close proximity to the surface (see Fig. 8.5) of the fin-loaded structure.

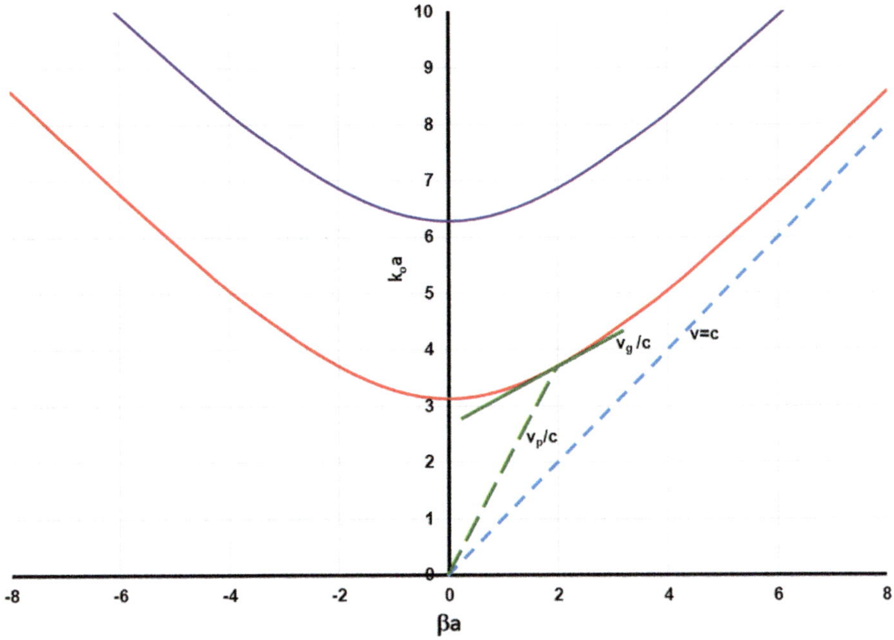

Fig. 8.4 Frequency/phase diagrams (ω–β diagrams) for a uniform rectangular waveguide—TE_{10} mode = red curve: TE_{20} mode = mauve curve

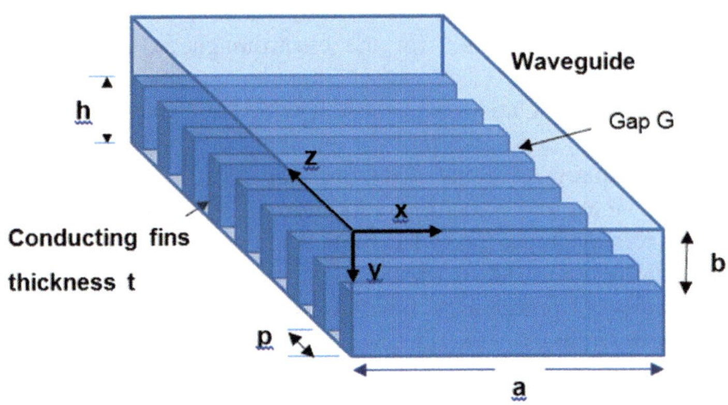

Fig. 8.5 Fin or iris loaded rectangular waveguide

8.3.2 Floquet Harmonics

Mathematically the condition of periodicity is expressed through Floquet's Theorem. This theorem can be succinctly stated as: *A steady-state modal solution to the electromagnetic boundary value problem, which accommodates a periodic structure, has the property that the fields in adjacent cells, distance p apart, are related by a complex constant.* For a *lossless* structure, it can be expressed in the form:

$$E_0(x, y, z+p) = E_0(x, y, z) \exp(-j\beta_0 p) \tag{8.7}$$

$E_0(x, y, z)$ represents a possible electromagnetic wave solution in the zero'th mode. In an infinitely long, closed and lossless, structure the solution is inevitably multi-modal. It is also a periodic function of z, the propagation direction in the waveguide, while β_o is the phase constant for the fundamental space harmonic of the zero'th mode. That $E_0(x, y, z)$ is a solution of the Maxwell equations which are applicable to the slow waveguide structure, we firstly represent it as follows:

$$E_0(x, y, z) = f(x, y, z) \exp(-j\beta_0 z) \tag{8.8}$$

where $f(x, y, z)$ is also a periodic function of z with period p. It follows therefore, referring to Fig. 8.5, that at location $z + p$ we must have:

$$E_0(x, y, z+p) = f(x, y, z+p) \exp(-j\beta_0(z+p)) \tag{8.9}$$

But, since $f(x, y, z+p) \equiv f(x, y, z)$ for a periodic function, then Eq. (8.9) becomes:

$$E_0(x, y, z+p) = f(x, y, z) \exp(-j\beta_0 z) \exp(-j\beta_0 p) \tag{8.10}$$

Hence, on combining Eqs. (8.10) and (8.8) we observe that the Floquet relationship [Eq. (8.7)] is confirmed for the electromagnetic solution of a closed periodic structure.

That the Floquet formulation is consistent with the Maxwell equations can be deduced by considering a parallel problem in signal processing theory, namely the representation of square or digital waveforms in the time domain by electrically more convenient sinusoidal waveforms of varying frequencies. The mechanism for generating this relationship is provided by Fourier series. The theory simply states that any periodic function that is finite, single-valued and continuous can be represented by a series of trigonometric functions. Good pictorial representations of the process are to be found on the internet [12]. Consequently, the periodic field function $E_0(x, y, z)$ can by expanded into a Fourier series in the space domain z (rather than time t as in signal processing) such that:

$$\begin{aligned} E_0(x, y, z) &= \sum_{r=-\infty}^{\infty} E_r(x, y) \exp\left(-\frac{j2\pi rz}{p}\right) \exp(-j\beta_0 z) \\ &= \sum_{r=-\infty}^{\infty} E_n(x, y) \exp(-j\beta_r z) \end{aligned} \tag{8.11}$$

Fourier theory dictates that the amplitudes of the r space harmonics $E_r(x, y)$ of the zero'th mode are given by,

$$E_r(x,y) = \frac{1}{p}\int_0^p E_0(x,y,z)\exp\left(\frac{j2\pi rz}{p}\right)dz \tag{8.12}$$

and

$$\beta_r = \beta_0 + \frac{2\pi r}{p} \tag{8.13}$$

is the phase coefficient for the rth space harmonic $(-\infty < r < \infty)$. Note that in the context of a fin loaded structure based on rectangular waveguide, as depicted in Fig. 8.5, the zero'th mode is essentially an evolution of the TE_{10} mode of the empty waveguide.

For higher order modes $E_{mn}(x, y)$ (m = 1, 2, 3,..., n = 0, 1, 2,...) the corresponding possible solution forms are

$$\begin{aligned} E_{m.n}(x,y,z) &= \sum_{r=-\infty}^{\infty} E_r(x,y)\exp\left(-\frac{j2\pi rz}{p}\right)\exp(-j\beta_0 z) \\ &= \sum_{r=-\infty}^{\infty} E_r(x,y)\exp(-j\beta_r z) \end{aligned} \tag{8.14}$$

where β_0 becomes the fundamental phase coefficient for the mnth mode, and the E_r represent the relevant harmonic magnitudes for this mode.

The question then remains as to whether or not the field forms represented by Eqs. (8.13) and (8.14) satisfy the electromagnetic wave equation as they must? For the electric field in z-directed waveguide this has the form (see Appendix B):

$$\nabla^2 E - \gamma^2 E = 0 \tag{8.15}$$

Consequently the postulated periodic structure solution $E_0(x, y, z)$ must satisfy:

$$\nabla^2 E_0(x,y,z) - \gamma^2 E_0(x,y,z) = 0 \tag{8.16}$$

A similar requirement applies to $E_{m,n}$ (x, y, z). Hence combining Eq. (8.16) with Eq. (8.11) we obtain:

$$\nabla^2 \sum_{r=-\infty}^{\infty} E_r(x,y)\exp(-j\beta_r z) - \gamma^2 \sum_{r=-\infty}^{\infty} E_r(x,y)\exp(-j\beta_r z) = 0 \tag{8.17}$$

which, given the linearity of the wave equation, leads to:

$$\sum_{r=-\infty}^{\infty} \left\{ \nabla^2 E_r(x, y) \exp(-j\beta_r z) - \gamma^2 E_r(x, y) \exp(-j\beta_r z) \right\} = 0 \tag{8.18}$$

Therefore, the postulated field solution $E_0(x, y, z)$ for the periodic structure will be a solution to the wave equation, and hence Maxwell's equations, if each space harmonic is also a solution. This will normally be the case since the space harmonics are generated by solving Maxwell's equations for individual periodic cells. The cells are generally represented by resonant structures exhibiting much simpler boundary conditions.

From an engineering perspective, the seemingly intractable complexity of potential solutions to the electromagnetic boundary value problem, as presented by a periodic structure, can be ameliorated significantly by noting that the structure is composed of a linear assembly of identical cells (Fig. 8.5). Actually, the resonant frequencies of these cells can be determined with little difficulty on an electromagnetic wave simulator such as HFSS [13] or COMSOL [14], and once known the formation of the phase (ω–β) diagram for the structure as a whole becomes possible. Furthermore, when the cells are at resonance, only two logical possibilities exist for neighbouring units. The fields either assume a symmetrical mode distribution with zero phase shift ($\beta p = 0$) from cell-to-cell, or they assume an anti-symmetrical form with π-phase shift ($\beta p = \pi$) from cell-to-cell. This behaviour can be observed in low frequency LCR filter circuits [5, 15]. In filter terms the first option represents a zero phase shift frequency cut-off condition while the second equates to a cut-off condition exhibiting π-phase shift.

8.3.3 Slow Waveguide Frequency Characteristics

For the fin-loaded rectangular waveguide the lowest possible operating frequency is determined by the fundamental (TE_{10}) mode cut-off in the empty waveguide, given by Eq. (8.6) with $\beta_{10}a = 0$. At this frequency $\lambda_{g10} \Rightarrow \infty$ which means that the gaps between the fins have effectively zero depth electrically and the modal fields are restricted to the empty region above the fins. That is, the periodic structure behaves electromagnetically like an empty rectangular waveguide of height b and replicates its frequency behaviour at cut-off (see Figs. 8.4 and 8.6) at $\beta p = 0$ and $\beta a = 0$ respectively). In cell modelling terms this resonance is generated when magnetic walls (H-walls) are located above neighbouring irises, as suggested in Fig. 8.7. The H-walls reflect the cell-to-cell zero phase shift requirement at this cut-off. The finite element simulation predicts, not unexpectedly, 6.55 GHz ($k_o a = \pi$) in accordance with Fig. 8.4 and with Eq. (8.6).

When an EM simulation is performed on the cell model with E-walls (Fig. 8.7b) the principal $\beta p = \pi$ frequency is predicted to be 9.87 GHz ($k_o a = 4.726$) with a full half cycle of TE_{10} mode field pattern formed between the base of the fin-gap and the top wall of the periodic structure. It is plotted on Fig. 8.6 as the first cut-off

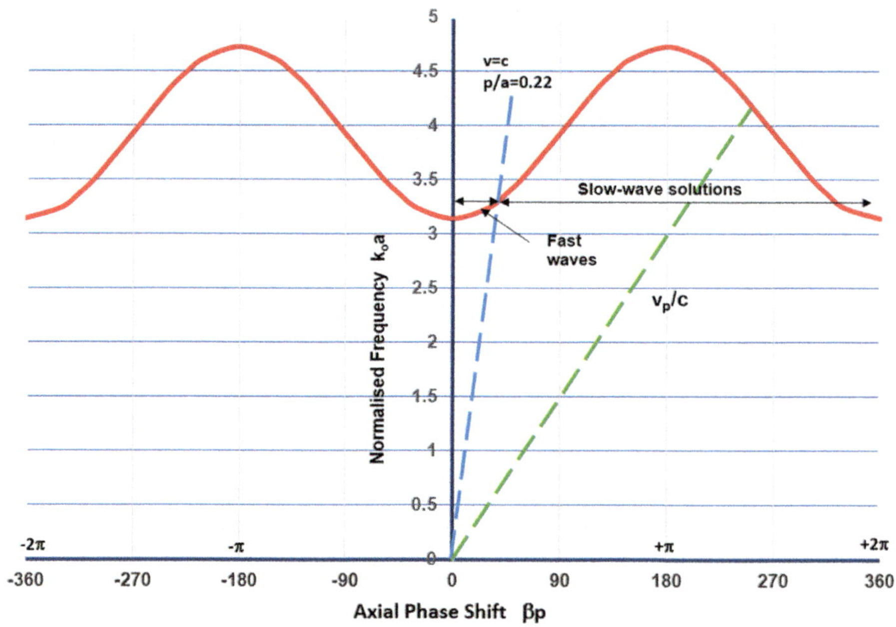

Fig. 8.6 Frequency/phase diagram (ω–β diagram) for rectangular waveguide periodically loaded with metal fins—showing fundamental plus first higher space harmonic (NB slow-wave solutions to right of v = c line: fast-wave solutions occur in region to the left of v = c line—note in an open structure (see Chap. 10) waves in the fast region become leaky)

maximum. Needless to say given the nature of the Floquet theorem, the $\beta p = 0$ frequency is repeated at $\beta p = \pm 2\pi, \pm 4\pi, \ldots \infty$, while the $\beta p = \pi$ frequency is repeated at $\beta p = \pm 3\pi, \pm 5\pi, \ldots \infty$.

The precise shape of the curve in Fig. 8.6 between the locations of key cut-off frequencies, which have just been identified, requires the solution of the Maxwell equations for the periodic structure as a whole. An investigator possessing electromagnetic simulation software such as HFSS [13] or COMSOL [14] could attempt a direct application of a finite element, or finite difference technique, although this route would demand considerable computational power to accommodate a model of the complexity presented by a periodically loaded waveguide. The computationally more efficient alternative would to set up a problem specific solution based on the moment method as outlined in Chap. 5, or perhaps on a variational method as advocated by Cairo and Kahan [16] or an alternative MoM [17]. Software such as MATLAB [18] can greatly facilitate the evolution of a suitable algorithm to achieve this, particularly in the case of the moment method.

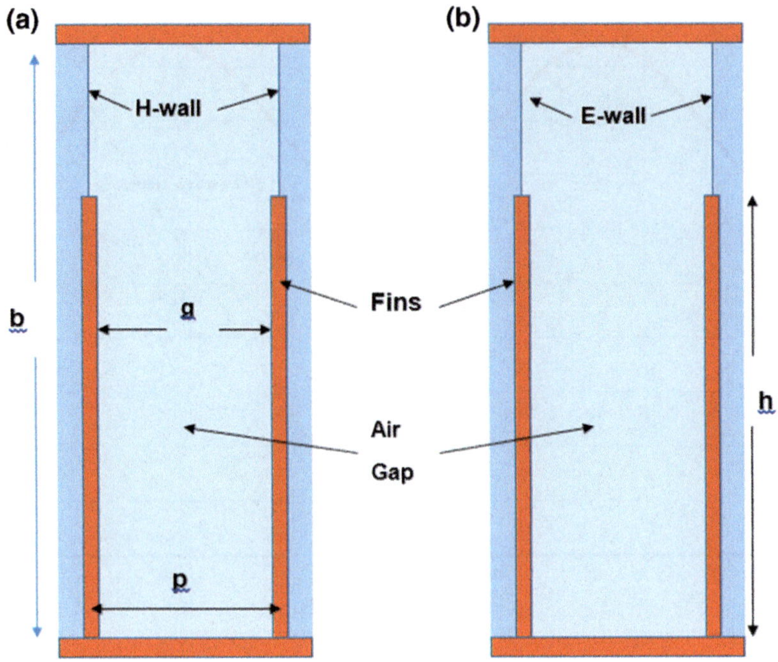

Fig. 8.7 Periodic cell models showing E-wall and H-wall inserts. **a** H-wall, **b** E-wall

8.3.4 Approximate Solution for Phase Coefficient

As we observed in Chap. 5 the moment method furnishes us with a mathematical procedure for solving electromagnetic boundary value problems generally comprising two coupled regions which are themselves fully described electromagnetically. In the slot radiator example the 'known' regions are empty rectangular waveguide and a half-space coupled through the 'uncharacterised' radiating aperture. Here, the coupled regions are the gaps in the periodic fin structure which behave electromagnetically like shorted low height waveguides, and the rectangular waveguide space located above the fins (see Fig. 8.5). The two regions are coupled through the 'apertures' representing the interface between the air gaps, between $y = b$ and $y = b + h$ (linked mathematically by means of the Floquet theorem), and the empty waveguide region. The unknown in the analysis is the axial propagation coefficient imposed by the periodicity. Needless to say, the literature presents a multitude of variational method, or moment method, solutions for this regular structure, and for a wide range of related structures [3, 6, 7, 11].

Alternatively, a sense of the nature of the solution to electromagnetic propagation in a periodically loaded waveguide can be achieved by following a heuristic methodology, as displayed in Chap. 2. There the approach aimed at formulating potential functions relevant to electromagnetic radiation [Eqs. (2.45), and (2.54)].

Here we wish to generate potential functions which are solutions of the electromagnetic wave equations within a fin loaded periodic waveguide (see Fig. 8.5). It is evident that Fig. 8.5 represents a rectangular waveguide, as discussed in Chap. 1, except that the lower broadwall presents an impedance other than zero to the electromagnetic waves. Consequently it as appropriate to evaluate waveguide propagation in this modified set-up.

In Chap. 1, it is demonstrated that the fundamental frequency characteristics of rectangular waveguide are contained in the propagation equation [Eq. (1.68)], which is also presented graphically in Fig. 1.6. For air filled waveguide with perfectly conducting walls this equation is:

$$\beta_{wmn}^2 = k_0^2 - k_{cmn}^2 = k_0^2 - \left(\frac{m\pi}{a}\right)^2 - \left(\frac{n\pi}{b}\right)^2 \tag{8.19}$$

for both TE and TM modes. If these are propagating solutions it is necessary that the propagation coefficient for the mnth mode in the waveguide, β_{wmn} is real, in which case:

$$k_0 > k_{cmn} \text{ and } \beta_{mn} < k_0$$

Since $\beta_{wmn} = \omega/v_{pmn}$ and $k_0 = \omega/c$ then clearly $v_{pmn} > c$. That is, the solutions are 'fast' with the phase velocity of the mnth mode travelling at a speed greater than that of light axially down the waveguide. So what conditions would be required for the propagating modes to be 'slow' with $v_{pm} < c$? Mathematically it would require:

$$\beta_m^2 = k_0^2 + k_{cm}^2 = k_0^2 - (k_{cy})^2 - \left(\frac{m\pi}{a}\right)^2 \tag{8.20}$$

'Slow' solutions seem intuitively more comprehensible in a world of physics which is used to the idea that nothing travels faster than light. The question, therefore, appears to be a not unnatural one. Note that since the x-directed geometry of the waveguide is unchanged by the fins (Fig. 8.5), the third term on the right of the equals sign in Eq. (8.20) must also remain unchanged. In the y-direction this is certainly not true, and therefore we introduce a k_{cy} parameter (to be determined) to reflect this. Hence we can write for the mth mode:

$$k_{cy}^2 = k_0^2 - \beta_m^2 - \left(\frac{m\pi}{a}\right)^2 \tag{8.21}$$

For slow-wave solutions we have already stipulated that $\beta_m > k_0$ and given that all the terms on the right side of Eq. (8.21) are real, k_{cy} must be imaginary (say $j\alpha_m$). This implies that in the fin loaded waveguide, in the space ($0 < y < b$) above the fins, valid fields solutions must display exponential decay in the −y direction.

Embracing the evident analogy between the fin loaded rectangular waveguide and its empty counterpart (Chap. 1) the following potential functions in phasor form can be constructed for TE and TM mode solutions. For the mth such mode:

$$
\begin{aligned}
A_{zm}^{TE} &= \sum_{r=-\infty}^{r=\infty} \frac{1}{k_{cu}^2}[A_r \sinh(\alpha_{mr}y) + B_r \cosh(\alpha_{mr}y)] \sin\frac{m\pi}{a} \exp -j\beta_{mr}z \\
A_{zm}^{TM} &= \sum_{r=-\infty}^{r=\infty} \frac{1}{k_{cu}^2}[C_r \sinh(\alpha_{mr}y) + D_r \cosh(\alpha_{mr}y)] \cos\frac{m\pi}{a} \exp -j\beta_{mr}z
\end{aligned}
\tag{8.22}
$$

summed over an infinite number of Floquet space harmonics. Note that for higher order space harmonics $\alpha_{cmr}^2 = \beta_{mr}^2 + \left(\frac{m\pi}{a}\right)^2 - k_0^2$. Also $k_{cu}^2 = k_0^2 + \alpha_{cmr}^2$ and $\beta_{mr} = \beta_{m0} + \frac{2r\pi}{p}$, while $p = G + t$.

On applying the Maxwell equations the following field expressions are generated for the mth mode in the rectangular space ($a \times b$) above the fins (denoted by subscript 'u'):

$$
\begin{aligned}
E_{ux} = &-\sum_{r=-\infty}^{\infty} \frac{1}{k_{cu}^2}\left\{\left(j\frac{m\pi}{a}\beta_{mr}A_{mr} + j\omega\mu_0\alpha_{cmr}D_{mr}\right)\sinh\alpha_{cmr}y\right. \\
&\left. + \left(j\frac{m\pi}{a}\beta_{mr}B_{mr} + j\omega\mu_0\alpha_{cmr}C_{mr}\right)\cosh\alpha_{cmr}y\right\} \times \cos\frac{\mathrm{m}\pi\mathrm{x}}{\mathrm{a}} \exp -j\beta_{mr}z
\end{aligned}
\tag{8.23}
$$

$$
\begin{aligned}
E_{uy} = &-\sum_{r=-\infty}^{\infty} \frac{1}{k_{cu}^2}\left\{\left(j\alpha_{cmr}\beta_{mr}A_{mr} + j\omega\mu_0\frac{m\pi}{a}D_{mr}\right)\cosh\alpha_{cmr}y\right. \\
&\left. + \left(j\alpha_{cmr}\beta_{mr}B_{mr} - j\omega\mu_0\frac{m\pi}{a}C_{mr}\right)\sinh\alpha_{cmr}y\right\} \times \sin\frac{\mathrm{m}\pi\mathrm{x}}{\mathrm{a}} \exp -j\beta_{mr}z
\end{aligned}
\tag{8.24}
$$

$$
E_{uz} = \sum_{r=-\infty}^{\infty} (A_{mr}\sinh\alpha_{cmr}y + B_{mr}\cosh\alpha_{cmr}y) \sin\frac{m\pi x}{a} \exp -j\beta_{mr}z \tag{8.25}
$$

$$
\begin{aligned}
H_{ux} = &-\sum_{r=-\infty}^{\infty} \frac{1}{k_{cu}^2}\left\{\left(j\omega\varepsilon_0\alpha_{cmr}A_{mr} + j\frac{m\pi}{a}\beta_{mr}D_{mr}\right)\cosh\alpha_{cnr}y\right. \\
&\left. + \left(j\omega\varepsilon_0\alpha_{cmr}B_{mr} + j\frac{m\pi}{a}\beta_{mr}C_{mr}\right)\sinh\alpha_{cmr}y\right\} \times \sin\frac{\mathrm{m}\pi\mathrm{x}}{\mathrm{a}} \exp -j\beta_{mr}z
\end{aligned}
\tag{8.26}
$$

$$
\begin{aligned}
H_{uy} = &-\sum_{r=-\infty}^{\infty} \frac{1}{k_{cu}^2}\left\{\left(j\omega\varepsilon_0\frac{m\pi}{a}A_{mr} + j\beta_{mr}\alpha_{cmr}D_{mr}\right)\sinh\alpha_{cmr}y\right. \\
&\left. + \left(j\omega\varepsilon_0\frac{m\pi}{a}B_{mr} + j\beta_{mr}\alpha_{cmr}C_{mr}\right)\cosh\alpha_{cmr}y\right\} \times \cos\frac{\mathrm{m}\pi\mathrm{x}}{\mathrm{a}} \exp -j\beta_{mr}z
\end{aligned}
\tag{8.27}
$$

$$
H_{uz} = \sum_{r=-\infty}^{\infty} (C_{mr}\sinh\alpha_{cmr}y + D_{mr}\cosh\alpha_{cmr}y) \cos\frac{m\pi x}{a} \exp -j\beta_{mr}z \tag{8.28}
$$

In the fin loaded region ($b < y < h + b$) electromagnetic TE and TM modes also exist in the gaps between the fins. This region will be denoted by subscript 'f' representing the 'fin' region. These gaps can sensibly be treated as short lengths (h) of low height rectangular waveguide of cross-sectional area $G \times a$, terminated in electrical short circuits. For a periodic structure with i fins and gaps ($0 < i < I$), the z-position of the ith gap z_i is given by $z_i = z - (i - 1)(G + t)$. Given that the mth mode in any prescribed gap displays the same x-functionality as the modes in the space above the fins, then for the mth mode we can write:

$$E_{fx} = -\sum_{n=0}^{\infty} \frac{1}{k_{cf}^2} \left(\frac{m\pi}{a} \frac{n\pi}{G} K_{mn} + j\omega\mu_0 \beta_{gmn} L_{mn} \right) \cos\frac{m\pi x}{a} \sin\frac{n\pi z_i}{G} \sin\beta_{gmn}(y - b - h) \tag{8.29}$$

where it is presumed that the Floquet periodicity is built into the constants K_{mn} and L_{mn}, for the mnth gap mode. This ensures that the gap modes are phase shifted from one gap to the next in sympathy with the eventual mnth modal solution (β_{m0}) for the periodic structure as a whole. Thus K_{mn} (or L_{mn}) at the ith slot (say) is related to K_{mn} in the first slot by,

$$K_{mn}^{(i)} = K_{mn} \exp(-ji\beta_{m0}(G + t)).$$

Also

$$E_{fy} = -\sum_{n=0}^{\infty} \frac{1}{k_{cf}^2} \left(\frac{n\pi}{G} \beta_{gmn} K_{mn} + j\omega\mu_0 \frac{m\pi}{a} L_{mn} \right) \sin\frac{m\pi x}{a} \sin\frac{n\pi z_i}{G} \cos\beta_{gmn}(y - b - h) \tag{8.30}$$

$$E_{fz} = -\sum_{n=0}^{\infty} K_{mn} \sin\frac{m\pi x}{a} \cos\frac{n\pi z_i}{G} \sin\beta_{gmn}(y - b - h) \tag{8.31}$$

$$H_{fx} = \sum_{n=0}^{\infty} \frac{1}{k_{cf}^2} \left(j\omega\varepsilon_0 \beta_{gmn} K_{mn} - \frac{m\pi}{a} \frac{n\pi}{G} L_{mn} \right) \sin\frac{m\pi x}{a} \cos\frac{n\pi z_i}{G} \cos\beta_{gmn}(y - b - h) \tag{8.32}$$

$$H_{fy} = \sum_{n=0}^{\infty} \frac{1}{k_{cf}^2} \left(-j\omega\varepsilon_0 \frac{m\pi}{a} K_{mn} + \frac{n\pi}{G} \beta_{gmn} L_{mn} \right) \cos\frac{m\pi x}{a} \cos\frac{n\pi z_i}{G} \sin\beta_{gmn}(y - b - h) \tag{8.33}$$

$$H_{fz} = \sum_{n=0}^{\infty} L_{mn} \cos\frac{m\pi x}{a} \sin\frac{n\pi z_i}{G} \cos\beta_{gmn}(y - b - h) \tag{8.34}$$

The phase coefficient β_{gmn} applies to the low height gap-waveguide and is given by the following equation (see Chap. 1):

$$\beta_{gmn}^2 = k_0^2 - \left(\frac{m\pi}{a}\right)^2 - \left(\frac{n\pi}{G}\right)^2 \tag{8.35}$$

and

$$k_{cf}^2 = k_0^2 - \left(\frac{n\pi}{G}\right)^2 \tag{8.36}$$

Note that for higher order modes in the gaps, the right-hand side of Eq. (8.35) can become negative, in which case the β_{gmn} terms in Eqs. (8.29) to (8.34) are replaced by $j\alpha_{gmn}$, when appropriate in the ensuing formulation. Also for n = 0, $\beta_{gm0} = \beta_{wm0}$.

8.3.5 Boundary Condition at Periodic Interface

The above partial field solutions, have been constructed by employing conventional waveguide analogies in the upper space (u) and the fin gap (f), by applying the electromagnetic field boundary conditions on the side walls at $x = 0$ and $x = a$, and on the fin surfaces within the gaps, and at the bottom of the gaps ($y = b + h$). Thus all are automatically satisfied. We are therefore left to ensure the boundary condition $E_{\tan} = 0$ is satisfied at $y = 0$, that is:

(1) $E_x = 0$ and $E_z = 0$ at $y = 0$
Furthermore the boundary conditions at the periodic surface at y = b must be observed. This implies that:

(2) $E_{\tan}$ and $H_{\tan}$ are continuous across the $y = b$ interface. It is sufficient to take care of the tangential field components only—since the Maxwell equations ensure that the normal components fall into line.

The application of condition (1) immediately requires that:

$$B_{mr} = C_{mr} = 0 \tag{8.37}$$

Condition (2) requires that $E_{ux} = E_{fx}$ and $E_{uz} = E_{fz}$ at $y = b$ in the range $0 < z < G + t$. This is automatically the case at the fin tips ($G < z < G + t$) where $E_{ux} = E_{fx} = E_{uz} = E_{fz} = 0$.

Continuity of E_z is enforced mathematically by performing an averaging process over one period of the periodic surface. That is:

$$\int_0^{G+t} E_{uz} \exp(j\beta_{mr}z)\, dz = \int_0^{G} E_{fz} \exp(j\beta_{mn}z)\, dz \tag{8.38}$$

On substituting Eqs. (8.25) and (8.31) into Eq. (8.38) and setting y = b yields:

$$(G+t)A_{m0} \sinh \alpha_{cm0}b = -\int_0^{G} \sum_{n=0}^{\infty} K_{mn} \sin \beta_{gmn}h \cos \frac{n\pi z}{G} \exp(j\beta_{gmn}z)\, dz \tag{8.39}$$

Note that on the left of the above equation the space harmonic summation disappears since the higher harmonics integrate to zero over the period p of the corrugated surface. A_{m0} represents the fundamental term in the space harmonic series for the mth slow-waveguide mode. This mode is neither TE nor TM. It is generally referred to as the HE_{11} hybrid mode. Equation (8.39) can be reconfigured to give:

$$A_{m0} = \frac{-G}{(G+t)\sinh \alpha_{cmr}b} \sum_{n=0}^{\infty} K_{mn}\beta_{m0} \sin(\beta_{gmn}h) \exp\left(\frac{j\beta_{m0}G}{2}\right) U(\beta_{m0}n) \tag{8.40}$$

Similarly enforcing continuity of E_x at the $y = b$ boundary gives:

$$\begin{aligned}&\frac{(G+t)}{k_{cu}G^2}\left[j\beta_{m0}\frac{m\pi}{a}A_{m0} + j\omega\mu_0\alpha_{cm0}D_{m0}\right] \sinh \alpha_{cm0}b \\ &= -\sum_{n=0}^{\infty} \frac{n\pi/G}{k_{cf}^2}\left[\frac{m\pi}{a}\frac{n\pi}{G}K_{mn} + j\omega\mu_0\beta_{gmn}L_{mn}\right] \sin(\beta_{gmn}h) \exp\left(\frac{j\beta_{m0}G}{2}\right) U(\beta_{m0}n)\end{aligned} \tag{8.41}$$

Continuity of H_z at $y = b$ generates:

$$\sum_{n=-\infty}^{\infty} j\left(\frac{n\pi}{G}\right) D_{mr} \cosh(\alpha_{cmr}b) \exp\left(-\frac{j\beta_{mr}G}{2}\right) V(\beta_{mr}n) = \frac{\xi_n L_{mn}}{2} \cos(\beta_{gmn}h) \tag{8.42}$$

Finally, continuity of H_x at $y = b$ gives:

$$\sum_{r=-\infty}^{\infty} \frac{\beta_{mr}}{k_{cu}^2}\left[j\omega\varepsilon_0\alpha_{cmr}A_{mr} + j\frac{m\pi}{a}\beta_{mr}D_{mr}\right]\cosh(\alpha_{cmr}b)\exp(-j\frac{\beta_{mr}G}{2})V(\beta_{mr}n)$$
$$= \frac{\xi_{\mathrm{n}}}{2\mathrm{k}_{\mathrm{cf}}^2}\left[j\omega\varepsilon_0\beta_{gmn}K_{mn} - \frac{m\pi}{a}\frac{n\pi}{G}L_{mn}\right]\cos(\beta_{gmn}h) \tag{8.43}$$

The following definitions apply to the above:

$$U(\beta_{mr}n) = \frac{\beta_{mr}}{\left(\beta_{mr}^2 - \left(\frac{n\pi}{G}\right)^2\right)}\begin{cases} \frac{\sin\psi}{\psi} & \text{m even} \\ +\frac{\mathrm{j}\cos\psi}{\psi} & \text{m odd}\end{cases} \tag{8.44}$$

$$V(\beta_{mr}n) = \frac{\beta_{mr}}{\left(\beta_{mr}^2 - \left(\frac{n\pi}{G}\right)^2\right)}\begin{cases} \frac{\sin\psi}{\psi} & \text{m even} \\ -\frac{\mathrm{j}\cos\psi}{\psi} & \text{m odd}\end{cases} \tag{8.45}$$

where $\psi_r = \frac{\beta_{mr}G}{2}$ and $\xi_n = \begin{cases} 2 & (n=0) \\ 1 & (\mathrm{n}>0)\end{cases}$.

8.3.6 Determinantal Equation

Relatively routine algebraic manipulation of Eqs. (8.40), (8.41), 8.42) and (8.43), leads to a pair of simultaneous equations for K_n and L_n (for the mth mode), which can be summarised as outlined below. Note that in terms involving *n,* which occur outside of the summations over *n*, it is convenient to replace the integer *n* with integer *s*.

$$\sum_{r=-\infty}^{\infty}\{W_{nrs}K_{ns} + X_{nrs}L_{ns}\} = 0$$
$$\sum_{r=-\infty}^{\infty}\{Y_{nrs}K_{ns} + Z_{nrs}L_{ns}\} = 0 \tag{8.46}$$

where

$$
\begin{aligned}
W_{nrs} &= \frac{\coth \alpha_{cmr} b}{\alpha_{cmr}} V(\beta_{mr} s) \sum_{n=0}^{\infty} \frac{\frac{m\pi}{a}\frac{n\pi}{G}}{j\omega\mu_0} \left[\beta_{mr}^2 \sin \beta_{gmn} h - \left(\frac{n\pi}{G}\right)^2 \frac{k_{cu}^2}{k_{cf}^2} \right] U(\beta_{mr} n) \\
X_{nrs} &= \delta_{ns} \left(\frac{G+t}{G}\right) \xi_s \frac{\cos \beta_{gms} h}{2} + \frac{\coth \alpha_{cmr} b}{\alpha_{cmr}} V(\beta_{mr} s) \sum_{n=0}^{\infty} \left(\frac{n\pi}{G}\right)^2 \beta_{gmn} \frac{k_{cu}^2}{k_{cf}^2} U(\beta_{mr} n) \\
Y_{nrs} &= \delta_{ns} \left(\frac{G+t}{G}\right) \xi_s j\omega\varepsilon_0 \beta_{gms} \frac{\cos \beta_{gms} h}{2k_{cf}^2} \\
&\quad + \beta_{mr}^2 \frac{\coth \alpha_{cmr} b}{\alpha_{cmr}} V(\beta_{mr} s) \sum_{n=0}^{\infty} \frac{j\omega\varepsilon_0}{k_0^2} \left[\beta_{w10}^2 \sin \beta_{gmn} h - \frac{\left(\frac{m\pi}{a}\right)^2 \left(\frac{n\pi}{G}\right)^2}{k_{cf}^2} \right] U(\beta_{mr} n) \\
Z_{nrs} &= -\delta_{ns} \left(\frac{G+t}{G}\right) \xi_s \frac{m\pi}{a} \frac{n\pi}{G} \frac{\cos \beta_{gms} h}{2k_{cf}^2} \\
&\quad + \beta_{m0}^2 \frac{\coth \alpha_{cmr} b}{\alpha_{cmr}} V(\beta_{mr} s) \sum_{n=0}^{\infty} \frac{j\frac{m\pi}{a}\frac{n\pi}{G} \beta_{mr} \beta_{gmn}}{k_{cf}^2} U(\beta_{mr} n)
\end{aligned}
\tag{8.47}
$$

The Kronecker delta δ_{ns} is introduced to ensure that modal orthogonality is maintained ($\delta_{ns} = 1$, when $n = s$, otherwise it equates to zero).

There is a non-trivial solution to Eqs. (8.46) only if the determinant of the matrix coefficients equates to zero. That is, if,

$$
\begin{vmatrix} W_{nrs} & X_{nrs} \\ Y_{nrs} & Z_{nrs} \end{vmatrix} = 0 \tag{8.48}
$$

The accuracy of any solution for the propagation behaviour of periodic waveguide, based on Eq. (8.48), is dependent on the number of terms in the infinite series that is retained in computations. Obviously there is a limit. Infinite series of similar character are reported in Ref. [19], which provides useful pointers to possible summation techniques based on the calculus of residues. However, in the present context we can test the validity of the heuristically deduced solution by noting that only the first passband for the periodic structure with a single field variation in the x-direction ($m = 1$) is significant. In this scenario drastic approximation can legitimately be made. In Eq. (8.48) the only term of any significance is Y_{nrs}, since this is the only term which does not equate to zero when n = 0. Consequently, Eqs. (8.48) can justifiably be approximated by setting $Y_{nrs} K_{ns} = 0$, to produce the following identity. It is pertinent to note that the W and Z terms in Eq. (8.48) emanate from the equations for E_x and H_z respectively. Thus equating them to zero to form (8.49) implies that $E_x = 0$ and $H_z = 0$ at $y = b$. In conjunction with the boundary conditions for E the implication is that $E_x =$ 0 everywhere inside the fin-loaded guide. While this is approximately in accord with reality for the fundamental mode, it is unlikely to remain valid for most high order modes in higher passbands.

$$0 = \sum_{r=-\infty}^{\infty} \left\{ \begin{array}{l} \delta_{ns} \frac{G+t}{G} \xi_s \frac{\beta_{gms} \cos \beta_{gms} h}{2k_{cf}^2} K_{ms} \\ \quad + \beta_{m0}^2 \frac{\coth \alpha_{cmr} b}{\alpha_{cmr}} V(\beta_{mr} s) \left[\frac{K_{m0} \beta_{w10}^2}{k_{cf}^2} \sin(\beta_{w10} h) U(\beta_{mr} 0) + \sum_{n=0}^{\infty} \frac{K_{mn} \beta_{gmn}^2}{k_{cf}^2} U(\beta_{mr} n) \right] \end{array} \right\} \tag{8.49}$$

Equation (8.49) has been programmed for three space harmonics above the corrugated surface and for three waveguide modes in the inter-fin gaps below the corrugated surface. The results for the phase coefficient (β_{11}) of the fundamental mode HE_{11} in the guide, as a function of frequency, are presented in Fig. 8.8. Two fin-loaded waveguide examples are examined, and when compared with measurements, performed carefully on precisely machined prototypes, the level of agreement between calculation and experiment is clearly extremely good. For the fundamental mode it is fair to conclude that the heuristically deduced solution, despite the evident approximations involved, is capable of providing a reliable source of design information for fin loaded slow-waveguide structures. Of course today, structures of this type can be comprehensively and accurately modelled by resorting to full-wave electromagnetic solvers [12–14]. Nevertheless such solvers have to be set up properly—a process which is open to errors—and the availability of accessible equations such as Eq. (8.49) can provide a very useful check on the modelling process.

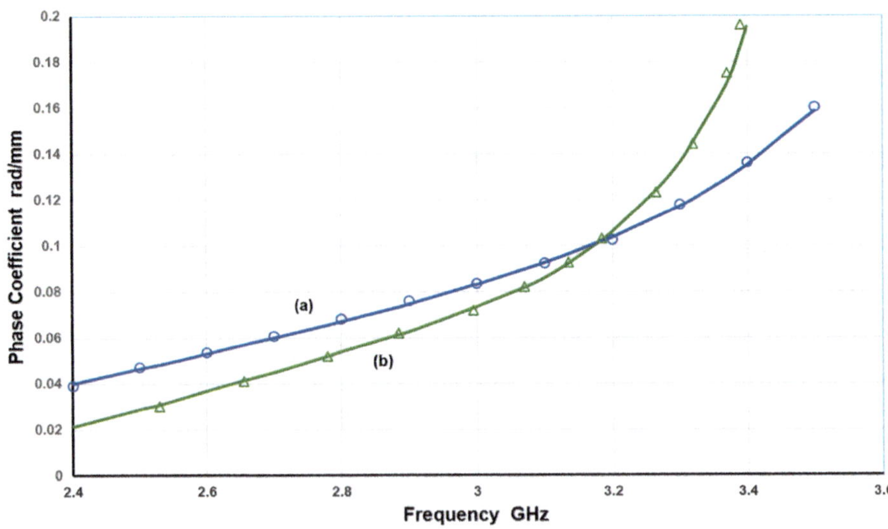

Fig. 8.8 Predicted and measured phase coefficients, as a function of frequency, for a fin-loaded rectangular waveguide **a** a = 72.14 mm, b = 12.24 mm, h = 21.8 mm, G = 4.76 mm, t = 1.58 mm, **b** a = 64.5 mm, b = 12.24 mm, h = 26.5 mm, G = 11.11 mm, t = 1.58 mm (curves = theory, points = measured)

8.4 Slot Radiators in a Periodic Waveguide Feed

The design of a scanned linear array antenna based on fin-loaded waveguide fed slot radiators requires knowledge of the radiation properties of such slots, which may be transversely or longitudinally directed, or inclined to the guide axis. Here we will limit our attention to the transverse slot radiator, as shown schematically in Fig. 8.9, which is fully representative of the computational process for slots in general. The application of a moment method to this problem leads, as we have seen for the conventional waveguide fed slot in Chap. 5, to a matrix equation of the form:

$$\begin{bmatrix} [A] & [\mathrm{C}] \\ [D] & [\mathrm{B}] \end{bmatrix} \begin{bmatrix} [\alpha] \\ [\alpha'] \end{bmatrix} = \begin{bmatrix} [h] \\ [0] \end{bmatrix} \tag{8.50}$$

The same equation applies to the periodic waveguide example in Fig. 8.9 but the matrix entries on the left of the equation require slight modification to reflect the waveguide change. The entries are:

$$A_{is} = -j\omega\varepsilon_0 \iint\limits_{S} f_i \hat{\mathbf{a}}_\eta \cdot \left\{ \hat{\mathbf{n}} \times \iint\limits_{\mathbf{S}} \left[\overline{\mathcal{G}}_{ma}(\mathbf{r}/\mathbf{r}_0) + \overline{\mathbf{G}}_{mb}(\mathbf{r}/\mathbf{r}_0) \right] \cdot f_s \hat{\mathbf{a}}_\xi dS_0 \right\} dS \tag{8.51a}$$

$$B_{is} = -j\omega\varepsilon_0 \iint\limits_{S'} f_i' \hat{\mathbf{a}}_\eta \cdot \left\{ \hat{\mathbf{n}}' \times \iint\limits_{\mathbf{S}'} \left[\overline{\mathbf{G}}_{mb}(\mathbf{r}/\mathbf{r}_0) + \overline{\mathbf{G}}_{mc}(\mathbf{r}/\mathbf{r}_0) \right] \cdot f_s' \hat{\mathbf{a}}_\xi dS_0 \right\} dS \tag{8.51b}$$

$$C_{is} = j\omega\varepsilon_0 \iint\limits_{S} f_i' \hat{\mathbf{a}}_\eta \cdot \left\{ \hat{\mathbf{n}} \times \iint\limits_{\mathbf{S}'} \left[\overline{\mathbf{G}}_{mb}(\mathbf{r}/\mathbf{r}_0) \right] \cdot f_s \hat{\mathbf{a}}_\xi dS_0 \right\} dS \tag{8.51c}$$

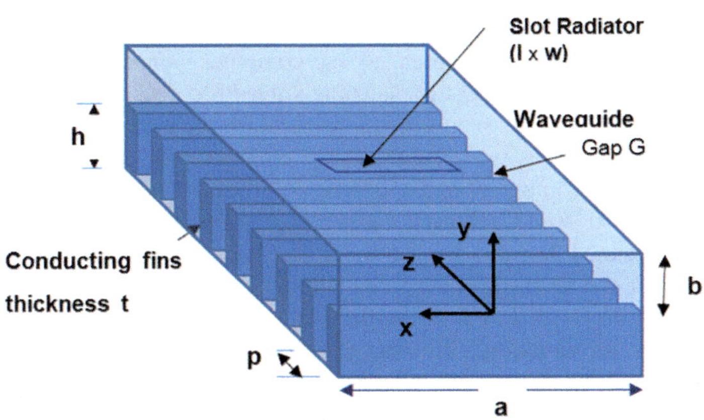

Fig. 8.9 Fin-loaded feed waveguide for transverse slot radiator in upper wall

$$D_{is} = j\omega\varepsilon_0 \iint_{S'} f_i \hat{\mathbf{a}}_\eta \cdot \left\{ \hat{\mathbf{n}}' \times \iint_{\mathbf{S}} \left[\overline{\mathbf{G}}_{mb}(\mathbf{r}/\mathbf{r}_0) \right] \cdot f_s' \hat{\mathbf{a}}_\xi dS_0 \right\} dS \quad (8.51\text{d})$$

$$h_i = -\iint_S f_i \hat{\mathbf{a}}_\eta \cdot [\hat{\mathbf{n}} \times \mathbf{H}_\mathbf{i}(\mathbf{r})] \, dS \quad (8.51\text{e})$$

The matrix entries A_{is}, B_{is}, C_{is}, D_{is} given mathematical expression in Eqs. (8.51a) are essentially electromagnetic coupling terms giving mathematical expression to the mutual coupling which occurs between the assumed current distributions ($f_{i,s}$) on the interior and external surfaces of the radiating aperture. For the 'slot-cavity' and for the exterior half-space the magnetic dyadic Green's functions $\overline{\mathrm{G}}_{\mathrm{mb}}$ and $\overline{\mathrm{G}}_{mc}$ respectively, remain unchanged from the conventional waveguide moment method derivation in Chap. 5. On the other hand, the magnetic dyadic Green's function $\overline{\mathcal{G}}_{ma}$ for the periodic feed waveguide clearly differs from $\overline{\mathrm{G}}_{\mathrm{ma}}$. Nevertheless it can readily be deduced from the plain waveguide form, as elaborated in Appendix C. For the transverse slot shown in Fig. 8.9 the matrix entries have the same meanings as for the slot radiator in conventional waveguide. These are:

A_{is} represents self-coupling between magnetic current elements on the inner slot surface S.
B_{is} represents self-coupling between magnetic current elements on the outer slot surface S′.
C_{is} represents mutual coupling to the inner slot surface (S) due to magnetic current elements on the outer surface (S′).
D_{is} represents mutual coupling to the outer slot surface (S′) due to magnetic current elements on the inner surface (S).

Once the integrals in Eqs. (8.51a) have been evaluated, which is mathematically routine but tedious [18, 20], the matrix Eq. (8.50) is amenable to solution by a range of matrix manipulation tools. These can be found in standard mathematical textbooks [21, 22] and in convenient to use computer software packages [12–14, 23]. The outcome of the process is plausible values for s, f_s, α'_s and α_s. These can then be used to compute the scattering parameters for the compact radiator much as outlined in Chap. 5.

Moment method computations have been performed on a fin-loaded waveguide fabricated in WG10 (a = 72.14 mm, b = 34.04 mm) designed to provide a feed for a frequency scanned linear array of transverse slots. These are summarised in Fig. 8.10, where the results are compared with an equivalent set-up in conventional waveguide (also WG10). In the figure the normalised power radiated by a centred transverse slot is presented as a function of frequency, for four different slot lengths of 30, 40, 42 and 46 mm, all at constant width (w = 4.5 mm). The solid (green) curves denote the conventional waveguide case, while the dashed (blue) traces

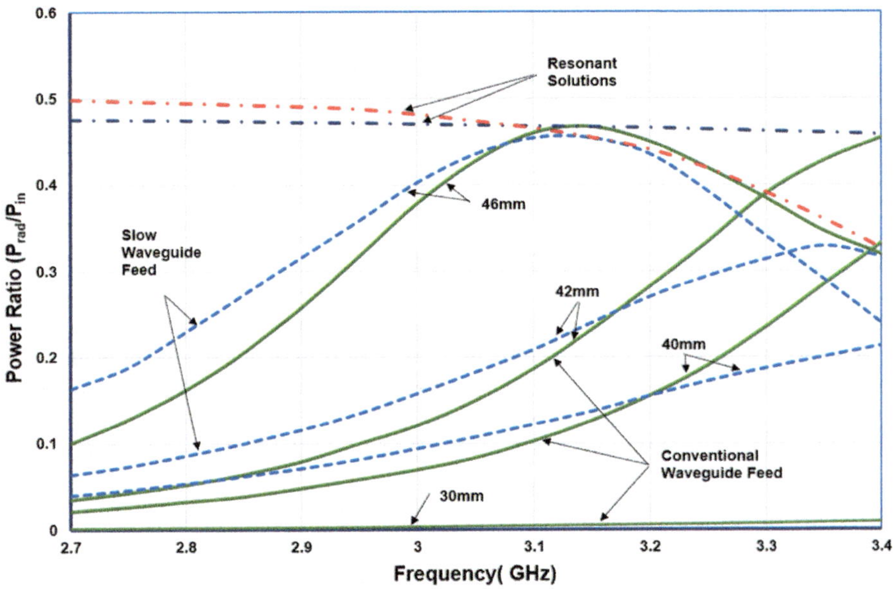

Fig. 8.10 Relative power radiated as a function of frequency for a centred transverse slot in a waveguide feed (solid curves) and a periodic waveguide feed (dashed curves) with slot length l as parameter (l = 30, 40, 42, 46 mm: w = 4.5 mm) Waveguide WG10: a = 72.14 mm, b = 34.04 mm Slow-waveguide: a = 72.14 mm, b = 12.24 mm, h = 21.8 mm, G = 4.7625 mm, t = 1.5875 mm

represent corresponding slow-waveguide calculations. At low frequencies it is clear that for the slow waveguide, with 21.24 mm deep fins separated by 4.76 mm, the normalised radiated power exceeds that of the equivalent slot in unmodified waveguide. This is largely because the height b (=12.24 mm) in the slow waveguide is made low enough to counteract the decay in fundamental mode field strength in the y-direction in the space above the periodic surface, thus enforcing good field coupling at the slot. This is emphasised by the curves depicting radiation from slots resonant at all frequencies (chain dotted) with the normalised power radiated deteriorating rapidly for the slow-wave case (red chain dotted) as the frequency extends beyond 3.15 GHz. As the frequency increases the decay phenomenon is enhanced [see Eq. (8.21)] so that between 3.1 and 3.2 GHz the plain waveguide begins to generate stronger radiation for the same feed conditions for all slot examples. For both feed types it is noticeable that for a 46 mm slot resonant at ~3.13 GHz, approximately 50% of the incident power is radiated and the operating bandwidths are comparable. These results confirm that the slow waveguide can provide an effective feed for a scanned linear array antenna.

8.4.1 Frequency Scanned Linear Slot Array Employing Periodic Waveguide Feed

The fin-loaded periodic waveguide has been shown to exhibit phase shift characteristics, as a function of frequency (see Fig. 8.8), which are comparable with the phase sensitivities manifested by dielectric loaded waveguide or sinuous waveguide. Furthermore slots located in the top wall opposite to the periodic surface (Fig. 8.9) can be excited to a level not unlike slots in a conventional waveguide feed (Fig. 8.10). Consequently, as a feed for a frequency scanned linear array antenna [10, 11, 24] the fin-loaded waveguide structure offers the advantages of much less weight than the dielectric loaded option and much less bulk than the serpentine waveguide alternative. Early attempts at forming fin-loaded waveguide feeds for a frequency scanning radar were derailed by heavy propagation losses [24]. However, it was eventually discovered that by adopting the highest accuracy and control of the manufacturing process the periodic feeder loss can be made comparable with a sinuous waveguide exhibiting similar scanning capability.

A slow-waveguide feed with the dimensions detailed in Fig. 8.8 [curve (a)] was used to create a scanned linear array comprising six slots on the broadface. The slots were of equal size, and with relatively low radiation per slot the field distribution across the aperture was close to uniform. The measured beam width and sidelobe levels for the array were consistent with this excitation set-up. The deviation of the peak of the main radiated beam has also been investigated. The predicted squint based of the theoretical phase coefficient change with frequency shown in Fig. 8.8 [curve (a)] was obtained using Eq. (8.1). It is plotted as a solid blue curve in Fig. 8.11, and compared with measured results (red diamonds). The degree of agreement further confirms the reliability of the developed formula [Eq. (8.49)] for the phase coefficient of the slow-wave feed.

8.5 Uniform Leaky-Wave Antennas

Careful and perceptive readers will have noted by now that electromagnetic waveguides display a degree of commonality in their frequency characteristics. This can largely be ascribed to the propagation equation which for conventional rectangular waveguide is given by Eq. (8.6). Standard waveguide text books [6, 7] confirm this by demonstrating that the same equation applies to empty waveguides of other cross-sections, such as circular, coaxial and elliptical. The only difference is that the transverse resonance terms, $(m\pi/a)$ and $(n\pi/b)$ in the rectangular case, must be replaced by Bessel function roots or elliptic function roots. Consequently, it is convenient to express Eq. (8.6) in a more general form, namely:

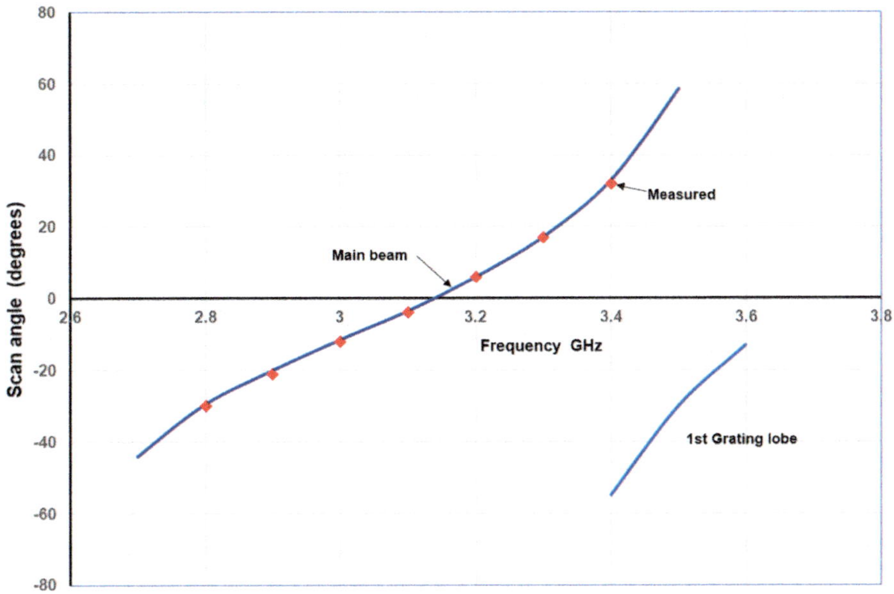

Fig. 8.11 Beam scan as a function of frequency for a slow-wave fin-loaded waveguide feed comprising six slots of the same size, equi-spaced by 65.4 mm (solid curves = theory, diamond markers = measurement)

$$k_0^2 = k_x^2 + k_y^2 + k_z^2 \quad (8.52)$$

where, for the rectangular waveguide: $k_x = \frac{m\pi}{a}$, $k_y = \frac{n\pi}{b}$, $k_z = \beta_{mn}$.

For slow rectangular waveguide we have observed that it is necessary for $k_z > k_0$ and this is constructed by making k_y^2 negative in which case k_y must be imaginary and equal to $j\alpha_m$ (say). That such solutions are possible is demonstrated in Sect. 8.3 for fin-loaded waveguide. Partial dielectric loading also generates solutions of this form.

Equation (8.52) points not only to trapped waves of the fast and slow variety, but also to leaky waves in an open structure. If such a wave propagates in the axial direction z, then we must have:

$$k_z = \beta_m - j\alpha_m \quad (8.53)$$

for the mth mode for which $k_x = \frac{m\pi}{a}$. β_m denotes the axial phase shift per unit distance in rad/m while α_m denotes axial attenuation rate in nepers/m. In this case Eq. (8.52) requires that:

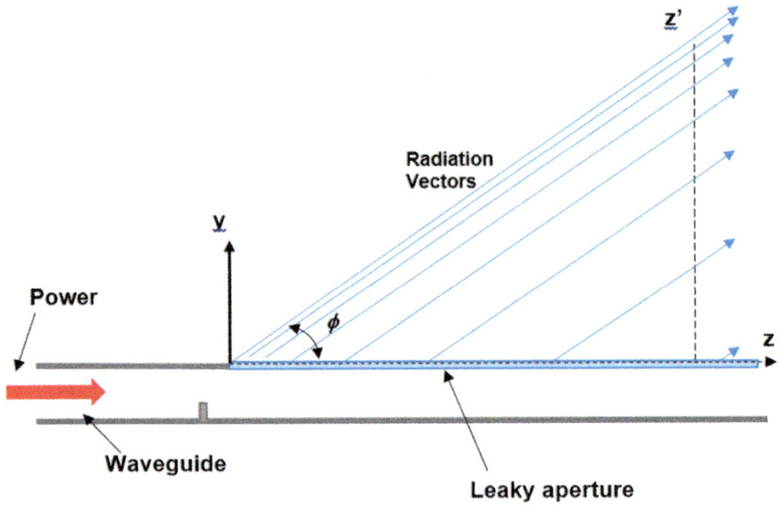

Fig. 8.12 Radiation from a leaky wave structure

$$k_y = \beta_{my} + j\alpha_{my} \tag{8.54}$$

if the space into which the electromagnetic waves leak is defined as y > 0 in Fig. 8.12. In this direction the leaky wave field strength has the functional form:

$$\exp(\alpha_{my} - j\beta_{my})y \tag{8.55}$$

The α_{my} term if positive implies, rather curiously that in the y-direction the leaky wave fields grow without limit. The consequent singular behaviour at infinity, means that mathematically, while it is a solution of the homogeneous field equations, the leaky wave cannot be a proper mode of the source free Maxwell equations. Nevertheless, despite this 'improperness' leaky waves can still be valid representations of the fields generated by an open structure in certain restricted regions provided a source is present [25].

The situation can perhaps be clarified by reference to Fig. 8.12 which depicts a leaky waveguide containing a long axially directed slit (commonly in the side wall [25]) fed from the left from a closed waveguide and an intervening matching transition. To the right, in the z-direction, the structure is assumed to be infinite. The slit is located at $y = 0$ and extends to infinity in the z-direction along with the waveguide. From Fig. 8.12 it is readily deduced that if the wave solution within the waveguide exhibits a dominant 'mode' phase coefficient β_0 then the wave component parallel to the waveguide and travelling along the outer surface must match it. This implies that:

$$\phi \approx \cos^{-1}\frac{\beta_0}{k_0} = \cos^{-1}\frac{\lambda_0}{\lambda_{g0}} \tag{8.56}$$

The 3 dB beamwidth $\Delta\phi$ can be deduced from array theory in Chap. 6 [Eq. (6.17)]. Thus for uniform excitation of the slit:

$$\Delta\phi \approx \frac{0.88\lambda_0}{L\sin\phi} \tag{8.57}$$

In Eq. (8.56) λ_{g0} is the wavelength of the dominant field solution in the guide. Note that it demands that $\beta_0 < k_0$. Consequently, leaky behaviour is linked to 'fast' waves and the radiation is launched into the forward quadrant. If the attenuation coefficient for this solution is α_0, then the field magnitude in the waveguide must diminish as z increases, which in turn implies that the power available for radiation decays. The resultant higher power density in the radiation fields at the input to the slit is represented in the diagram by closely spaced rays, with the spacing growing with increasing z. Consequently, it is not difficult to appreciate that if leaky radiation occurs the radiation intensity, at any position z', grows in the y-direction up to the point $y = z' \tan\phi$. This growth is in accordance with Eq. (8.55). In practical terms the leaky wave radiation is confined to the region $y < z\tan\phi$.

The key to long-slot leaky-wave antenna design in which the rectangular waveguide feed is modified by a long axial slit, through which the modal energy in the waveguide dissipates, lies in determining the propagation and attenuation coefficients for the adapted waveguide. For long leaky slits which can be presumed, with little error mathematically, to be of infinite extent, an electromagnetic boundary problem is formed. The resultant boundary value problem is readily amenable to solution by an electromagnetic solver either based on the moment method, or on the finite element method [13, 14]. As we have noted previously, while such solvers can be very accurate, if set up properly, it is generally good engineering practice to anticipate their use by developing simple analogues which are open to approximate field solutions. In this way a thorough insight into the antenna characteristics becomes available, while at the same time furnishing a method of checking the predictions of an electromagnetic solver if adopted.

For the long slot leaky-wave antenna with the aperture located in the sidewall of the rectangular feed waveguide, an elegant and surprisingly accurate field solution, using approximate methods, is provided by Goldstone and Oliner [25]. The analysis employs a solidly established transverse resonance technique which recognises that the non-axial propagation coefficients (k_x, k_y) in Eq. (8.52) represent the transverse resonances of all possible modes. Transmission line techniques are used to model the resultant field interactions in the transverse directions, including the slit discontinuity, represented as a complex impedance, between the transmission line analogue and a half-space. For this class of leaky-wave antenna it is recognised that the leaky-wave solution is essentially a perturbation of the electromagnetic mode existing in the waveguide in the absence of the slit. Consequently, perturbation theory is used to establish 'closed' form equations to represent the real and imaginary parts of the propagation coefficient in the leaky-wave structure.

For a rectangular waveguide perturbed by a slit running along the length of a sidewall Eq. (8.52) can be expressed as:

$$k_0^2 = (k_x + \Delta\kappa)^2 + k_y^2 + k_z^2 \tag{8.58}$$

where $\Delta\kappa$ is the perturbation on the closed waveguide value for the x-directed mode number, and is assumed to be complex, which means that k_z is complex and given by:

$$k_z = \beta - j\alpha \tag{8.59}$$

Hence, some routine algebra leads to:

$$\begin{aligned} \frac{\beta}{k_0} &= \frac{\lambda_0}{\lambda_g} \approx \frac{\lambda_0}{\lambda_{go}}\left(1 - \frac{k_x\lambda_{go}^2}{4\pi^2}\Delta\kappa_r\right) \\ \alpha\lambda_0 &\approx \frac{\lambda_0\lambda_{go}}{2\pi}k_x\Delta\kappa_j \end{aligned} \tag{8.60}$$

where λ_{go} is the guide wavelength in the unperturbed waveguide, while $\Delta\kappa_r$ and $\Delta\kappa_j$ are the real and imaginary parts of $\Delta\kappa$. Transmission line analysis of the transverse field relations in the perturbed waveguide yields the following approximate equations [25]:

$$\Delta\kappa \approx \frac{j}{a}\frac{\left(Y_T'(\kappa_0) - j\cot\kappa_0 a\right)}{\csc^2\kappa_0 a} \tag{8.61}$$

Here Y_T' denotes the normalised terminating admittance of the transmission line analogue. Expressions for the terminating impedance or admittance, where the slit as 'seen' from the transverse propagation perspective can be viewed as a capacitive iris, are available in many standard texts on electromagnetism [6, 7, 26]. Not surprisingly they confirm that this admittance is strongly dependent on slit width d. Reference [25] gives:

$$\begin{aligned} G_T' &= \frac{G_T}{Y_0} \approx 0.285\left[\frac{\pi^2 b\left(1 - \left(\frac{d}{b}\right)^2\right)^2}{16d\cos^2\left(\frac{\pi d}{2b}\right)}\right]\frac{2\pi}{\kappa d} \\ B_T' &= \frac{B_T}{Y_0} \approx -0.156\left[\frac{\pi^2 b\left(1 - \left(\frac{d}{b}\right)^2\right)^2}{16d\cos^2\left(\frac{\pi d}{2b}\right)}\right]\frac{2\pi}{\kappa d} \end{aligned} \tag{8.62}$$

with $\kappa = \sqrt{k_0^2 - \left(\frac{\pi}{a}\right)^2 - \beta^2}$.

The phase shift constant β expressed as λ_0/λ_g, and the attenuation coefficient αa are plotted in Fig. 8.13 as functions of the slit width d. The plots show clearly the

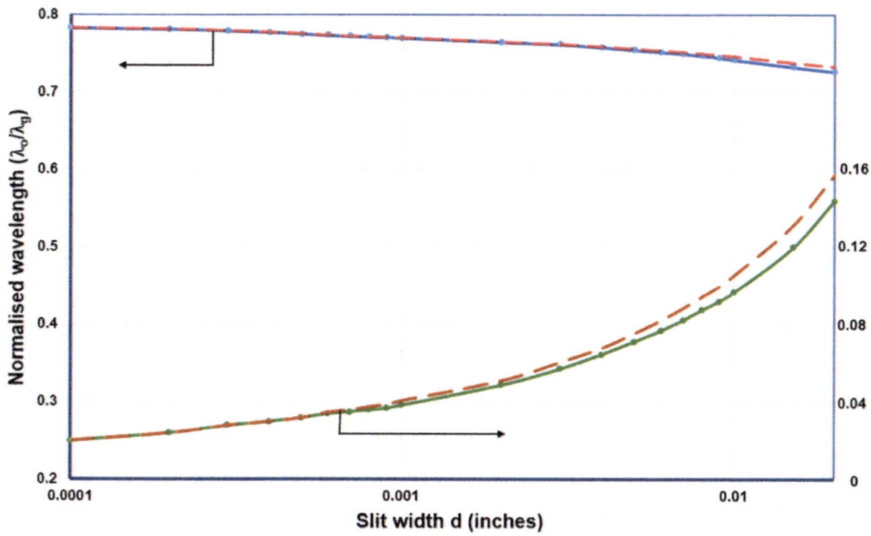

Fig. 8.13 Guide wavelength and attenuation coefficient as a function of slit width d in a rectangular waveguide with a = 22.86 mm and b = 10.16 mm at a frequency of 9.4 GHz. (Dashed curves = approximate solution: Solid curves = exact solution)

dependence of α and β on the slit width, with the influence on the attenuation rate more evident. However, more importantly the curves show minimal deviation from the approximate closed form solutions (Eq. (8.54); dashed curves) from super-positions which are nominally 'exact'. From an engineering view-point it cannot be too often stated that the availability of easily understood and easily implemented solutions for a given class of antenna is immensely helpful to the design process.

Once α and β have been ascertained to a requisite degree of accuracy, it becomes a relatively straightforward matter to determine the radiation characteristics of the leaky-wave antenna. Such characteristics include beam direction, beamwidth, radiation efficiency and scan sensitivity to frequency. Since, in this section scanning antennas are the focus of our attention we will limit ourselves to the scan sensitivity issue. For long-slot antennas which are largely uniform in cross-section, and unchanging geometrically in the axial direction, the scan behaviour usually depends on whether or not the feed waveguide is air filled or partially dielectric filled.

Generally, the air filled option provides a superior scanning antenna. This is explicable by directing one's attention toward transverse wave numbers (k_x, k_y) which in the air filled case are independent of frequency [see Eq. (8.52)]. The implication is that the beamwidth of the radiation pattern of the leaky-wave antenna does not change as the beam is scanned with frequency. This scan behaviour is summarised in Eq. (8.56), which dictates that the beam scan angle ϕ progresses from broadside ($\beta/k_0 \Rightarrow 0$:$\phi = 90°$) near cut-off, to endfire at high frequencies in the direction of propagation within the feed (+z-direction: $\beta/k_0 \Rightarrow 1$). Furthermore, the application of elementary trigonometry to (8.56) yields:

$$\sin^2\phi = 1 - \cos^2\phi = 1 - \frac{\beta^2}{k_0^2}$$

But $\beta^2 = k_0^2 - k_c^2$, consequently for an air filled feed $1 - \left(\frac{\beta}{k_0}\right)^2 = \left(\frac{k_c}{k_0}\right)^2$, and therefore:

$$\phi = \sin^{-1}\frac{k_c}{k_0} \tag{8.63}$$

Substituting into (8.57) then reveals that:

$$\Delta\phi \approx \frac{5.53}{k_c L} \tag{8.64}$$

Note that (8.64) confirms that the beamwidth is independent of frequency for the air filled feed, as indicated above.

In the partially dielectric loaded feed increasing the frequency causes the field distribution of the internal mode to become more and more confined to the dielectric, weakening the slit excitation level. This process significantly degrades the beamwidth which is inevitably considered to be a disadvantage. On the other hand, it is well known that the phase shift coefficient in a dielectric loaded waveguide exhibits an increasing magnitude with frequency, which is significantly larger than for an equivalent empty waveguide. The consequence is, that with dielectric loading, the uniform leaky waveguide antenna can be designed to scan over a larger range of angles than the air filled alternative, for a prescribed frequency range.

8.6 Periodic Leaky-Wave Antenna Arrays

While leaky-wave antennas evidently manifest themselves in a wide variety of geometries, which are uniform rather than periodic in nature [1], nevertheless we will direct our attention to the antenna array with periodically and closely spaced slot radiators, as depicted in Fig. 8.14b. This is consistent with the slot antenna emphasis of the book, and in this section, it is the evolution of linear slot arrays into structures incorporating overt slow-wave characteristics [1, 11, 25], which is of primary interest. This means that in a rectangular waveguide fed travelling-wave array, formed from transverse slots in the broadwall, if the slots become so closely spaced ($S \ll \lambda_o$; see Fig. 8.14b) that mutual coupling makes a significant contribution to the array behaviour, then the slot array as a whole is considered to represent periodic loading of the waveguide. The periodically loaded waveguide

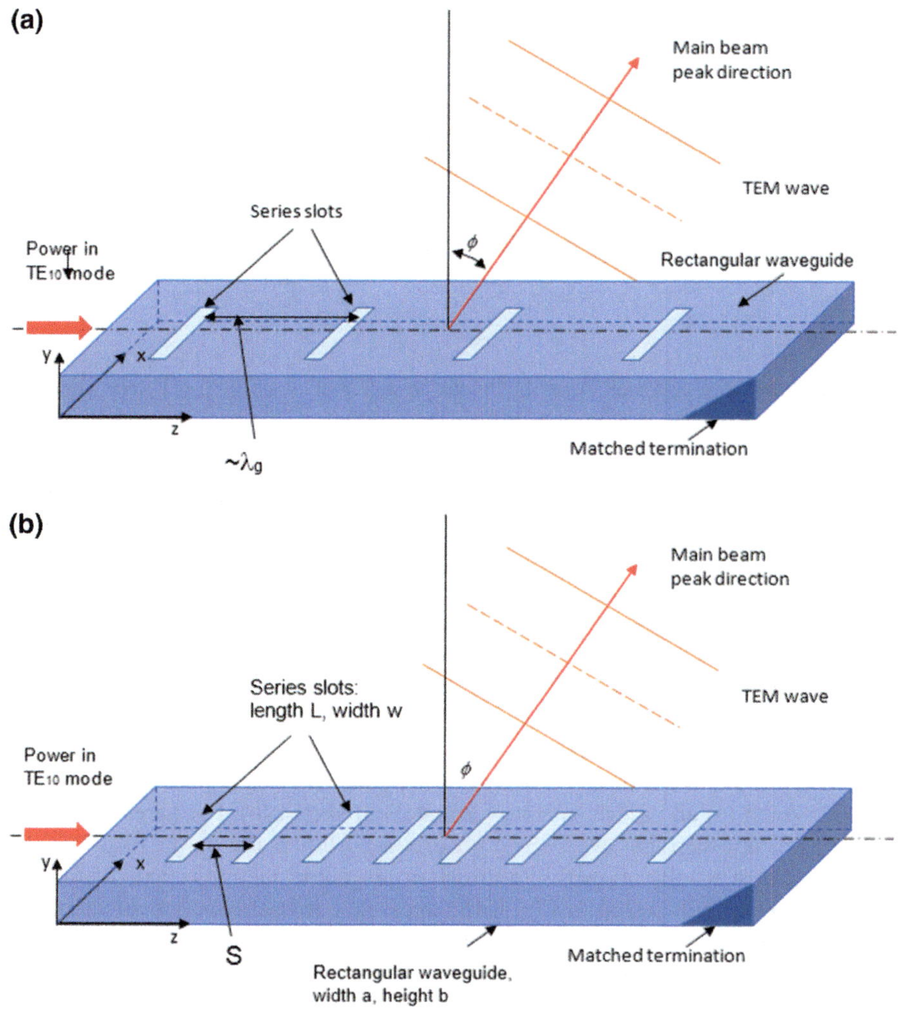

Fig. 8.14 Waveguide linear array antennas employing transverse (series) slots. **a** Travelling-wave array, **b** leaky wave array

structure must then be treated as a single electromagnetic boundary value problem, the solutions for which are rather different from those of the original empty waveguide as we have seen. Electromagnetically, the transitioning from uniform rectangular waveguide to a periodically loaded version has been addressed in Chap. 7, Sect. 7.3 and this material provides a helpful route into the analysis of the closely spaced transverse slot array.

The reader will be aware at this juncture that the travelling-wave array antenna in a rectangular waveguide feed is largely realised using longitudinally directed (shunt) slots. Nevertheless alternative slot arrangements for travelling-wave arrays

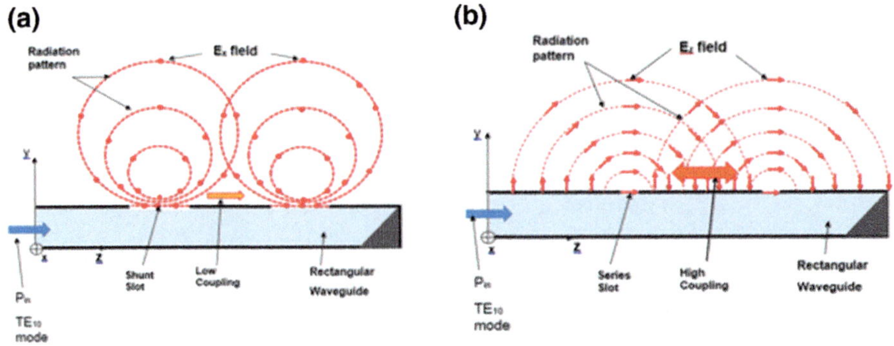

Fig. 8.15 Pictorial representation of the mutual coupling mechanism for **a** shunt slots and **b** transverse or series slot in a linear waveguide array antenna

have been tried, most commonly in the form of inclined slots and transverse (series) slots. The transverse slot in rectangular waveguide as shown in Fig. 8.14 most clearly illustrates the difficulty of adopting slot forms other than the shunt variety, so the following exposition is focused upon it.

The overwhelming reason for the low adoption of non-longitudinal slots in waveguide arrays of this type is mutual coupling between neighbouring slots, which is strongly influenced by slot orientation and positioning. The electromagnetic field mechanism underpinning divergent levels of mutual coupling between notionally similar slots in waveguide is readily explained by reference to Fig. 8.15. The figure depicts a y-z plane cut of a rectangular waveguide feeding a pair of longitudinal slots in sketch (a) and a pair of transverse slots in sketch (b). For narrow slots it is now well established that the electric field induced in the slot by the exciting TE_{10} mode is almost wholly aligned in a direction normal to the major axis of the slot. So, for the longitudinal slots in Fig. 8.15a the E-field in each slot is x-directed. This means that in the y-z plane containing the slot the radiated E-field is also x-directed, exhibiting the typical (dipole) pattern shape with nulls at the conducting surface of the waveguide where the x-directed electric field is 'shorted'. This pattern dictates that at the neighbouring slot located $d \sim \lambda_o/2$ away the E-field is weak and mutual coupling is very low. Similar observations apply to the internally scattered fields. As a consequence shunt slots in waveguide represent almost ideal array elements and shunt slot arrays can be investigated easily, as we have seen, using the conventional array theory presented in Chap. 6.

On the other hand, the electromagnetic interaction between a neighbouring pair of transverse (series) slots in the broadwall of a rectangular waveguide is by no means insignificant, as Fig. 8.15b purports to demonstrate. In this case, with the major axis of each slot aligned in the x-direction the induced E-field in both is z-directed. Thus, in any y-z plane intercepting the slots the radiated E-field forms omni-directional, semi-circular, equi-field contours as shown. This results in the radiation power density propagating over the guide surface (z-direction) matching the normally directed radiation (y-directed). Furthermore, the E-field for this z-directed surface radiation is normal to the waveguide wall. This behaviour means that the electromagnetic coupling between slots is no longer weak unless the slots are separated by more than the free space wavelength. Actually to form a broadside mainbeam with an array of series slots, with no possibility of a π-phase shift adjustment, as with shunt slots, slot separation is required to be $\sim \lambda_{g10}$. For an air filled waveguide feed this equates to a slot interval $d > \lambda_o$. While this is advantageous in lowering mutual coupling, it has the effect of permitting the formation of undesirable grating lobes (see Chap. 6). It can be solved by dielectric loading of the waveguide which has the effect of shrinking λ_{g10}. However, such a change simply negates the attempt to reduce mutual coupling. Actually, dielectric loading is seldom an acceptable solution due to weight and power loss penalties. It is perhaps pertinent to note that conventional array theory (Chap. 6) can continue to be applied to transverse slots afflicted by mutual coupling provided the radiation pattern of each slot element in the array is not modelled as an isolated slot as in Chap. 5. Rather the presence of coupled neighbouring slots is modelled by assuming that the active slot is embedded in a ground plane of passive slots arranged in a periodic grid representative of the array antenna design. Note that in planar slot arrays transverse slot mutual coupling is largely unavoidable. Not surprisingly, for inclined slots mutual coupling strength is a function of the inclination angle [1].

The consequence of the above is that transverse slot arrays can be realised only if the mutual coupling between slots is viewed as instrumental to the array operation rather than as disruptive to it. This means that the feed is no longer just a support structure whereby the TE_{10} mode illuminates a sequence of slot radiators (Fig. 8.14a), but forms an integral part of a slow-wave structure created by the periodic array of slots. In addition to the TE/TM modes of rectangular waveguide the ladder structure (Fig. 8.14b) can also support surface wave and leaky-wave modes, which significantly contribute to the radiation characteristics of the antenna.

For the waveguide periodically loaded with slot radiators (Fig. 8.14b) the leaky phenomenon, as we have seen, results in complex propagation coefficients for the fundamental space harmonic of each mode—α and β (say) for the dominant mode [27]. For a typical structure with $b/a = 0.466$, $L/a = 1.0$ theoretical estimates of these parameters, normalised to α/k_o and β/k_o, are presented in Figs. 8.16 and 8.17. In a manner reminiscent of the uniform leaky-wave antenna, radiation beamwidth and radiation efficiency are influenced by the magnitude of α, so that operation near the waveguide cut-off, where α is large, results in a wide primary beam, which narrows as the frequency is raised (see Fig. 8.16). Above cut-off α rises rapidly to a peak once finite power is able to enter the guide. It then falls away with increasing frequency as the slot radiators, with $L = a$, become increasingly non-resonant. The relationship between attenuation constant and frequency can obviously be tailored by dimensional adjustments as is hinted at in Fig. 8.16.

Generally the 'leaky' slotted ladder in the broadwall of waveguide, as suggested in Fig. 8.14, does not impose heavy loading on the waveguide mode. Consequently the phase coefficient for the modified guide deviates little from the closed waveguide prediction, as indicated in Fig. 8.17. The noteworthy observation, which can be extracted from Fig. 8.17, is that the closed waveguide β-curve represents the

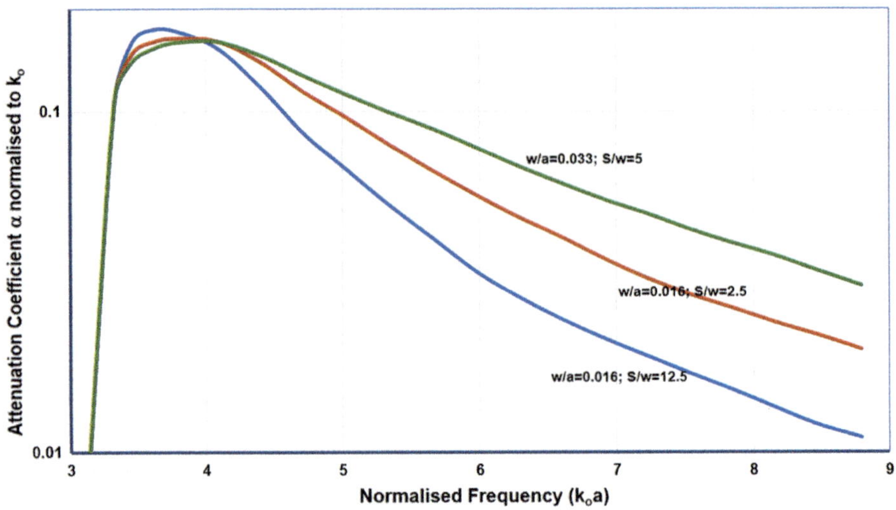

Fig. 8.16 Attenuation coefficient as a function of frequency for the dominant mode of a leaky ladder of slot radiators (with L = a)

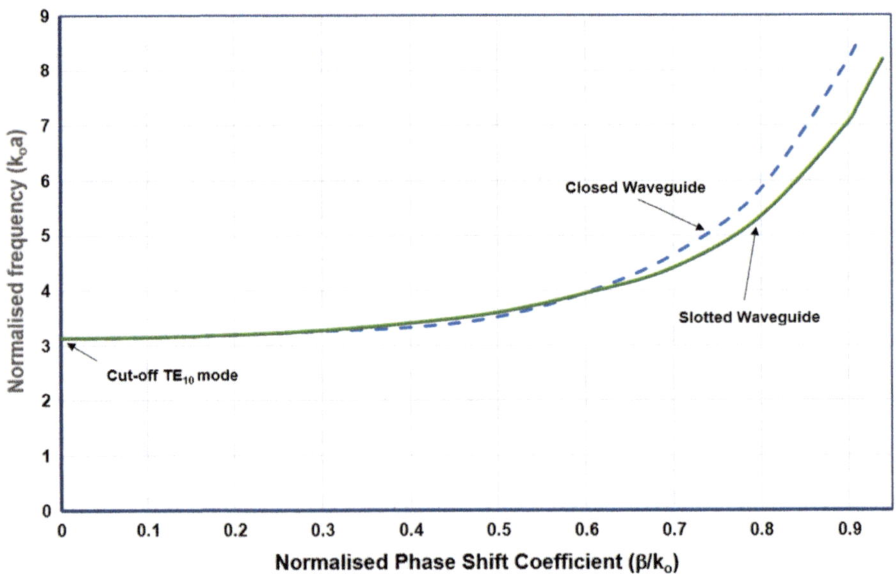

Fig. 8.17 Brillouin diagram for the dominant mode of a leaky ladder of slot radiators: w/a = 0.0016, S/w = 5

border between fast wave and slow wave solutions. Where the leaky wave curve falls below this line, at approximately $\beta = 0.6k_o$, the solution ceases to be leaky and reverts to a trapped surface wave. Therefore, for the specified antenna geometry, leaky wave performance is available only between cut-off at $k_oa = \pi$ and $k_oa = 4$.

Finally, it is pertinent to note here that an important difference between the uniform and the periodic leaky-wave antenna is that the dominant mode in the former is a fast wave so that the structure radiates whenever it is coupled to free-space. On the other hand, the dominant mode for a periodic leaky-wave antenna is a slow-wave that is intrinsically non-radiating. However, 'fast' space harmonics of a higher order mode can exist, and one of these can be a source of leaky-waves in an unclosed structure such as Fig. 8.14b. Note that if a harmonic is 'leaky' the mode as a whole is 'leaky'. Since the engineering aim is to create an antenna that forms a single radiating beam the structure must be designed to favour the mode which is potentially leaky. Sophisticated design of a periodic leaky-wave antenna, combining a slow-wave structure with a leaky ladder of slots, can achieve very large scan range, potentially over the angular range $-90° < \phi < 90°$ by operating over a frequency range which transitions from a backward space harmonic to a forward space harmonic. This is not possible with the uniform slit based alternative.

8.7 Chapter Summary

Linear slotted array antennas implemented in regular rectangular waveguide are generally considered to be restricted in their use because of 'squinting' which links the beam angle to frequency (see Chap. 7). However, as is often the case in engineering developments, disadvantages of a technology in one context can be a boon in another.

Array antennas, which are designed to enhance beam scanning for 3-D radar applications, have been examined in this chapter. Several different forms are considered including the sinuous feed, the dielectric loaded feed, and the periodic waveguide feed. The latter relies on the enhanced frequency dependency of the phase shift constant of the fundamental mode in the periodically loaded waveguide. The key to the development of scanned antenna applications is precise knowledge of the modal characteristics of the waveguide. This is particularly so for periodic waveguide. The required knowledge is provided in this chapter, while relevant theoretical methods for estimating frequency scanning capability are described.

When the periodicity is created by the slot array itself, it is demonstrated that this leads to leaky-wave behaviour. Leaky-wave antennas exist in two basic forms, namely uniform versions and periodic versions. These are compared in this chapter. It is shown that scan range is optimal in the leaky-wave antenna array format, provided full cognizance of the complex modal structure is exercised in the design process.

References

1. R. Johnson, *Antenna Engineering Handbook* (McGraw-Hill, New York, 1993)
2. R.C. Hansen, *Microwave Scanning Antennas, Vol III*, (Academic Press Inc., New York, 1966), N.A. Begovich, Chap. 2
3. J.C. Slater, *Microwave Electronics* (Van Nostrand, 1950)
4. R.M. Bevensee, *Slow-wave Structures* (Wiley, New York, 1964)
5. W. Walkinshaw, Theoretical design of linear accelerators for electrons. Proc. Phys. Soc. **61**, 246–254 (1948)
6. R.E. Collin, *Field Theory of Guided Waves* (McGraw-Hill, New York, 1960)
7. J.A. Stratton, *Electromagnetic Theory* (Wiley, New Jersey, 2007)
8. S.B. Cohn, Analysis of a wideband waveguide filter. Proc. IRE **37**, 651–656 (1949)
9. P.J.B. Clarricoats, P.K. Saha, Propagation and radiation behaviour of corrugated feeds. Proc. IEE **118**(9), 1167–1186 (1971)
10. K. Milne, The combination of pulse compression with frequency scanning for three dimensional radars. Radio Electron. Eng. **26**(2), 89–97 (1964)
11. A.W. Rudge, K. Milne, A.D. Olver, P. Knight, *The Handbook of Antenna Design*, vol. 2 (Peter Peregrinus Ltd., London, 1983)
12. http://mathworld.wolfram.com/FourierSeriesSquareWave.html
13. HFSS, *High Frequency Structure Simulator*, Ansoft Corp
14. COMSOL, https://www.comsol.com/comsol-multiphysics
15. R.F. Harrington, *Moment Methods*
16. MATLAB, Mathworks Ltd., https://uk.mathworks.com/products/matlab.html
17. A.J. Sangster, B. Maher, Higher order evanescent modes on slow wave structures. Electron. Lett. **8**, 608–609 (1972)
18. R.W. Lyon, *An Investigation of Apertures in Dual Mode Rectangular Waveguide for Variable Polarisation Arrays,* Ph.D. Thesis, Heriot-Watt University, 1979
19. R.A. Hurd, The propagation of an electromagnetic wave along an infinite corrugated surface. Can. J. Phys. **32**(12), 727–734 (1954)
20. R.W. Lyon, A.J. Sangster, Efficient moment method analysis of radiating slots in thick walled rectangular waveguide. IEE Proc. **128**(4), 197–205 (1981)
21. M.G. Salvadori, M.L. Baron, *Numerical Methods in Engineering* (Prentice-Hall International, London, 1961)
22. T.J. Akai, *Numerical Methods* (Wiley, Canada, 1993)
23. R.S. Esfandiari, *Numerical Methods for Engineers and Scientists using MATLAB* (CRC Press, Boca Raton, USA, 2013)
24. A.J. Sangster, D.C. Hawkins, in *IEEE Conference Publication*. A slotted waveguide circularly polarised linear array for a 3D antenna system, No. 105, pp. 94–97, Oct 1973
25. L.O. Goldstone, A.A. Oliner, Leaky-wave antennas – part 1: rectangular waveguides. IRE Trans. Antennas Propag. **AP-7**, 307–319 (1959)
26. N. Marcuvitz, *Waveguide Handbook* (M.I.T. Radiation Laboratory Series, McGraw-Hill Book Co., Inc., New York, 1951)
27. R.F. Hyneman, Closely spaced transverse slots in rectangular waveguide. IRE Trans. Antennas Propag. **AP-7**, 335–342 (1959)

Chapter 9
Compact Planar Reflectarrays

9.1 Introduction

It is appropriate at this juncture in our examination of slot antennas to apply what we have learned so far, on slots and radiation, to the examination of planar arrays, and in particular planar slot arrays. Chapters 2–7 have furnished the reader with the electromagnetic 'tools' required to evaluate radiation from antennas of the slot type, particularly when excited by the most commonly employed electromagnetic feed structure, namely rectangular waveguide. This propagation format is almost invariably adopted wherever significant power delivery and/or low loss are desirable. However, to be an effective antenna in most practical scenarios the slot radiator cannot act on its own. It inevitably needs to form an element in an array. The 'tools' needed to evaluate the radiation behaviour of a range of possible linear array configurations are provided in Chaps. 6 and 7. It is pertinent to note that 'full blown' planar phased array systems, where each array element is backed by a self-contained transmit/receive device, remain very expensive. Consequently, they are generally limited to military installations such as long range radar, and to 'big science' applications such as astronomy. Needless to say, a 'half-way house' planar array antenna can readily be formed from the paralleling of linear arrays fed from branching networks of couplers. This is particularly true of waveguide systems where the linear slotted waveguide arrays are excited by end-fed, or centre fed, waveguide distribution systems. However, such antenna developments tend to be equally large being aimed at providing long range 3-D radar installations in the lower microwave frequency bands (L-band to X-band). They remain costly despite the technological simplifications which can be adopted by comparison with maximally versatile phased arrays. An illustrative example is shown in Fig. 8.3.

At the time of writing this text, it is probably fair to say that commercial applications of planar arrays have tended to concentrate on formats which offer specified functions designed to deliver highly specialised aims at affordable cost levels. Such systems predominantly provide compactness, conformability and

A. J. Sangster, *Compact Slot Array Antennas for Wireless Communications*, Signals and Communication Technology, https://doi.org/10.1007/978-3-030-01753-8_9

lightness based on the adoption of advanced fabrication methods, such as 3-D printing. They find applications in mobile and satellite communications systems and satellite sensing and radiometry, generally in operating bands at the mm-wave and low infrared sections of the electromagnetic spectrum, competing with slotted waveguide array antennas as indicated in Chap. 7, Sect. 7.2.5. To be consistent with the thrust of this book, it is appropriate here to direct our attention toward these planar slot array antennas, which fall into the compact category. The availability of increasingly accurate fabrication methods enables the realisation of the high precision periodic surfaces which are in evidence in planar arrays operating at mm-wave and low infra-red frequencies. Typical examples which we will explore in this Chapter are the reflectarrays. Antennas in the Fabry-Perot cavity resonator (CRA) class, which offer similar sensing solutions, are discussed in Chap. 10.

In essence the reflectarray [1, 2] represents an attempt by antenna designers, at least at the outset, to replace ubiquitous and very traditional parabolic reflector antennas in satellite communications roles, by planar slot/patch array technology which offers potential improvements such as lower weight, less bulk, increased ease of assembly and installation (in space), and a more cost effective alternative to the parabolic dish. The latter is particularly true at mm-wave frequencies.

9.2 Passive Arrays

While both the reflectarray and CRA antennas presents an electromagnetic boundary value problem which is straightforward enough to permit direct analysis by commercially available full wave finite element based solvers, such as HFSS or COMSOL, significant computing power remains necessary to simulate, with meaningful detail and accuracy, a high gain structure. On the other hand, by generating an approximate analysis constructed from established field equations and antenna analysis techniques, basic scattering relationships can be deduced while at the same time procuring valuable insight into the fundamental nature of the passive array antenna. Such an analysis can usefully double up as a check on simulation predictions where these are pursued.

It is perhaps not too surprising that the passive array can be analysed using antenna array theory (Chap. 6) together with aperture theory (Chap. 3) and plane wave spectrum theory (Chap. 3). It can comprise many elemental scatterers (patches [3] or grooves [4] or dielectric protrusions [5]), and to some extent the most appropriate analysis technique is dependent of the array size and hence the number of scatterers. For small arrays an element-to-element or mutual impedance approach is preferred [6] whereby the interactions between every element in the array is tabulated leading to a potentially large set of simultaneous equations. While the method is intrinsically simple the computation of the mutual interactions between realistically modelled elements is not. Consequently even for relatively small arrays computations can be restrictively time consuming.

The alternative approach, which is more relevant to multi-element arrays, embraces the argument that in a large array the vast majority of elements, remote from the array periphery, are embedded in an environment which is largely that of an infinite array. The adoption of an infinite array analogy introduces the significant mathematical advantage of permitting the array to be treated as a periodic surface. This facilitates [7] the application of plane wave expansion or spectral techniques to the passive reflectarray geometry.

In Chap. 3 the essence of antenna analysis is demonstrated to reside in the integral equations (see Table 3.1):

$$\mathbf{A}(\mathbf{r},t) = \frac{\mu_0}{4\pi}\iiint\limits_V \frac{\mathbf{J}(\mathbf{r}',t)}{|\mathbf{r}-\mathbf{r}'|}dV' \tag{9.1}$$

$$\mathbf{A}_m(\mathbf{r},t) = \frac{\varepsilon_0}{4\pi}\iiint\limits_V \frac{\mathbf{J}_m(\mathbf{r}',t)}{|\mathbf{r}-\mathbf{r}'|}dV' \tag{9.2}$$

As we have observed in earlier chapters for wire, dipole and patch antenna elements on which an electric current density $\mathbf{J}$ exists (Chap. 2), the starting point for a radiation calculation is Eq. (9.1), while for antenna elements of the aperture type (slots, holes, slits and patch/slots) where radiation emanates from an E-field usually referred to, by analogy, as the magnetic current density ($\mathbf{J}_m$), Eq. (9.2) represents the appropriate commencement of any calculation. It is, hopefully, also clear from Chaps. 2, 3 and 4 that these equations lead to the powerful observation that for any wire antenna or aperture antenna, if the current densities on a specified antenna geometry are known and integrable, and if the dyadic Green's functions (see Table 4.1) for the coupled spaces are available, then near and far field radiation patterns can be computed.

The introduction of passive arrays, such as reflectarrays and frequency selective surfaces, into the analysis mix requires the application of Eqs. (9.1) and (9.2) to a space, or spaces, which are not accommodated by the traditional antenna analysis tools. A procedure for accommodating radiators, which form elements of a periodic surface, is required [7]. We commence the process by replacing Eq. (9.1) by its arguably more familiar usage as it applies to the Hertzian current element (see Chap. 2). It has the form:

$$\mathbf{A} = \frac{\mu_0 I_0 \mathbf{dl}}{4\pi r}\exp(j(\omega t - k_0 r)) \tag{9.3}$$

We could have commenced with Eq. (9.2) and proceeded in a similar manner by applying it to an isolated magnetic current element (i.e. an elemental slot). However, we shall persevere with Eq. (9.3) since periodic surfaces used in antenna pattern control are more commonly fabricated from multiple scatterers of the 'wire' based variety (see Fig. 9.1). Note that in Eq. (9.3) the term

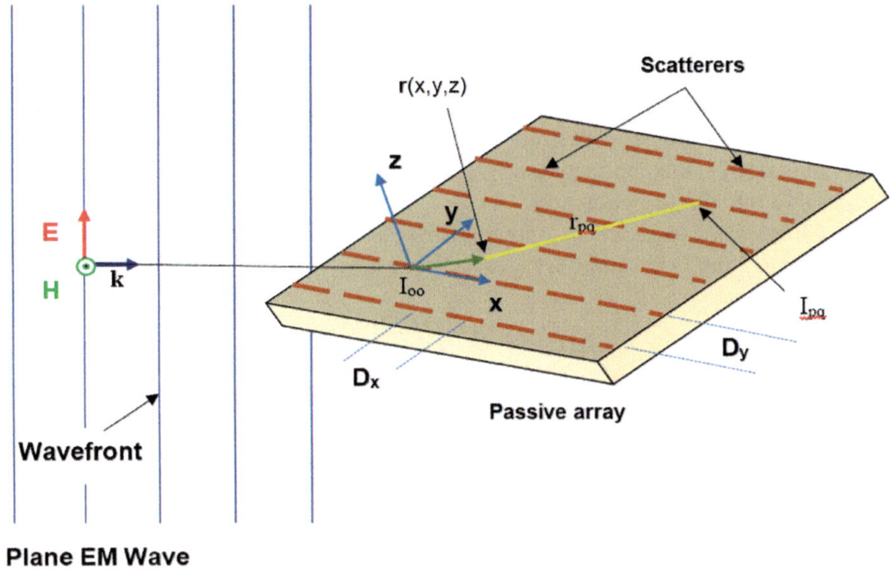

Fig. 9.1 Plane wave illumination of passive periodic array of current elements

$$\frac{1}{4\pi r}\exp(j(\omega t - k_0 r))$$

represents in essence the scalar Green's function for free-space in phasor form. The vector form is given in Table 4.1.

Generally speaking, the incorporation of frequency selective surfaces or periodic surfaces into antenna technology aims to influence radiation characteristics by making use of the controlled and adjustable scattering behaviour of the surface, when illuminated by a plane electromagnetic wave. There are few antenna situations where plane wave illumination cannot be applied and therefore we concentrate on this aspect. In Fig. 9.1 a plane electromagnetic wave is presumed to propagate from the left, in direction $\mathbf{k}(= \hat{\mathbf{a}}_k k_0)$, where $\hat{\mathbf{a}}_k$ is the unit vector and $k_0 = \omega/c$, and illuminates an inclined flat surface supporting a regular lattice of dipole like scatterers. The periodic surface lies in the *x-y* plane of the *x*, *y*, *z* Cartesian system, with x-directed metallic dipoles spaced D_x apart in the x-direction and D_y apart in the y-direction. A more general system would have the dipoles directed at angle ψ to the x-axis, but in such a case the dipole can be resolved into x-directed and y-directed portions, so it is not too restrictive to limit consideration to the less complex x-directed case. In the x-direction the elements are numbered from $-\infty < p < \infty$ while in the y-direction they are numbered from $-\infty < q < \infty$, so we can identify the current on a typical element as $\hat{\mathbf{a}}_\mathbf{x} I_{pq}$.

At the reference element I_{oo} the plane wave arrives in the direction.

$$\mathbf{k} = \hat{\mathbf{a}}_x k_x + \hat{\mathbf{a}}_y k_y + \hat{\mathbf{a}}_z k_z \tag{9.4}$$

Thus, if the incident plane wave **k** induces a current I_{oo} at the reference element, then at a typical element located at p, q, where p is the element count in the x-direction and q is the count in the y-direction, we must have, on applying the Floquet theorem:

$$I_{pq} = I_{00} \exp(-jpk_x D_x) \exp(-jqk_y D_y) \tag{9.5}$$

The exponential factors express the plane wave phase shift between element 0, 0 and element p, q. Note that the currents I_{pq} are also secondary sources of radiation, and if the periodic surface is immersed in an incident plane wave of infinite extent, the elements will radiate equally strongly.

To assess the scattering from the passive array as a whole, we first need to evaluate the radiation at an observation point $\mathbf{r}(x, y, z)$ exterior to the array surface from each current element p, q. Relative to this observation point the distance to the element p, q is given by (see Fig. 9.1):

$$r_{pq} = z^2 + (pD_x - x)^2 + (qD_y - y)^2 \tag{9.6}$$

Consequently, by analogy with Eq. (9.3) it is possible to expect that the magnetic vector potential at $\mathbf{r}$, due to I_{pq}, is $\mathbf{A}_{pq}$ and that it can be expressed as:

$$\mathbf{A}_{pq} = \frac{\mu_0 I_0 \mathbf{dl}}{4\pi r_{pq}} \exp(-jk_0 r_{pq}) \tag{9.7}$$

Given that all of the current elements are aligned in the direction $\mathbf{dl}(= \hat{\mathbf{a}}_\mathbf{x} dl)$ then the magnetic vector potential $\mathbf{A}_p$ for the entire linear array at p, with $-\infty < q < \infty$ can be expressed as:

$$\mathbf{A}_p = \frac{\mu_0 I_0 \mathbf{dl}}{4\pi} \sum_{q=-\infty}^{\infty} \frac{I_{pq} \exp(-jk_0 r_{pq})}{r_{pq}} \tag{9.8}$$

This can be formulated in terms of I_{oo} by substituting Eq. (9.5) into Eq. (9.8), to give:

$$\mathbf{A}_p = \frac{\mu_0 I_{00} \mathbf{dl}}{4\pi} \exp(-jpk_x D_x) \sum_{q=-\infty}^{\infty} \frac{\exp(-jqk_y D_y) \exp(-jk_0 r_{pq})}{r_{pq}} \tag{9.9}$$

Generally the infinite sum in the equation is convergent, except where $k_y D_y$ is a multiple integer of π—a grating lobe condition. Unfortunately, the summing convergence rate is very slow [6]. Relatively routine mathematical manipulation of the summation in Eq. (9.9) can generate a more amenable version [6] which can be expressed as:

$$\mathbf{A}_p = \frac{\mu_0 I_{00}\mathbf{dl}}{4jD_y}\exp(-jpk_xD_x)\sum_{n=-\infty}^{\infty}\exp\left(-jk_0y\left[\frac{k_y}{k_o}+\frac{n\lambda_0}{D_y}\right]\right)\mathrm{H}_0^{(2)}(k_0as_n) \tag{9.10}$$

where

$$a^2 = z^2 + (pD_x - x)^2 \tag{9.11}$$

and

$$s_n = \left[1-\left(\frac{k_y}{k_0}+\frac{n\lambda_0}{D_y}\right)^2\right]^{1/2} \tag{9.12}$$

In (9.10), $\mathrm{H}_0^{(2)}$ is the Hankel function of order zero and of the second kind. While Eq. (9.10) is useful, here it represents a step toward a solution for the infinite planar array, since it is the adoption of planar periodic surfaces, rather than passive linear arrays, that provide routes toward developing compact, electronically controllable antennas. For the infinite planar array formed from an array of y-directed linear arrays (q), it is clear that:

$$\mathbf{A} = \sum_{\mathrm{p}=-\infty}^{\infty}\mathbf{A}_q \tag{9.13}$$

On substituting Eq. (9.10) into Eq. (9.13) we obtain:

$$\mathbf{A}_p = \frac{\mu_0 I_{00}\mathbf{dl}}{4jD_y}\sum_{n=-\infty}^{\infty}\exp(-jk_0y\left[\frac{k_y}{k_o}+\frac{n\lambda_0}{D_y}\right])\sum_{p=-\infty}^{\infty}\exp(-jpk_xD_x)\mathrm{H}_0^{(2)}(k_0as_n) \tag{9.14}$$

To improve the rate of convergence the p-summation can again be manipulated, in much the same way as for the q summation. This is expounded in detail in [7]. The result is:

$$\mathbf{A} = \frac{\mu_0 I_{00}\mathbf{dl}}{4jD_y}\sum_{n=-\infty}^{\infty}\exp\left(-jk_0y\left[\frac{k_y}{k_o}+\frac{n\lambda_0}{D_y}\right]\right)\frac{2\pi}{D_x}\sum_{i=-\infty}^{\infty}\exp\left(-jk_0x\left[\frac{k_x}{k_0}+\frac{i\lambda_0}{D_x}\right]\right) \times\frac{\exp\left(-jk_0z\left[1-\left(\frac{k_x}{k_0}+\frac{i\lambda_0}{D_x}\right)^2-\left(\frac{k_y}{k_0}+\frac{n\lambda_0}{D_y}\right)^2\right]^{1/2}\right)}{\left[1-\left(\frac{k_x}{k_0}+\frac{i\lambda_0}{D_x}\right)^2-\left(\frac{k_y}{k_0}+\frac{n\lambda_0}{D_y}\right)^2\right]^{1/2}} \tag{9.15}$$

Equation (9.15) can be written more compactly as:

$$\mathbf{A} = \frac{\mu_0 I_{00}\mathbf{dl}}{2jk_0D_xD_y}\sum_{n=-\infty}^{\infty}\sum_{i=-\infty}^{\infty}\frac{\exp(-jk_0\mathbf{r}\cdot\hat{\mathbf{a}}_{r\pm})}{s_z} \tag{9.16}$$

where $\mathbf{r}(x, y, z)$ is the vector denoting the reference position (Fig. 9.1) while the unit vector is given by:

$$\hat{\mathbf{a}}_{r\pm} = \hat{\mathbf{a}}_x\left[\frac{k_x}{k_0} + \frac{i\lambda_0}{D_x}\right] + \hat{\mathbf{a}}_y\left[\frac{k_y}{k_o} + \frac{n\lambda_0}{D_y}\right] \pm \hat{\mathbf{a}}_z\frac{k_z}{k_0} \tag{9.17}$$

Also

$$\frac{k_z}{k_0} = \left[1 - \left(\frac{k_x}{k_0} + \frac{i\lambda_0}{D_x}\right)^2 - \left(\frac{k_y}{k_0} + \frac{n\lambda_0}{D_y}\right)^2\right]^{1/2} \tag{9.18}$$

By analogy with Eq. (9.3) it is evident that the Eq. (9.16) represents, in effect, the free-space Green's function for a space bisected by a plane periodic surface. Furthermore, relatively straightforward applications of the Maxwell equations (see Chap. 4) to Eq. (9.16) generates the corresponding expressions for electric field and magnetic field, namely:

$$\mathbf{E} = \frac{\eta_0 I_{00}\mathbf{dl} \times \hat{\mathbf{a}}_{r\pm} \times \hat{\mathbf{a}}_{r\pm}}{2D_xD_y}\sum_{n=-\infty}^{\infty}\sum_{i=-\infty}^{\infty}\frac{\exp(-jk_0\mathbf{r}\cdot\hat{\mathbf{a}}_{r\pm})}{s_z} \tag{9.19}$$

$$\mathbf{H} = \frac{I_{00}\mathbf{dl} \times \hat{\mathbf{a}}_{\mathbf{r}\pm}}{2D_xD_y}\sum_{n=-\infty}^{\infty}\sum_{i=-\infty}^{\infty}\frac{\exp(-jk_0\mathbf{r}\cdot\hat{\mathbf{a}}_{r\pm})}{s_z} \tag{9.20}$$

where $\eta_0 = \sqrt{\mu_0/\varepsilon_0}$. Note that Eqs. (9.16), (9.19) and (9.20) apply to both the forward and backward scattering into the neighbouring half-spaces. A thorough assessment of the infinite array Green's function issue is presented in Ref. [7].

The nature of the solutions contained in Eqs. (9.16), (9.19) and (9.20) is depicted schematically in Fig. 9.2. The plane passive periodic surface created by an array of current elements (dipoles, patches) is supported by a dielectric sheet (yellow) which lies in the x-y plane of the Cartesian coordinate system. The sheet is illuminated by an incident plane electromagnetic wave propagating in the direction $\hat{\mathbf{a}}_k$ illustrated by the large blue arrow.

The aligned scatterers in the uniform passive array are equally excited by the incoming plane wave and they adopt its phase profile over the array surface. The initial response electromagnetically is specular scattering of a single reflected wave into the incidence half-space ($z < 0$), and a single transmitted wave into the reverse space ($z > 0$). This behaviour matches that of plane wave incidence on a smooth flat semi-transparent screen. The reflected wave (red arrow) is formed in the direction $\hat{\mathbf{a}}_{r-}$, while the transmitted wave (green arrow) is directed along $\hat{\mathbf{a}}_{r+}$. For a plane screen this is the only scattering which occurs, whereas the periodic surface generates additional outcomes depending on the magnitudes of the ratios D_x/λ_o and D_y/λ_o.

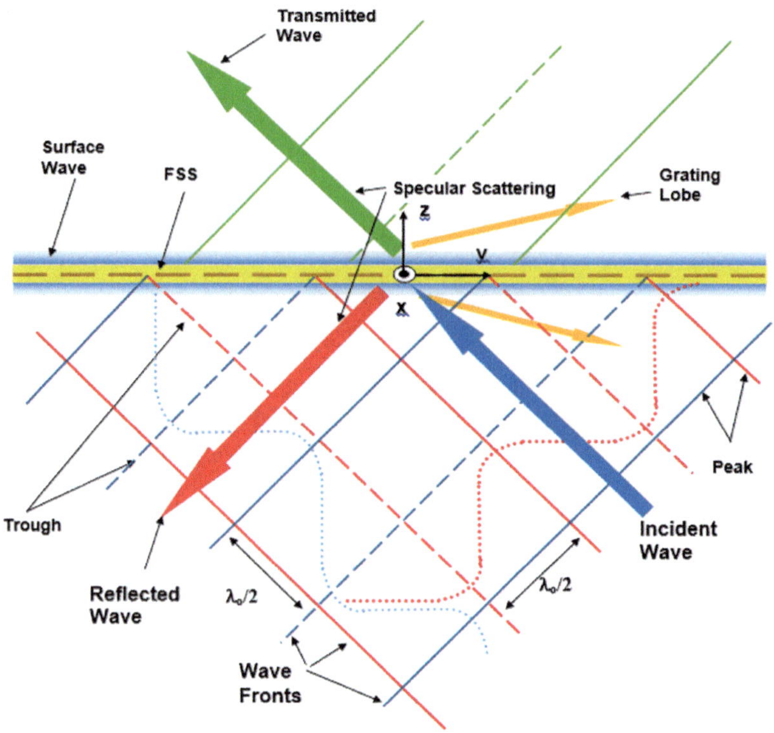

Fig. 9.2 Electromagnetic wave scattering for FSS illuminated by a plane wave

The key equations are (9.17) and (9.18). When $i = n = 0$ in these equations, the fundamental non-periodic scattering is being implemented. The equations reduce to:

$$\hat{\mathbf{a}}_{r\pm} = \hat{\mathbf{a}}_x \frac{k_x}{k_0} + \hat{\mathbf{a}}_y \frac{k_y}{k_0} \pm \hat{\mathbf{a}}_z \frac{k_z}{k_0} \tag{9.21}$$

indicating, as anticipated, that two plane waves are scatterered in the directions $\hat{\mathbf{a}}_{r\pm}$, namely the transmitted wave $(\hat{\mathbf{a}}_{r+} k_0)$ and the reflected wave $(\hat{\mathbf{a}}_{r-} k_0)$. Note that these directions are independent of the nature of the surface periodicity. On the other hand, for large enough values for D_x and D_y it is possible for other values of i and n to generate more real solutions for $\hat{\mathbf{a}}_{r\pm}$ in which case additional scattered waves can appear. These are the grating lobes (see Chap. 6) as shown schematically by the orange coloured arrows in Fig. 9.2, approximating the particular case of $i = 0$ and $n = -1$. Furthermore, it is clear from Eq. (9.18) that for high enough values of i and n the propagation ratio k_z/k_0 can be imaginary $(= jk_z/k_0)$, whereas k_x and k_y remain real. This implies that above $(+z)$ and below $(-z)$ the periodic surface, waves can exist which propagate freely in the x and y directions, but decay exponentially in the z-directions. These solutions are termed surface waves, slow waves or trapped waves. They are depicted in Fig. 9.2 as blue bands above and below the array which fade monotonically with distance $|z|$. While in radiation terms these surface waves are insignificant in magnitude a finite distance away from the array, they represent vital stored energy in the array and are very influential to the radiation control which the passive array surface provides.

9.3 Reflectarrays

The reflectarray provides a good example of a flat periodic surface being used to control the scattering of incident electromagnetic waves. There are three basic options (1) all metallic, (2) patch, (3) printed. The design principles underpinning the reflectarray concept are fully explored in Refs. [1, 2] and are well rehearsed in [3–6, 8–11]. Obviously, the essence of focusing electromagnetic waves with a reflector is well understood and can readily be explained by reference to Fig. 9.3a. In antenna terms, the requirement of the reflector is to reflect the spreading (un-focused) rays from the source (usually a horn) so that waves emanating from the source arrive at a plane surface ahead of the reflector with the same phase, i.e. focused. This surface is termed the equiphase surface (Chap. 3). It is usually chosen to coincide with the extremities of the reflector and is then viewed as the radiating aperture of the antenna as a whole. For curved reflector systems the route to forming an equiphase surface is to ensure that all rays from the horn or feed to this 'aperture' surface are equal in electrical length. This in turn dictates that the reflector is parabolic in shape.

For the reflector in Fig. 9.3a, at the equiphase front, the total phase delay from the horn phase centre is

$$-k_0(r_0 + r_1) = \text{constant} \tag{9.22}$$

for all possible r_o and r_1 values, where $k_o = 2\pi/\lambda_o$ is the free space propagation coefficient. So provided the reflector has a parabolic profile a focused beam is formed in the direction normal to the reflector (z-direction).

The reflectarray has been proposed as a means of producing focused reflection using flat, low cost microstrip or stripline substrates. More recently reflectarrays formed using all metallic grooved plates or 3-D dielectric printed protrusions on a metal base have been advocated. In the microstrip option, it usually comprises a periodic array of square resonant patches or narrow resonant slots, as suggested in Fig. 9.3b. This patch/slot dual interpretation is possible since the radiation mechanism for a resonant patch is associated with the fields in the 'open-circuited' edges (edge slots) of the patch (see Fig. 1.4). However if the patches are all the same each patch reflects in mirror fashion so that the array as a whole acts like a flat mirror, reflecting the radiated wave from the horn back towards it without modifying the spreading unfocused pattern as Eqs. (9.17) and (9.21) predict.

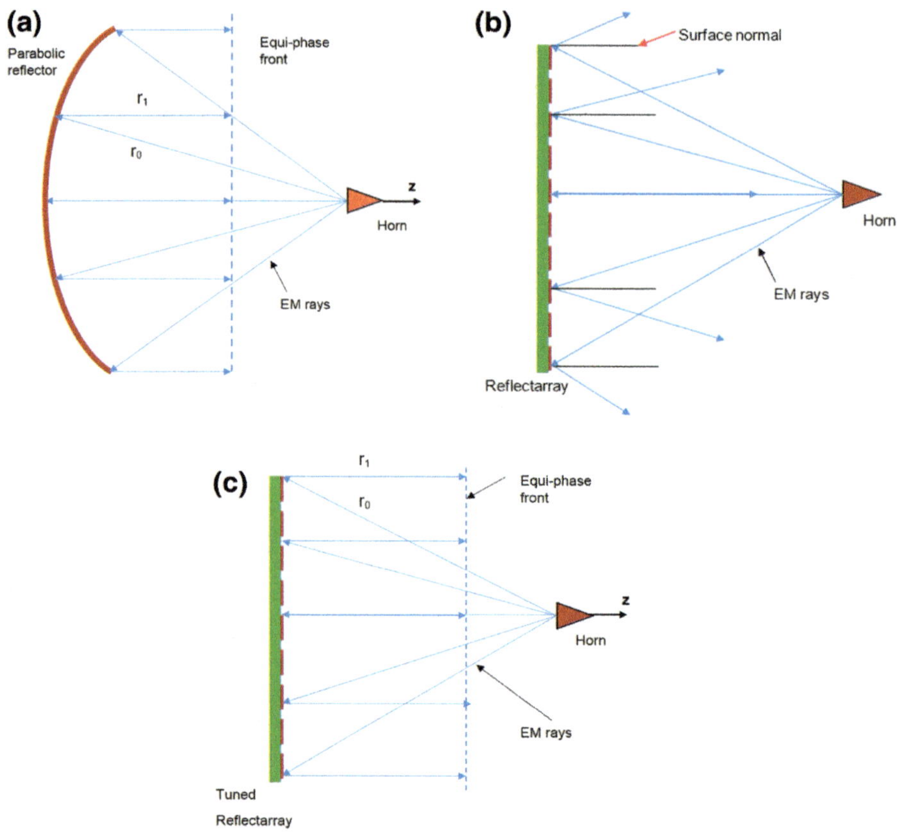

Fig. 9.3 **a** Focused reflector antenna, **b** unfocused slot/patch array, **c** reflectarray focusing

To procure focusing with a flat reflectarray an adjustment mechanism not unlike that distilled in Eq. (9.22) is required where a phase altering procedure can be incorporated, not by path length, but by reflection phase modifications at the individual patches [1, 2]. If this patch phase adjustment for the *mn*th patch is ϕ_{mn}, then for the reflectarray in Fig. 9.3c to be focused, we require:

$$\phi_{mn} = k_0(r_{0mn} + r_{1mn}) - (2N + 1)\pi \tag{9.23}$$

where ϕ_{mn} counteracts the physical length differences for the rays from the horn to the equiphase surface. Note that the $2N\pi$ factor incorporates the arbitrary and electrically meaningless full cycle phase differences between well separated rays. These are irrelevant to the design calculation since electrically $2\pi = 360° \equiv 0°$. The additional $-\pi$ accommodates the reflection phase shift for a resonant patch which is 180°.

To get a better appreciation of the dimensional constraints it is generally quite helpful to consider a real example such as a 12 GHz design [3]. In this design the

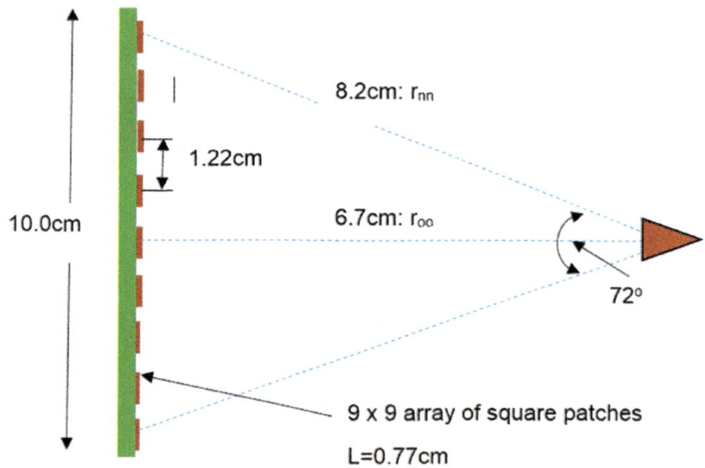

Fig. 9.4 Basic dimensions for 10 cm × 10 cm reflectarray

flat array is required to subtend an angle of 72° at the horn phase centre to achieve a compromise between excessive aperture blockage (high sidelobes) and excessive distance (r_{oo}) which results in an unwieldy array and lowered bandwidth. The square array dimensions are set at 10 cm × 10 cm. The angle choice immediately leads to the dimensions for r_{oo} and r_{nn} as shown in Fig. 9.4. For an array with a period of 1.22 cm the reflectarray in this case will comprise a 9 × 9 grid of patches.

When Eq. (9.2) is applied to this geometry we obtain, at 12.25 GHz, the following reflection phase requirements for a central patch and an edge patch. The free-space wavelength at 12.25 GHz is $\lambda_o = 2.45$ cm. Therefore, for the 6.7 cm ray there are 6.7/2.45 = 2.75 wavelengths between the horn phase centre and the central patch of the array. The 2 full wavelengths can be ignored (2*N* term). Thus for the central patch a reflection phase of $\phi_{oo} = 0.75\lambda_o = 0.75 \times 360° = 270°$. From this we subtract 180° to allow for the reflection phase of a resonant patch. For the 8.2 cm long ray to an edge-patch a similar calculation yields $\phi_{mn} = 0.347\lambda_o - 180° = -55.1°$. The establishment of the reflection phase settings for all patches, consistent with forming a focused beam, is relatively routine (see Fig. 9.5) requiring only a simple algorithm to step through all ray possibilities. This procedure is particularly required for much larger arrays.

The phasing over the surface of a 30 × 30 element front fed reflectarray of square patches, as computed by a commercial mathematics package, is depicted in Fig. 9.5. The square elements of the array are colour coded to display the reflection phases required by every element in order to produce a focused beam at broadside to the reflectarray surface. The right hand scale demonstrates the link between the colour range and reflection phase, with dark blue representing 0° and dark red/brown representing 360°. A typical rippling colour pattern is depicted as the

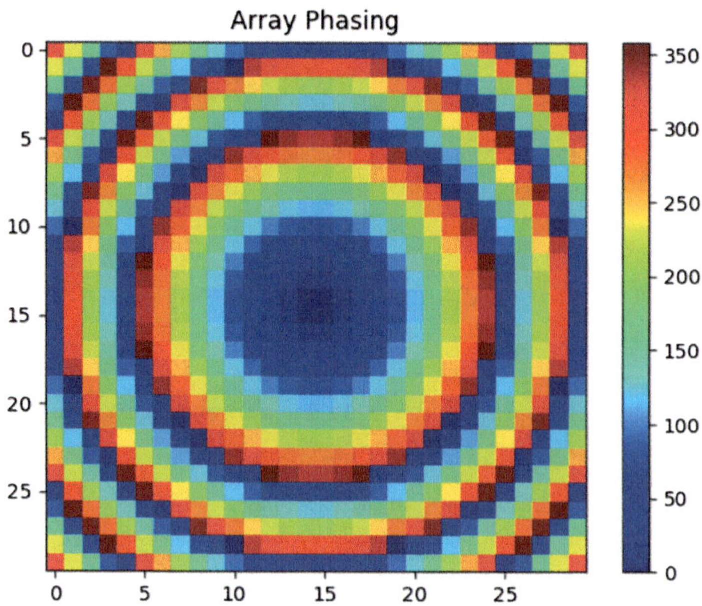

Fig. 9.5 Array phasing to form a broadside beam for a 30 × 30 square reflectarray (0° = dark blue, 360° = brown)

element phases compensate for increasing ray path length between the centre and the periphery of the reflectarray surface. Note that the proximity of deep red and dark blue patches does not indicate a large phase jump, since 360° equates to 0°.

9.3.1 Reflection Phase Computation

Once the element phase requirements are known it becomes necessary to implement these phases within the hardware. In the case of a patch array the patch size is used with a reflection phase of 180° at resonance. In all metal arrays the groove depth is tailored to the phase requirement, while in 3D printed arrays the protrusion heights provide phase control. Note that the computation of the reflection phase shifts of elements in an array cannot be determined from calculations on an isolated element, because the mutual interactions from neighbouring elements in the array contribute significantly to the result. However, for arrays of the order of 10 × 10 elements, or more, the finite array can generally be replaced by an infinite array with seriously influencing the behaviour of individual elements. This, in turn, means that periodic array techniques can be adopted to determine reflection phase. The most efficient route to a solution is by means of the moment method [6, 7] applied to an integral equation for the unknown currents in the patch element, or the unknown fields at the groove surface in the case of all metal arrays (see Chap. 5). Of course the

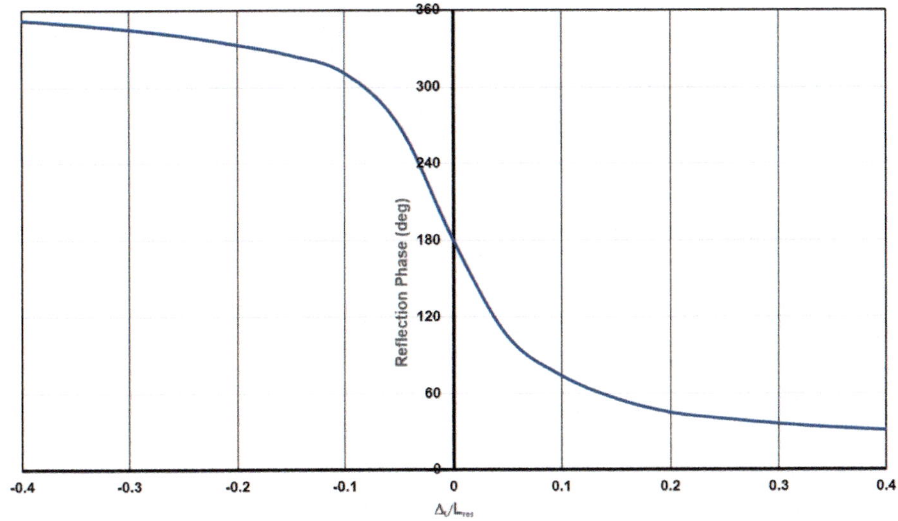

Fig. 9.6 Typical reflection phase versus patch size for 28 GHz reflectarray [3]

external region must be represented by the periodic surface Green's function (Sect. 9.1) which consists of a double summation of Floquet modes. In Ref. [3], Pozar used a simulation package based on the moment method to determine the reflection phases of patches within a 28 GHz reflectarray fabricated on a 0.02 in. thick substrate with $\varepsilon_r = 2.95$. A typical result from the Pozar analysis is presented in Fig. 9.6.

The square patch used in the derivation of Fig. 9.6 had a resonant length $L_{res} = 0.2756$ cm, and the curve shows the change in reflection phase from $\sim 345°$ well below resonance to $\sim 30°$ well above it. This is a sufficiently large phase swing to enable an effective reflectarray to be formed. Note that the unavoidable 'mirror' reflection phase shift of π at the surface of each reflectarray patch is incorporated within the total reflection phase of the detuned patch. Hence the resonant patch at $\Delta_L/L = 0$ in Fig. 9.6 has a phase shift of 180°. Δ_L is the incremental size change from the resonant case.

In simulation terms, a patch element, a metal reflectarray groove, or printed array protrusion, forming a segment of an infinite array, can be modelled quite readily, by any modern finite element based simulation package such as COMSOL or HFSS. This is achieved by embedding the element of interest in a 'square waveguide' whose 'walls' are periodic boundaries, as suggested in Fig. 9.7. Periodic boundaries are standard features in most packages and replicate the role of the Green's function for the periodic surface as required in moment method calculations. The results obtained from these calculations are similar to those displayed in Fig. 9.6.

Once the element phasings have been established, the radiation characteristics of the reflectarray antenna as a whole can be investigated. Generally, this takes the form of an iterative process towards satisfying a predetermined radiation pattern

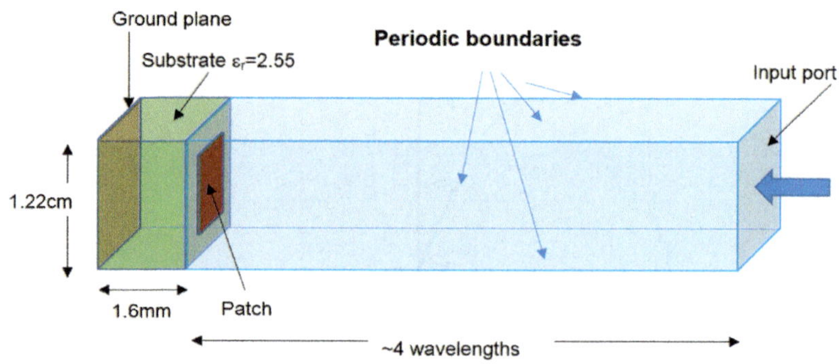

Fig. 9.7 COMSOL model to determine the patch reflection phase when embedded in an array

specification. Given that the relative position of the feed horn and the reflect array are defined in order to compute the element phases, the precise illumination of the array elements by the chosen feed, entails a determination of its radiation pattern. The techniques for doing this are provided in Chap. 3. With this information, backed by geometric optics, it becomes possible to compute the magnitude and phase of the incident radiation at every element. Each element, of the now finite reflectarray, can be viewed as a radiator with known excitation phase and signal level. Hence, the radiation pattern of the reflectarray antenna as a whole can be constructed from the array theory developed in Chap. 6. If the pattern fails to meet the desired specification on an initial pass then further iterations will be required. This computational procedure will generally lead to acceptable design [3]. Computed E-plane and H-plane patterns of just such an exercise are presented in Fig. 9.8. These were calculated, as described above, for a 77 GHz reflectarray, which is defined in Ref. [3]. The differences between the traces can be attributed to, firstly, pattern differences in the E and H planes for the feed and, secondly, to the patch elements which radiate differently in the E and H planes near endfire. For the front mounted feed, blockage can differ between the two principal planes and may be reflected, in addition, in the overall patterns. Comparisons with measurements, which are generally excellent, are provided in the original paper.

9.4 Reflectarray—State of the Art

At the time of compiling this text, in 2017, with the primary aim of collecting, in a single volume, a compendium of design techniques which are generally now available for creating antennas employing passive arrays of slots, the reflectarray was featuring strongly in the literature. Some versions of the reflectarray naturally emphasise the use of slots, while others based on patches, for example, can be viewed as slot dependent insofar as patch radiation can be attributed to fields in the edge ‘slots’ (see Chap. 1, Fig. 1.4). Consequently, it seems apposite at this point to

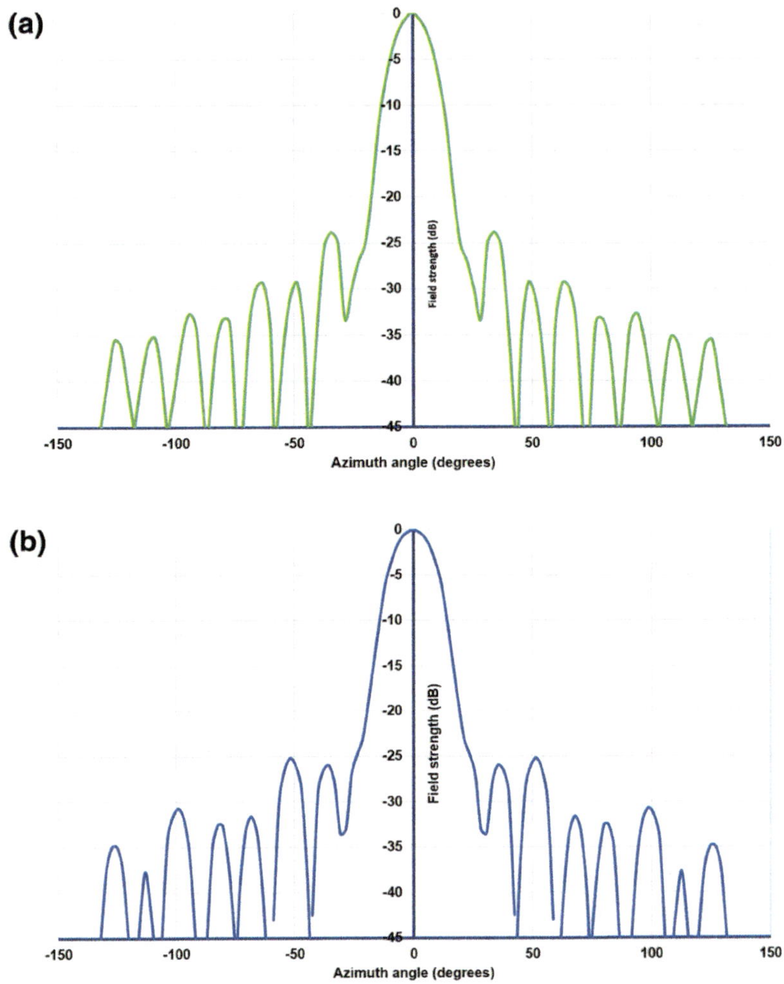

Fig. 9.8 Typical H-plane **a** and E-plane **b** radiation patterns for a 77 GHz, 6 in. square reflectarray, formed from 0.005 in. thick Duroid substrate with 5776 linearly polarized patch elements

review recent publications on this topic while placing these contributions in the context of earlier work on slot array antennas described in previous chapters.

9.4.1 Microstrip Reflectarrays

The design principles underpinning the reflectarray concept are fully explored in Ref. [3] and are well rehearsed in Refs. [8–11], which are summarized in Sect. 9.2.

The Pozar paper [3] provides three examples of antennas designed for use as high frequency space-based sensors. These would in the past have been implemented using parabolic dish technology, but have instead been designed around planar reflectarrays, formed from microstrip patches. Such arrays are potentially more convenient to transport and to install. Simple square patches are used in all three designs with the required reflection phase adjustment at every patch being achieved by dimensional modifications. The operational conditions for individual patches may thus move from below resonance to above resonance, thus providing the required reflection phase ranging from −180° to +180°. The examples presented in [3] are of millimetre-wave designs—three operating at 28 GHz and two at 77 GHz. Of the 28 GHz designs two are directed at similar 6 in. square reflectarrays but fabricated onto different substrates—0.01 in. thick Taconic, and 0.02 in. thick RT Duroid. The former required 528, 0.609 cm × 0.667 cm, off-square patches, while the latter was formed from 784, 0.544 × 0.544 square patches. In reported antenna modelling outcomes, and in comparisons with measurement, the reflectarrays are illuminated from the boresight direction from suitably designed circular corrugated horns. The differences in performance between the two antennas are shown to be small, and measurements are demonstrated to be in very good agreement with theory, confirming the accuracy of the analysis method outlined in Sect. 9.2.

Comparisons are reported to have been made with an equivalent parabolic reflector antenna, and generally indicate that any differences in performance are slight, except for bandwidth, which is at best 5% for the reflectarray, much less than for a parabolic reflector antenna which is limited mainly by the feed. The source of this limited bandwidth is the resonant patches themselves which display impedance bandwidths in the 5% range. Nevertheless, for antennas operating in the millimetre-wave bands, a 5% bandwidth still provides considerable gross frequency range, in GHz terms, and for many applications, it seems to be not too restrictive. On the other hand, in the 12–14 GHz frequency band, for antennas in satellite communications roles 5% is likely to represent a distinct disincentive to the adoption of reflectarray technology.

The 77 GHz reflectarray [3] is presented both in prime focus, and in Cassegrain, formats. The microstrip reflectarray itself is 6 in. square, fabricated on 5 thousandth's of an inch thick RT Duroid substrate, and required 5776 patches. Both antenna types are shown to exhibit high gain, but the Cassegrain version is less efficient largely because of higher 'blockage' generated by the sub-reflector. The antennas are reported to exhibit −1 dB gain bandwidths of 2.5%. A typical example of a prototype microstrip reflectarray in a Cassegrain geometry, and aimed at operation in W-band (94 GHz), is shown in Fig. 9.9.

The bandwidth limitation inherent to the single layer microstrip patch reflectarray is tackled in three ways in the literature. Firstly, non-linear changes in reflection phase when presented as a function of patch size (see Sect. 9.2) is considered to be a major factor [3] in narrowing the bandwidth in patch based reflectarrays, and it has been suggested that much more linear phase change could be secured by using line length [8]. The proposed X-band array in this submission [8] employs equi-size square patches and each patch is coupled through ground

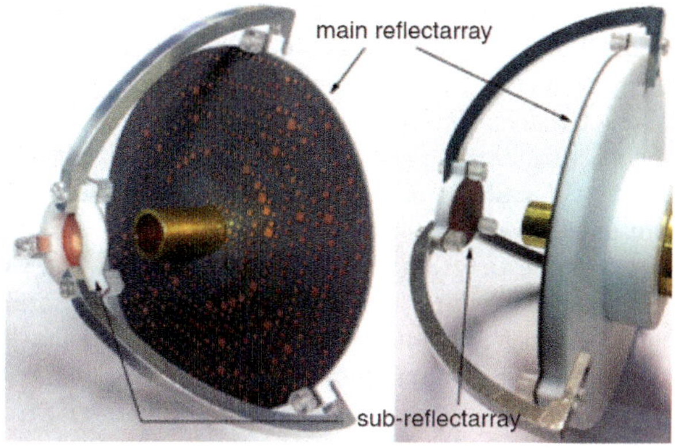

Fig. 9.9 Square patch reflectarray fabricated in microstrip and illuminated by a Cassegrain sub-reflector (see Yoon [13])

plane slots to a microstrip circuit on the reverse of the ground plane. This circuit contains couplers and junctions linked to tailored line-lengths which establish the required reflection phases at each patch in accordance with beam focusing requirements. Needless to say, with the introduction of a microstrip circuit to control reflection phases it is a small step to consider introducing active elements [9] to enhance antenna performance. Such solutions are inevitably complex. From the perspective of securing low costs in satellite communications and sensing systems, there seems, at least in 2017, little advantage in taking this route.

Secondly, multiple substrate layers have been employed for bandwidth enhancement in other patch related applications with adjacent patches on different layers tuned to slightly different frequencies. The approach is advocated by Encinar and Zornoza [11] who describe two-layer and three-layer printed reflectarrays. In Ref. [10] two 40 cm circular reflect arrays with slightly different patch size patterns are described. The paper observes that for a single layer reflectarray a reflection phase range >300° can be secured by adopting as thin a substrate as possible, but that this means that the phase, as a function of patch size, becomes highly non-linear, thus limiting bandwidth. Modelled and measured results for patch reflectarrays fabricated on two layers, show that with the outer patch of the co-located patch pair is set smaller than the inner patch by 30%. It is observed that reflection phase as a function of patch size can be linearised without degrading phase range. A designed 40 cm reflectarray, front fed from an 11.95 GHz corrugated horn used in TV receivers, displays a creditable −1.5 dB gain bandwidth of 16.7%. However, when the two layer technology is applied to a 150 cm off-circular reflectarray the bandwidth decreases significantly to ~4% [9]. This is largely attributed to insufficient reflection phase variation in order to accommodate the much larger range of ray lengths from horn to patch. By introducing a third layer [9, 10] the reflection phase range can be increased from 400° to 800° [9]. With the introduction of a three layer patch array the 1.5 m diameter reflectarray antenna is shown to exhibit a credible 10% operating bandwidth [10] but obviously at the expense of fabrication complexity.

An alternative scheme to achieve enhanced gain-bandwidth in a single-layer microstrip reflectarray is reported by Yoon [12, 13]. The proposal comprises an array formed from patch elements of differing geometries and differing frequency characteristics. The recommended elements are the standard patch (SP) the hollow patch or square ring (SR), and the ring loaded patch (RLP), which is a combination of the SP and SR. Curves of reflection phase as a function of element size are presented to demonstrate how the elements can be adjusted to accommodate different phase ranges between −300° and zero. A 15 GHz reflectarray on a 2.4 mm substrate and with a diameter of 25.5 cm is described. It is claimed to possess a −1 dB gain bandwidth of 11.7%. In Ref. [12] the Yoon multi-element concept is developed to a further level with the range of element geometries increased to nine. These geometries are the square, round, oval and triangular patch, and the square, circular, elliptical and triangular metallised track or ring. The ninth form is a square track with a T-shaped insert. To realise the reflectarray these elements are etched onto a 1 mm thick RT Duroid 5580 substrate, to a predetermined software generated pattern. Interestingly, the prototype 10 GHz antenna incorporating the reflectarray is illuminated by a fixed beam feed which is itself an array, comprising sixteen patch elements, and having a 17 dB directivity. The overall outcome is a prototype reflectarray antenna exhibiting an impressive −3 dB measured gain/bandwidth of 24.73%.

The above scheme has also been applied to a 95 GHz Cassegrain antenna in which both the main reflector and the subreflector are implemented by reflectarrays of this multi-element type [13]. A Cassegrain antenna for remote sensing of the atmosphere (radiometry) at 95 GHz is also reported in Ref. [4]. It uses a conventional parabolic main reflector but a reflectarray for the subreflector. A 3% bandwidth is claimed.

9.4.2 All-Metal Reflectarrays

Conceptually, the all-metal reflectarray is arguably a less complex entity than the patch versions described in the previous section [14–19]. As with the patch based reflectarray, formed from a flat microstrip substrate, the all-metal alternative is configured from a flat metal sheet. In order to replicate the beam forming characteristics of a parabolic reflector it must adhere to the reflection phase requirements at the elements as represented by Eq. (9.23). Inevitably these adjustments at the element level are achieved rather differently in the all metal alternative. There are actually three reflection phase altering mechanisms which have been thoroughly explored in the literature. These are (a) a flat metal plate with (usually) square grooves of varying depth, (b) a flat conducting sheet with raised metallic 'pill-boxes' hollowed to form elemental shorted waveguides, and (c) a pixelated or stepped structure without the hollowing. These alternatives are illustrated schematically in Fig. 9.10.

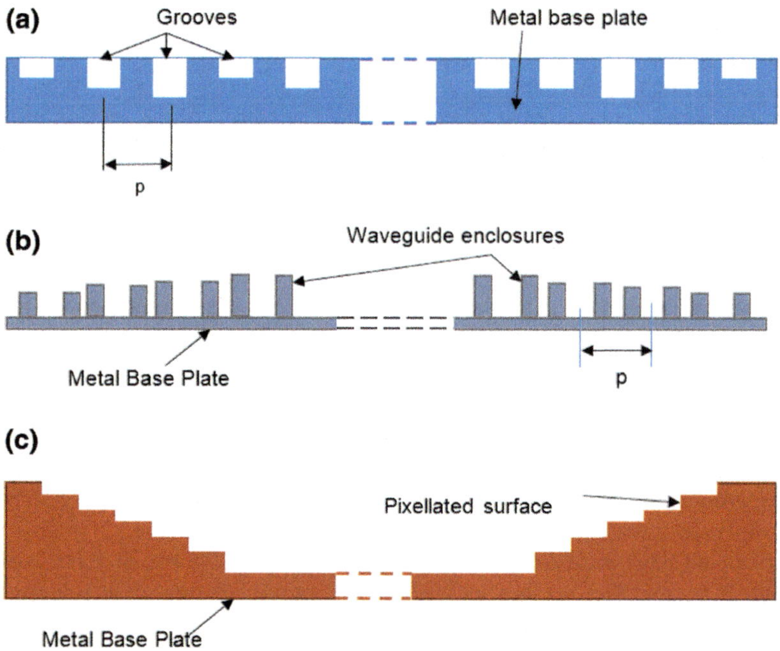

Fig. 9.10 2-D schematics of possible all-metal reflectarray antennas **a** grooved metal plate, **b** hollowed 'pill-boxes' raised from surface of metal backing plate, **c** stepped or pixelated metal plate

Case (*a*) The periodic surface in this option is strictly flat and comprises an array of small rectangular recesses of varying depth. Radiation from the individual elements of the array emanates from a magnetic current density $\mathbf{J}_\mathrm{m}$ at the outer surface of the groove created by the incident illumination reflecting off its base. That is, the periodic surface can be viewed as a conducting plane supporting a lattice of magnetic current radiators. The arrangement is in electromagnetic terms the dual of the patch reflectarray studied in Sect. 9.2, where the dielectric surface of the substrate supports a lattice of electric current radiators. A solution paralleling the procedure outlined for the microstrip reflectarray in Sect. 9.2 is evidently available. A Green's function which is in effect the dual of Eq. (9.15) has been formulated elsewhere for the grooved flat metal plate (see Eqs. (9.9) and (9.10) in Ref. [20]). This Green's function is defined as applicable to the 'semi-infinite' array which means an infinite array curtailed in size to the required reflectarray area but without accommodating edge effect. This is not unreasonable for large arrays where edge elements represent a very insignificant contribution to the overall array behaviour. The availability of a suitable Green's function means that it is not too difficult to formulate an integral equation for the unknown aperture magnetic current, and the moment method can, as described in Chap. 5 and in Sect. 9.2, be employed to develop depth profiles for the recess elements, which acknowledges that they are embedded within a periodic surface.

Actually, much of the literature directed at the all metal reflectarray, in the case (a) configuration, is noticeably exercised by the limitations of the semi-infinite array model [15–17]. The alternative solution accommodates edge effect by adopting a modelling technique (T-block Method) which is largely borrowed from a deconstruction procedure widely used in dispersion calculations for ridged metal waveguide [5]. It is claimed to offer more rapid computations than earlier solutions. The introduction of the T-block concept into the reflectarray analysis is depicted schematically in Fig. 9.11. In the central region, elemental recesses in the metal ground plane can be viewed as residing in a semi-infinite periodic surface, for which the half-space Green's function exists. Array scattering computations for this portion of the array can proceed as outlined above. For the edge portions of the periodic array an alternative procedure is required. In the T-block method each groove on this part of the array is viewed as a separate element in a finite conducting ground plane. Scattering solutions can be formulated for this geometry by using diffraction theory for an aperture in a finite ground plane [21]. This is performed for all edge grooves with their relative positions in the periodic lattice are guaranteed by the T-block process [15–17]. T-block superposition is used subsequently to determine the scattering for the periodic surface as a whole. In some contributions, this problem specific analysis is backed up by a full wave electromagnetic simulation based on the finite element technique, mainly as a check on accuracy.

Beam focusing by means of the plane all-metal reflectarray antenna requires reflection phase adjustment from individual grooves paralleling the microstrip based reflectarray summarised in Eq. (9.23). The desired phase condition is illustrated in Fig. 9.12 where the reflection phase from each groove is adjusted using its depth ($d^{(n)}$) to achieve a condition which matches that of the equivalent parabolic reflector antenna. Relative to the reflectarray coordinate system at (0, 0) the focus of

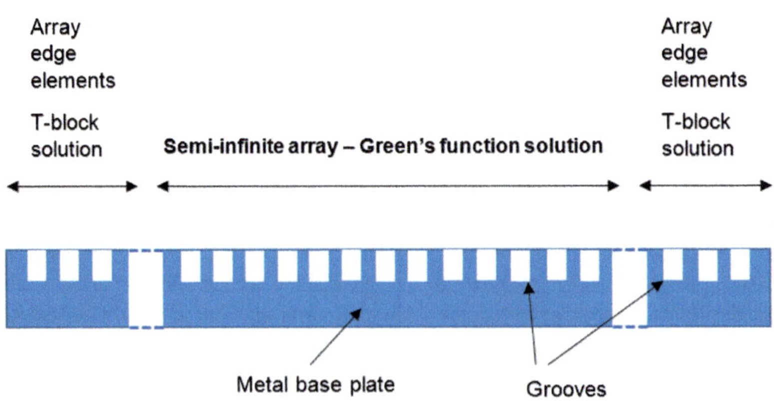

Fig. 9.11 Multiple groove flat metal-only reflectarray—solution options

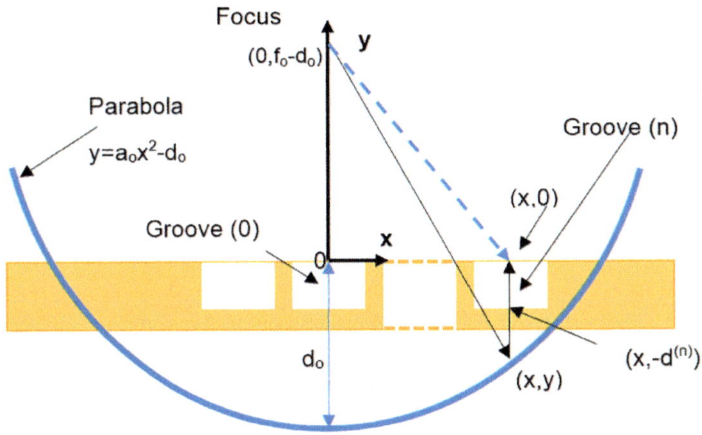

Fig. 9.12 Phase matching between all-metal reflect array and equivalent parabolic reflector

the parabolic curve is at $(0, f_o - d_o)$ where d_o is the distance from (0, 0) to the lowest point on the parabola represented mathematically by the relation:

$$y = a_0 x^2 - d_0 \tag{9.24}$$

Hence, using Fig. 9.12 phase equality requires that:

$$2\beta_{10} d^{(n)} + k_0 \left[x^2 + (f_0 - d_0)^2\right]^{1/2} = k_0 \left[f_0 + a_0 x^2 + |y|\right] \tag{9.25}$$

where β_{10} is the phase coefficient for the lowest mode in the groove waveguide, k_0 is the free-space phase coefficient, $d^{(n)}$ is the depth of the nth groove, while f_0 and d_0 are defined on the diagram. The nth groove, in T-block notation, is defined as being located distance $T^{(n)}$ from the origin. Consequently, if we take $a_0 = 1/(4f_0)$, which is not untypical of reflectors for very high frequency sensor systems, then the groove depth requirement for a focused reflectarray is given by [5, 15]:

$$d^{(n)} = \left[d_0 + f_0 - \sqrt{(T^{(n)})^2 + (f_0 - d_0)^2}\right] \frac{\lambda_{g10}}{2\lambda_0} \tag{9.26}$$

The reflection phase as a function of groove depth, normalised to the deepest case, is presented as a chain-dotted orange curve in Fig. 9.13. There it is compared with the equivalent patch array curve normalised to the patch dimension at resonance. It is evident that the grooved reflectarray provides a much more linear relationship between reflection phase and element size. This is responsible for much improved bandwidth for the antenna. That the curve for the grooved all-metal reflectarray is not more linear than is depicted in Fig. 9.13 can be attributed both to mutual coupling between grooves and edge effect.

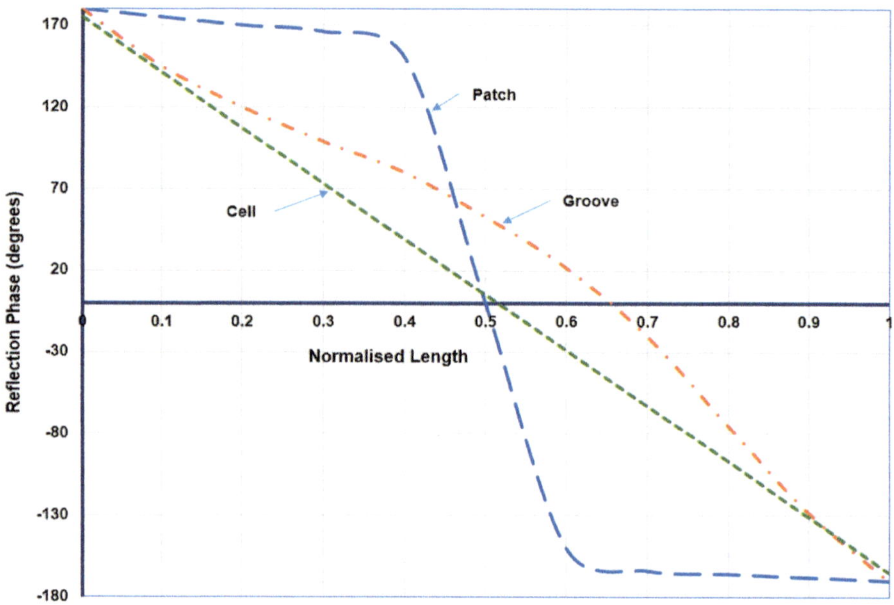

Fig. 9.13 Reflection phase computations for three all-metal reflectarray options

Fig. 9.14 Prototype all-metal reflectarray formed from 5961 rectangular grooves machined into a 30 cm diameter base plate (from [5])

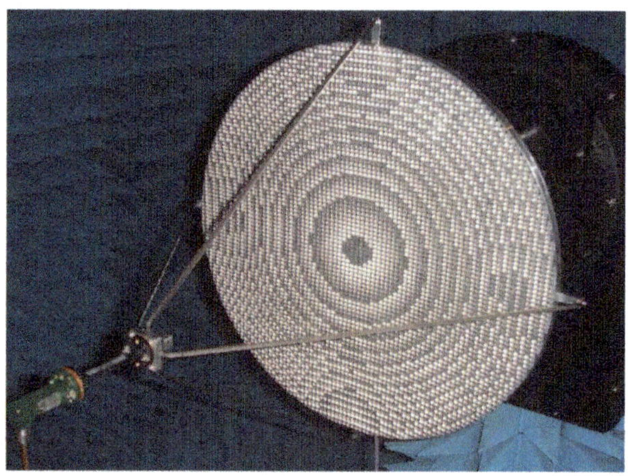

A prototype of a front illuminated grooved all-metal reflectarray [5], fabricated from a 1 cm thick, 30 cm diameter, flat metal plate, is presented in Fig. 9.14. It comprises 5961 rectangular grooves and is fed from a pyramidal horn designed to operate at 78.5 GHz. At this frequency a gain of 42 dBi is reported for the antenna as a whole.

Case (*b*) An alternative to the tuned recesses of Ref. [5] is to introduce raised metallic inserts above a flat conducting plate (see Fig. 9.10). Such a scheme is described by Chou et al. [14] where the variable height metal protrusions essentially enclose shorted cylindrical waveguides, which operate not unlike the recesses in [5]. The claimed advantage of this approach is that the reflectarray can be manufactured using a low cost die-cast process. However, this is off-set by the fact that the array surface is no longer planar and consequently the semi-infinite array approach to reflectarray analysis, employed earlier, is less valid.

A 48 cm × 48 cm square reflectarray operating at 12.45 GHz and fed from an off-set circular corrugated horn is described in [14]. A measured gain of 33 dB together with a −3 dB gain/bandwidth of ~3% is demonstrated.

Case (*c*) The elements for an all-metal reflectarray described by Lee et al. [18, 19] take the form of *non-resonant* square cross-section metal posts or cells rising above the flat metal base plate of the reflector. The lattice of protrusions of varying height results in a quasi-curved reflecting surface which generates a high gain beam when illuminated from a horn feed. In effect what is created here is a parabolic reflector with a 'pixellated surface' (Fig. 9.10). It can be designed using geometric optics. It is not really a reflectarray in the conventional 'flat plate' sense. The design, fabrication and measurement of a precision machined 7.2 cm × 7.2 cm square reflector operating at 94 GHz is described in the published papers. Reflection phase calculations emanating from the surface of each step are simple. For a step of height d above a reference ground level it is given by:

$$\angle\rho = -\frac{4\pi d}{\lambda_0} + \pi \tag{9.27}$$

Clearly the reflection phase is a linear function of height d, and this is shown in Fig. 9.13 (green dashed trace). However, it is also a linear function of frequency, which inevitably curtails the available bandwidth, when compared with the equivalent parabolic reflector.

A gain of ~35 dB for the front fed antenna is reported in the literature [18]. The −3 dB bandwidth is not specified but it is likely to closely match that of an equivalent parabolic dish version of this mm-wave antenna. In fact, a Cassegrain version of this antenna [19], which displays a slightly higher gain of 36 dB, has been fabricated and measured and shown to have a creditable −1 dB gain/bandwidth of 10.7%. A photograph of this reflectarray is reproduced in Fig. 9.15.

9.4.3 3D-Printed Reflectarrays

The growing interest in reflectarrays at the beginning of this millennium, which is revealed by the antenna literature, shows a predominant emphasis in mm-wave systems largely for radiometry. Around the same time polymer jetting 3D printing technology was moving from a novel laboratory curiosity to a serious contender for the manufacturing of complex devices with sub-millimetre dimensions. Inevitably, at the frequencies involved, from 100 GHz to 1 THz, the very small array

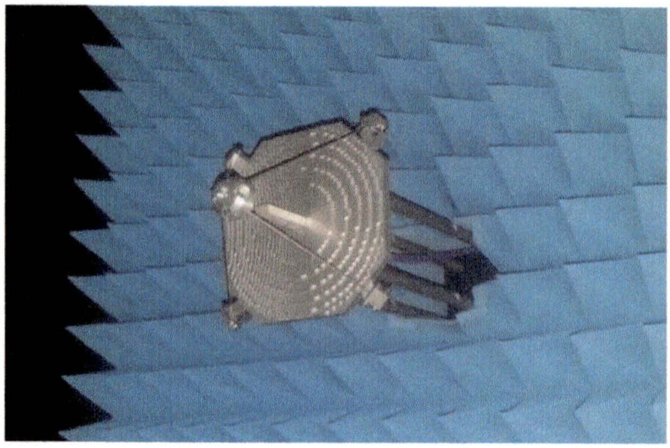

Fig. 9.15 Cassegrain illuminated all-metal reflectarray with pixelated or stepped periodic surface (from [19])

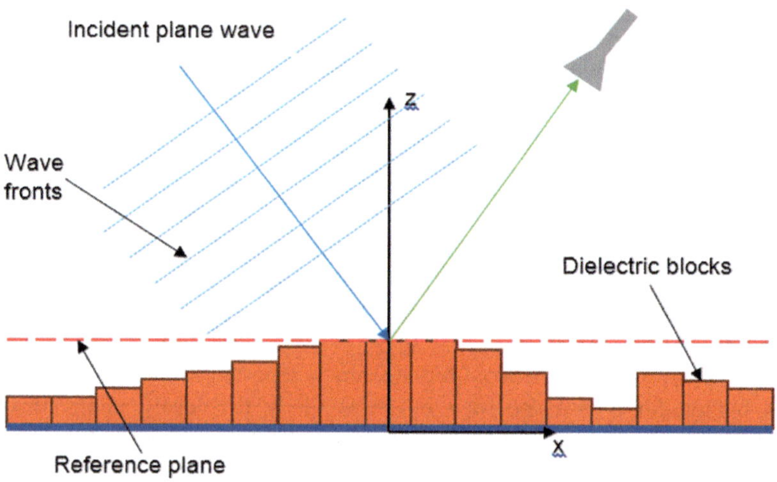

Fig. 9.16 Plane wave illumination of 3D-printed reflectarray in off-set receiver geometry

dimensions and element sizes of reflectarray solutions, was deemed to favour this evolving technology [6]. The paper describes the printing of array elements in the form of variable height dielectric blocks onto a flat conducting surface. With the blocks occupying the whole periodic cell—an option rather than a necessity—the process results in a stepped surface formed from dielectric blocks of square cross-section and varying height. Each step in the non-flat periodic surface acts as a dielectric resonator if tuned to the array operating frequency with height used to procure reflection phase adjustments, as suggested in Fig. 9.16. The reference plane for phase determination is adjusted to coincide with the tallest block, and all blocks are then adjusted by the same height in order to ensure that the reflection phase from the tallest element at the centre of the array is 360°.

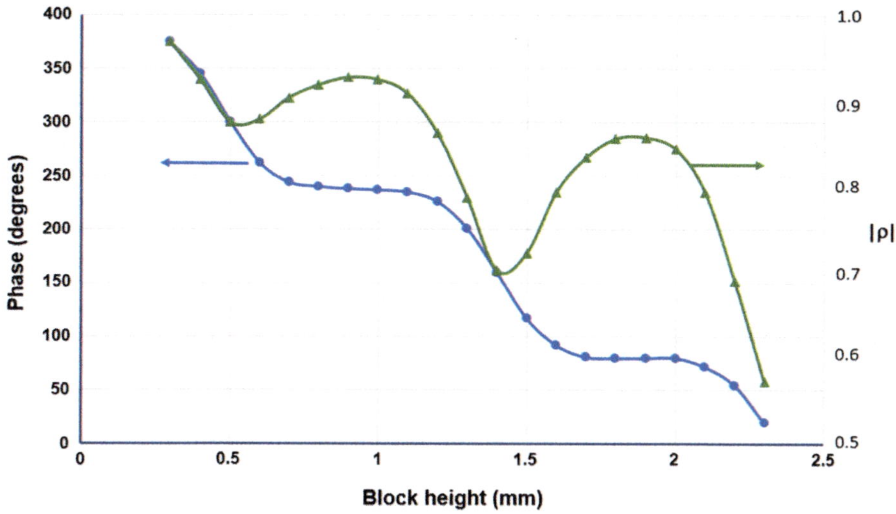

Fig. 9.17 Phase and magnitude of reflection coefficient (ρ) for a dielectric block in 3D-printed reflectarray

A drawback for the printed reflectarray is that unless the dielectric used is very dense, and therefore heavy, only partial reflection occurs at the top surface of each block. This is illustrated in Fig. 9.17. The transmitted component reflects off the conducting base plate creating a resonant condition within the bloc. The re-transmitted portion then interferes with the direct reflection from the block resulting in the typical oscillatory reflection pattern (see Fig. 9.17) for the reflection magnitude (green curve, triangles) and phase (blue curve, circles) as block height changes. Interestingly, to correct this difficulty, the paper suggests that in the future dielectric constant variations, rather than cell height, could represent a better method of phase correction.

Three 100 GHz reflectarrays are reported to have been designed and fabricated using different criteria for determining the dielectric resonator patterns. These arrays are horn fed from a 30° off-set position. Gains of ~25 dB and encouragingly high −1 dB gain/bandwidth levels ranging from 14 to 20% are claimed.

9.5 Chapter Summary

A comprehensive examination of the reflectarray is furnished in this chapter and it is shown to provide an instructive insight into the design of planar passive periodic structures aimed at specialist antenna applications. It is clear from the literature that the recent surge in the development of reflectarrays has been driven by growing interest in atmospheric monitoring from space at mm-wave frequencies and this has inevitably influenced developments.

The relevant technologies which have evolved to date are as follows.

1. Double layer microstrip array as described in [22].
2. Variably shaped reflecting patch elements in single layer microstrip reflectarray [4].
3. Die-cast metal reflectarray as described in [6].

The analysis of arrays of this description is dependent of the formulation of a general mathematical representation of radiation in the presence of a passive periodic surface. The basic steps towards establishing such a formulation are presented in Sect. 9.1. By commencing with the Hertzian potential function for radiation from a current element, derived in Chap. 2, and by assuming, not unreasonably, that the non-edge elements of a finite multi-element passive array are effectively embedded in an infinite array, it is demonstrated that a Green's function for an element of the periodic array radiating into the surrounding space can be formed. It then becomes possible to synthesise radiation patterns for typical passive arrays such as the reflect-array. The phase compensation calculations required to procure parabolic reflector type focusing from a 'flat' array are also studied.

The chapter is completed with a resume of the state-of-the-art in reflect-arrays.

References

1. D.G. Berry, R.G. Malech, W.A. Kennedy, The reflectarray antenna. IEEE Trans. Antennas Propag. **11**(10), 645–651 (1963)
2. J. Huang, J.A. Encinar, *Reflectarray Antennas* (Wiley-IEEE, Hoboken, NJ, 2008)
3. D.M. Pozar, S.D. Targonski, H.D. Syrigos, Design of millimetre wave microstrip reflectarrays. IEEE Trans. Antennas Propag. **45**(2), 287–296 (1997)
4. Hu Wenfei et al., 94 GHz dual-reflector antenna with reflectarray subreflector. IEEE Trans. Antennas Propag. **57**(10), 3043–3050 (2009)
5. Y.H. Cho, W.J. Byun, M.S. Song, High gain metal-only reflectarray antenna composed of multiple rectangular grooves. IEEE Trans. Antennas Propag. **59**(12), 4559–4568 (2011)
6. P. Nayeri et al., 3D printed dielectric reflectarrays: low cost high-gain antennas at sub-millimetre waves. IEEE Trans. Antennas Propag. **62**(4), 2000–2008 (2014)
7. B.A. Munk, *Frequency Selective Surfaces* (Wiley, New York, 2000)
8. A.W. Robinson, M.E. Bialkowski, H.J. Song, A passive reflectarray with dual-feed microstrip patch elements. Microw. Opt. Tech. Lett. **23**(7), 295–299 (1999)
9. A.W. Robinson, M.E. Bialkowski, H.J. Song, A 137-element active reflectarray concept. Microw. Opt. Tech. Lett. **26**(3), 147–151 (2000)
10. J.A. Encinar, Design of two layer printed reflectarrays using patches of variable size. IEEE Trans. Antennas Propag. **49**(10), 1403–1410 (2001)
11. J.A. Encinar, J.A. Zornoza, Broadband design of three-layer printed reflectarrays. IEEE Trans. Antennas Propag. **51**(7), 1662–1664 (2003)
12. J.H. Yoon et al., Single-layer reflectarray with combination of element types. Electron. Lett. **50**(8), 574–576 (2014)
13. J.H. Yoon et al., Axially symmetric dual-reflectarray antennas. Electron. Lett. **50**(13), 908–910 (2014)

14. H.-T. Chou, C.-Y. Lin, M.-H. Wu, A high efficiency reflectarray antenna consisted of periodic all-metallic elements for the Ku-band DTV applications. IEEE Antennas Wirel. Propag. Lett. **14**, 1542–1545 (2015)
15. Y.H. Cho, W.J. Byun, M.S. Song, Metallic-rectangular-grooves based 2-D reflectarray antenna excited by an open-ended parallel plate waveguide. IEEE Trans. Antennas Propag. **58** (5), 1788–1792 (2010)
16. Y.H. Cho, Transverse magnetic plane-wave scattering equations for infinite and semi-infinite rectangular grooves in a conducting plane. IET Microw. Antennas Propag. **2**(7), 704–710 (2008)
17. Y.H. Cho, TM plane-wave scattering from finite rectangular grooves in a conducting plane using overlapping T-block method. IEEE Trans. Antennas Propag. **54**(2), 746–749 (2006)
18. W. Lee et al., Non-resonant conductor reflectarray element for linear reflection phase. Electron. Lett. **51**(9), 669–671 (2015)
19. W. Lee, Y.J. Yoon, A broadband dual-metallic reflectarray antenna for millimetre wave applications. IEEE Antennas Wirel. Propag. Lett. **15**, 1–4 (2016)
20. J.P. Skinner, P.J. Collins, A one-sided version of the Poisson sum formula for semi-infinite array Green's function. IEEE Trans. Antennas Propag. **45**(4), 601–607 (1997)
21. R.E. Collin, F.J. Zucker, *Antenna Theory* (McGraw-Hill Book Co., New York, 1969)
22. A.P. Feresidis, J.C. Vardaxoglou, High gain planar antenna using optimised partially reflective surfaces. IEE Proceedings-Microwave, Antennas Propag. **148**(6), 345–350 (Dec. 2001)

Chapter 10
Compact Planar Resonator Arrays

10.1 Introduction

Arguably the search for low cost antenna solutions, which embody aspects of phased array technology, has underpinned recent interest in compact array antennas (CRA) excited by the modal fields in a planar resonant cavity. This technology also offers the intriguing possibility of securing what has been termed 'configurability', which simply implies an ability to achieve some level of electronic positioning or adjustment of the radiated beam. Electronic beam control is a technology which has traditionally been restricted to full phased arrays. Intrinsically high gain, and efficiency, represent a further attraction for CRA's, as does the possibility of low fabrication costs for antennas in the millimetre-wave range of the electromagnetic spectrum.

In its original form, as described by Trentini [1], the planar resonator array was referred to as a partially reflective surface (PRS) antenna and is founded on an essentially rather simple concept. It comprises a Fabry-Pérot cavity resonator created by the sandwich formed by the electrically conducting (perfect electric conductor—PEC) antenna ground plane and an electromagnetically ('porous') periodic surface (PRS) located approximately half of a wavelength above it. In principle the structure, which can realistically be viewed as a parallel plate waveguide, may be unconfined in the transverse directions; that is, parallel to the reflecting planes (x-y plane in Fig. 10.1). An elementary feed antenna, which may be a microstrip patch antenna or a ground plane slot, excites the system [1–4]. A weakly resonant electromagnetic mode (usually TE_{10}) is set up in the cavity due to the multiple reflections between the reflecting layers. Curiously, antennas of this type have been simultaneously and independently studied from the viewpoint of the leaky-wave antenna (LWA), rather as if an alternative technology was being pursued [5–8]. However, this is strictly not true since the antennas in question are also based on the Fabry-Perot resonator. They differ from the PRS class by trapping the electromagnetic resonance between a high impedance surface (HIS) or an artificial

A. J. Sangster, *Compact Slot Array Antennas for Wireless Communications*, Signals and Communication Technology, https://doi.org/10.1007/978-3-030-01753-8_10

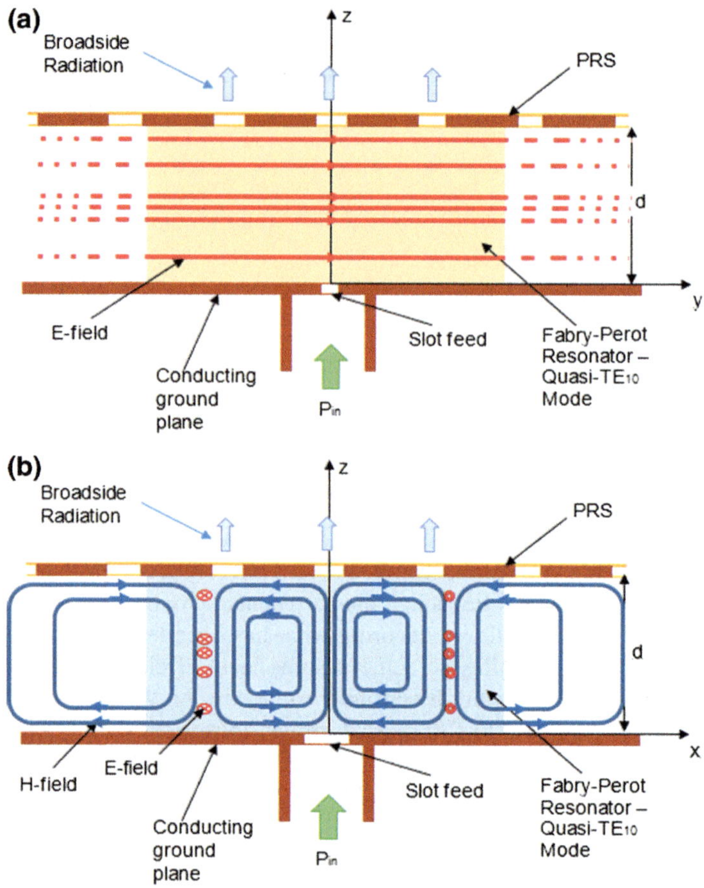

Fig. 10.1 Radiation mechanism for a partially reflective surface on an infinite ground plane **a** E-plane (y-z plane); **b** H-plane (x-z plane)

magnetic conductor (AMC), replacing the PEC in the PRS version. Electrically the operating principles are interchangeable so that the PRS antenna is actually a special case of the LWA. Consequently simplicity is served by viewing all antennas of this type as cavity resonance antennas (CRA).

10.2 Cavity Resonator Arrays

In order to realise a CRA it is essential that the PRS is only partly reflective to the cavity mode fields, resulting in electromagnetic coupling to the external space through periodically spaced apertures, or radiating elements, embedded in the PRS. Depending in the rate of power leakage the finite volume containing meaningful stored energy in the Fabry-Perot resonator will rise or shrink to fulfil the dictates of power conservation. This is depicted by the shaded areas in Fig. 10.1a, b. High leakage will result in minimal or weak mode formation and hence low directivity with only a small area of the PRS illuminated by the minimally established mode. On the other hand a highly reflective surface, offering a low level of PRS transmission, permits high energy accumulation before the power radiated from a now larger area of the PRS balances the input power. This condition equates to high directivity for the antenna as a whole [1, 3].

10.3 PRS Antenna

The original parallel plate cavity resonator, on which the PRS antenna is based, is essentially a 2-D resonator trapping a TE_{10} mode (one half-cycle E-field variation in the z-direction, invariant in the y-direction). If the waveguide is cut-off at the resonant frequency the fields decay monotonically and rapidly in the x-direction and contribute little to the resonance condition (c.f. Figure 10.1a). For the PRS therefore, it is clear from Fig. 10.1 that the essentially two dimensional Fabry-Perot cavity will support quasi-TE_{m0} modes which could be either y-polarised as shown, or x-polarised. With a preselected reflector spacing of $d \sim \lambda/2$ the quasi-TE_{10} mode, as depicted, will be formed. Note that while the quasi-TE_{10} mode produces a largely monotonic field variation (in y) due to power loss through the PRS, as suggested in Fig. 10.1a, for an infinite structure the x-directed propagation must display a travelling wave behaviour [1] as shown in Fig. 10.1b. The same observation is true for the orthogonal TE_{10} mode if excited by an omni-directional source. However, for $d \sim \lambda/2$ the mode is very close to cut-off and the guide wavelength for the fundamental mode, $\lambda_y = \lambda_{g10}$, will be very large. Thus from the slowly varying excitation field, exhibiting a very long wavelength, a broadside pencil beam is formed by the PRS for both modes. Clearly, in the Fig. 10.1a representation, coupling apertures on the PRS will be excited in phase, resulting in an intrinsically broadside mode of operation for the antenna in both planes. One the other hand, the aperture field distribution forming the radiation pattern is not a uniform one, since the mode, as explained above is concentrated in the central region of the resonator. This is also true in the H-plane (Fig. 10.1b), if λ_{g10} is long enough, in which case essentially only one cycle is effective.

For a Fabry-Perot resonator operating in the $n = 1$ mode (quasi-TE_{10} mode) as depicted in Fig. 10.2 the plate separation d dictates the internal reflection angle θ in accordance with the relationship:

$$\frac{2\pi d}{\lambda_z} = \frac{2\pi\sqrt{\varepsilon_r}}{\lambda_0} d \cos\theta = \pi \tag{10.1}$$

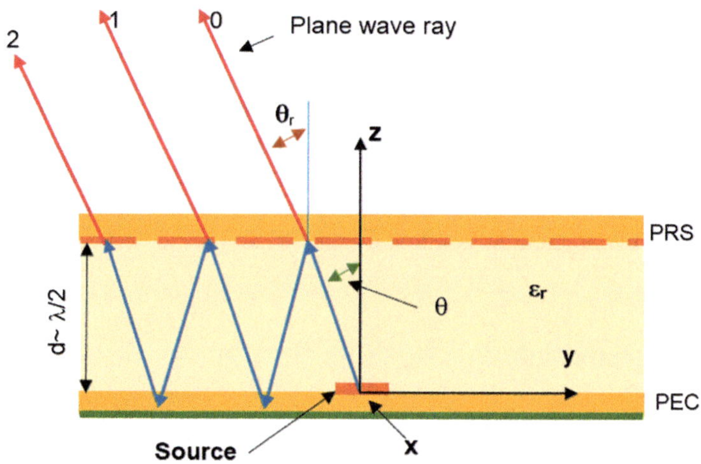

Fig. 10.2 Ray optics representation of PRS cavity antenna

If the cavity is dielectric filled, with a relative permittivity of ε_r, the radiation angle θ_r is given by Snell's laws and hence for the dielectric filled cavity:

$$d = \frac{0.5\lambda_0}{\sqrt{\varepsilon_r - \sin^2\theta_r}} \tag{10.2}$$

Clearly for a broadside beam we require $d \approx \lambda/2$. The dimensions of the cavity (d) and the relative permittivity of the dielectric spacer (ε_r) require to be adjusted to ensure higher order mode radiation lobes (secondary lobes) are suppressed particularly where scanning of the main beam is anticipated. Generally the scanning requirement is that the main (n = 1) beam can be located anywhere between $\theta_r = 0$ and $\theta_r = \pi/2$ without the n = 2 mode forming a lobe in real space. The n = 1 mode can be located at endfire when:

$$d = \frac{0.5\lambda_0}{\sqrt{\varepsilon_r - 1}} \tag{10.3}$$

and the n = 2 mode will be present, with a lobe located at broadside, if:

$$d = \frac{\lambda_0}{\sqrt{\varepsilon_r}} \tag{10.4}$$

Equating (10.3) and (10.4) leads to the restriction that for beam scanning without secondary lobes the relative permittivity must be greater than 1.333. The maximum n = 1 beam scan angle, which can be secured without degrading the radiation pattern with a higher order mode lobe, is then given by:

$$\theta_r = \sin^{-1}\frac{\sqrt{3\varepsilon_r}}{2} \tag{10.5}$$

This condition should not be confused with the occurrence of grating lobes (see Chap. 6) which is dictated by the element separation in the periodic surface of the PRS antenna.

Scanning in the z-x plane is possible [5, 6] for a PRS antenna by frequency adjustment. For a periodic element separation, or pitch, in the x-direction of magnitude p, array theory in Chap. 6 informs us the when the main (n = 1) beam is scanned to endfire in the forward direction a grating lobe at backward endfire is avoided if $p/\lambda_0 < 0.5$. This generally represents the normal mode of operation of a PRS.

10.3.1 Radiation Pattern Estimation

The excitation of the PRS array is mainly shaped by the modal concentration, and significantly it is only very slightly influenced by the feed. A radiation pattern with a tendency toward low sidelobes is the result [1]. This can be illustrated very effectively by performing a simple ray analysis of the parallel plate geometry. It is a generally accepted principle in electromagnetic wave systems, as we have already observed, that any legitimate field pattern can be resolved into a spectrum of plane waves (see Chap. 3). In a rectangular box, such as a conventional waveguide, it is shown in Chap. 1 that the waveguide mode patterns can be broken down into plane wave components. For the parallel plate waveguide, which is just a rectangular waveguide minus the top and bottom walls, the TE_{10} mode (see Fig. 10.1) is formed from a pair of y-polarised plane waves repeatedly reflecting on the side walls at z = 0 and z = d. However, for the PRS antenna the wall at $z = d$ is only partially reflective with a reflection coefficient $\rho(\theta)\angle\phi(\theta)$ which is a function of the angle of incidence θ of the constituent plane waves. This is illustrated in Fig. 10.2. Transmission through the wall at $z = d$ can therefore be expressed as:

$$\tau_o = \left[1 - \rho^2\right]^{1/2} \tag{10.6}$$

at the location of ray zero, for example.

For ray 1, the incident field amplitude at the PRS has, ignoring secondary effects, the magnitude $\rho\angle(\phi - \pi - \frac{4\pi d}{\lambda_o}\cos\theta)$ for an air filled structure. Therefore, at the position of ray 1:

$$\tau_1 = \rho\left(1 - \rho^2\right)^{1/2}\angle(\phi - \pi - \frac{4\pi d}{\lambda_o}\cos\theta) \tag{10.7}$$

and so on toward the more remote of the n rays.

Hence, the radiated field pattern for the PRS antenna can be crudely expressed as:

$$E_{rad} = \sum_{n=0}^{\infty} E_0 g(\theta)\rho^n (1-\rho^2)^{1/2} \exp(j\Phi_n) \tag{10.8}$$

where the angles Φ_n represent the range of possible reflection angles at the PRS for the n possible rays. In this equation $g(\theta)$ is the element pattern for the source at (0, 0, 0) in Fig. 10.2, and

$$\Phi_n = n\Theta = n\left(\phi - \pi - \frac{4\pi d}{\lambda_o}\cos\theta\right) \tag{10.9}$$

Since the reflection coefficient magnitude for the PRS is by definition less than unity it follows that:

$$\sum_{n=0}^{\infty} (\rho \exp j\Phi_n)^n = \frac{1}{1-\rho \exp j\Theta} \tag{10.10}$$

On inserting Eq. (10.10) into Eq. (10.8) an equation for the radiated power density ($P(\theta)$) emerges [1]:

$$P(\theta) = \frac{(1-\rho^2(\theta))}{1+\rho^2(\theta) - 2\rho(\theta)\cos[\phi(\theta) - \pi - \frac{4\pi d}{\lambda_0}\cos\theta]} g^2(\theta) \tag{10.11}$$

Equation (10.11) emphasises the earlier qualitative observation that the larger the magnitude of ρ the sharper is the primary beam resulting in high directivity.

The radiation mechanism could be viewed as electromagnetic seepage from a standing wave (in the z-direction) within the parallel plate housing. For a patch based PRS this mechanism has been illustrated in reference [4] by modelling the near field 2 mm above the PRS (see Fig. 10.3). It is important to note that with a dual polarised feed (square waveguide or square patch the modal patterns in Fig. 10.1 will occur with Fig. 10.1a representing the y-z plane, and Fig. 10.1b representing the x-z plane (TE_{01} mode replacing TE_{10}). If these orthogonal modes are excited simultaneously and in phase then a high gain, linearly polarised and symmetrical pencil beam is formed at broadside. The leaky energy dissipation rate results in an aperture field shape which is virtually free of sidelobes (see Chap. 6, Sect. 6.3.2). If the feed modes are in quadrature phase the radiation will be circularly polarised for a PRS with dual-polarisation capability—for example a periodic surface of square patches [4] or square apertures [3] or cross-slots [6].

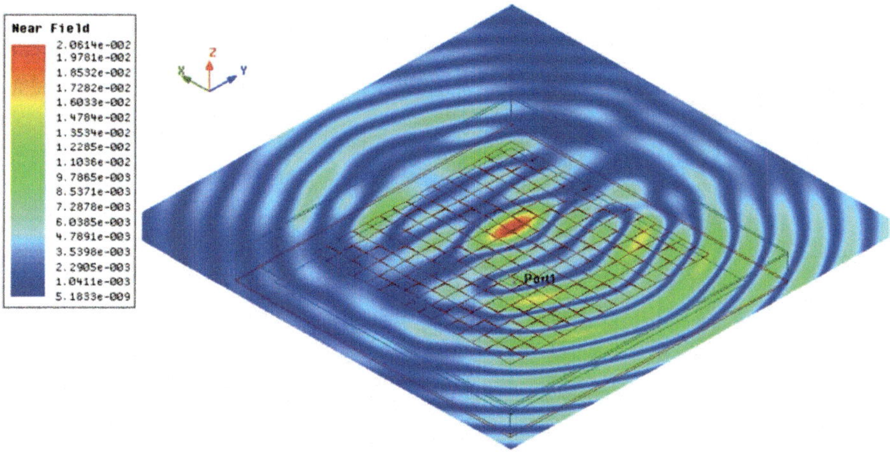

Fig. 10.3 PRS antenna near-field 2 mm above periodic surface. Note the concentration of energy towards the mid-region (gridded) for a surface modelled as unrestricted in x and y (from [4])

10.4 Leaky-Wave Fabry-Perot Antenna

In parallel with innovative developments on the PRS antenna described above, research has simultaneously been proceeding strongly on the rather similar planar leaky-wave array antenna which has also evolved from the Fabry-Perot cavity. It is probably fair to say the PRS version represents a special case, or particular embodiment, of the planar leaky-wave antenna. The latter is discussed in references [5] through [9].

The primary features which distinguish the PRS antenna from the leaky wave alternative are depicted in Fig. 10.4. Both antennas are derived from planar Fabry-Perot cavities but, while in the PRS antenna the cavity is formed by a periodic surface (PRS) in parallel with, and separated by $\lambda/2$, from a perfectly conducting ground plane (PEC), in the LWA the ground plane is replaced by a high impedance surface (HIS), or perfect magnetic conductor (PMC), and the separation distance reduces to $\sim\lambda/4$, or less. In principle the LWA makes possible a much slimmer package.

10.4.1 Reciprocal Periodic Moment Method for CRA

It is apparent from just a cursory look at Fig. 10.4b that any serious theoretical treatment of the cavity resonator antenna has to accommodate the presence of one or more periodic surfaces. As we have noted in earlier chapters, while it is possible with today's 'full-wave electromagnetic solvers' based on either finite element or finite difference techniques to 'crunch out' a solution to any given CRA geometry,

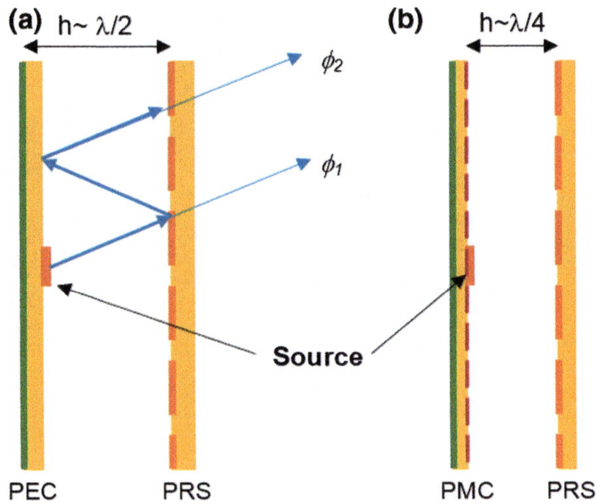

Fig. 10.4 Sketch depicting the fundamental difference between a PRS antenna and a leaky-wave antenna **a** PRS antenna with Fabry-Perot cavity formed by a PRS in parallel with a perfect electrical conductor (PEC) or ground plane and separated by a distance of $\lambda/2$ **b** LWA employing a Fabry-Perot cavity formed from a PRS above a high impedance surface (HIS) and separated by a distance $< \lambda/2$ ($h = \lambda/4$ if the HIS functions as a perfect magnetic conductor (PMC))

insight and understanding of the electromagnetic wave interactions within the CRA usually entails modelling, typically based on the moment method. In Chap. 9 the issue of electromagnetic scattering, and the formulation of a moment method model in the presence of a periodic surface, also arose in connection with planar reflectarray antennas. However, from a periodic surface perspective, while superficially similarities exist between a CRA and a reflectarray, electromagnetically they call for rather different solutions. With the passive reflectarray, scattering at the periodic surface is made accessible to the moment method, by adopting the notion of the infinite array. This then permits a degree of mathematical simplification by enabling the imposition of Floquet's theorem to the electromagnetic field representations and hence the formulation of a Green's function for an exterior space bounded by a periodic surface. For relatively large (in wavelength terms) reflectarray surface areas this is actually a rather insignificant approximation, as we have observed.

In contrast to the reflectarray, the CRA is an active antenna, for which the Fabry-Perot resonator acts as an electromagnetic field transition between the elementary source, whether slot, patch, dipole or probe, and the antenna aperture formed by the PRS (see Fig. 10.4a). Consequently, modelling for the CRA can follow more conventional antenna array and moment method techniques as outlined in Chaps. 3, 5 and 6. The essence of such a moment method analysis is detailed in Chap. 5 Sect. 5.2. In simplistic terms it entails the computation of the coupling between two electromagnetic systems such as a feed system (waveguide in Chap. 5) and free space usually modelled as a half-space. The unknown quantity, which needs to be established, is the field distribution in the coupling mechanism, generally a radiating aperture or slot. This is accomplished by applying field boundary conditions at the interface. The resultant coupled equations are then solved by means of the moment method. Once the aperture fields are established it becomes possible following conventional antenna theory to determine the scattering parameters and the antenna radiation pattern through the plane wave spectrum technique applied to the already computed aperture field distribution.

However, with the CRA, the open microstrip substrate, bounded as it is on one side by a PEC and on the other by a PRS (Fig. 10.5a), presents a complication to the direct application of a moment method solution both because of its 'open' geometry and also because of the periodicity of the PRS. The solution to this dilemma is presented in references [5, 6]. Well established aperture antenna theory (Chap. 3) dictates that the radiation pattern for such an antenna can be computed by representing the radiation as a spectrum of plane waves. Consequently, by invoking the reciprocity, which exists between an antenna in transmission and an antenna in receive mode, and provided that the moment method calculations are performed in the spectral domain, it is feasible to calculate an antenna radiation pattern by illuminating it with plane waves from a testing current element in the radiation space. For the CRA the far-field pattern computation then reduces to the calculation of the field in the interior of the substrate due to incident plane waves (Fig. 10.5a).

The CRA analytical model for a reciprocity computation of the antenna far-field pattern is depicted in Fig. 10.5. The remote testing dipole is directed along $\hat{\mathbf{a}}_\theta$, although it could equally be aligned with $\hat{\mathbf{a}}_\phi$. The computation endeavours to equate the testing field, emanating from the remote dipole, to the field E_x at the location of the CRA source (at the PEC) as generated by the illuminating plane wave from the test dipole. In short, the radiation pattern computation for the CRA reduces to the substrate pattern due to an incident plane wave. As with conventional aperture antennas such as slot radiators (Chap. 5) it is appropriate to adopt the equivalence principle which permits the CRA slots to be replaced by magnetic currents $\mathbf{J_{mx}}$.

The direction vector and the field vectors for plane wave incidence on a surface in the x-y plane are shown in Fig. 10.6. For excitation of a substrate mode it is the plane wave component tangential to the surface which is relevant. In the figure the x-direction is highlighted, but the y-direction could equally have been selected. The direction of the incoming wave is denoted by k_o while the tangentially directed wave component lies along k_x (or k_y). Consequently for propagation in the x-z plane, we can write:

$$k_x = k_0 \sin\theta_r \tag{10.12}$$

$$\text{and} \quad k_0^2 = k_x^2 + k_z^2 \tag{10.13}$$

The incident electric field and magnetic field components associated with the k_o directed plane wave are E_{inc} and H_{inc}. In the left hand diagram H_{inc} is entirely transverse to k_x and is labelled as TM incidence, while in the right hand diagram E_{inc} is entirely transverse to k_x generating TE incidence.

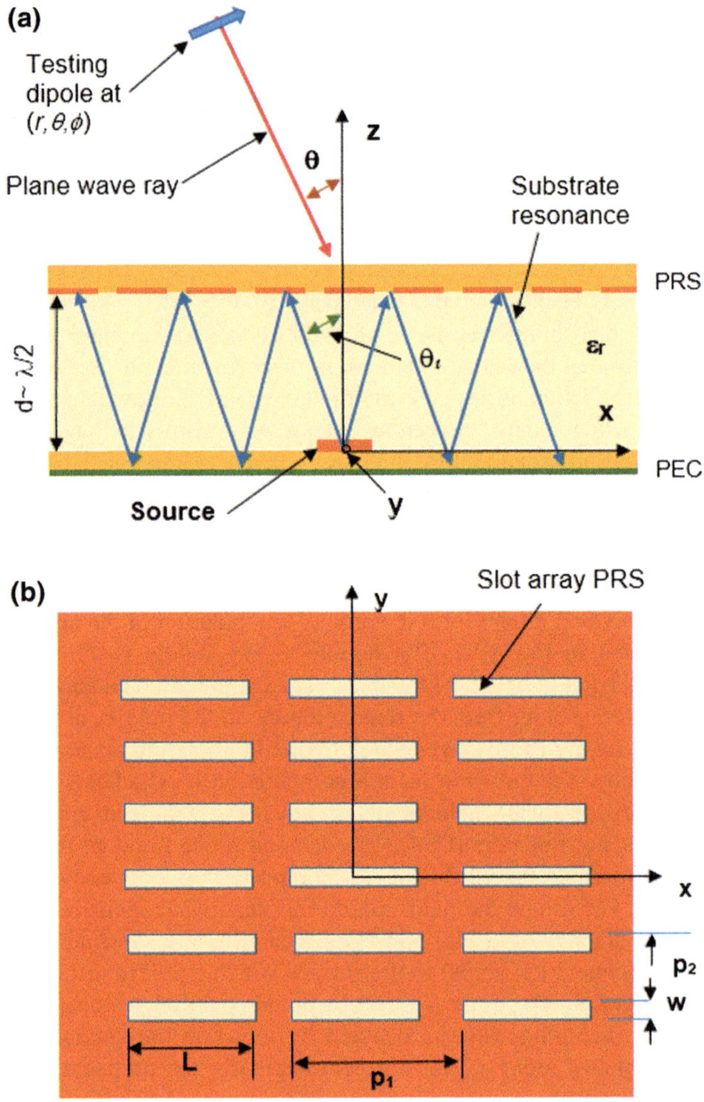

Fig. 10.5 Fabry-Perot resonator antenna **a** illuminated by a plane wave from a distant test dipole at r, θ, ϕ **b** PRS comprising a periodic array of slots

For a closed PRS, that is a surface with no slots, the plane wave reflects totally on the outer surface and hence for TE incidence (see Fig. 10.6):

$$H_{xp}^{+} = 2H_x^{inc} \tag{10.14}$$

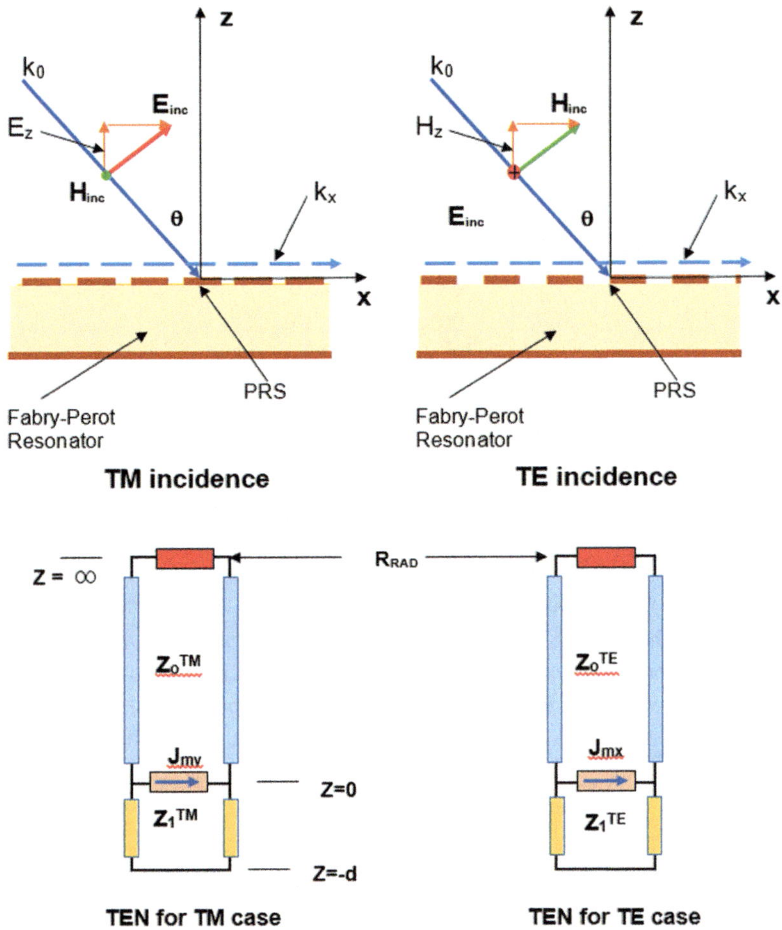

Fig. 10.6 Electromagnetic wave field components at CRA surface when illuminated by a plane wave and transmission-line equivalent circuits

where H_{xp}^{+} denotes the x-directed magnetic field tangential to the outer surface of the plate. H_x^{inc} is the x-directed magnetic field component of the incident plane wave. Furthermore for this surface wave directed along k_x, E and H are related through:

$$E_y^{inc} = Z_0^{TE} H_x^{inc} \tag{10.15}$$

A similar relationship exists for the TM case. Given that the plane wave at the CRA emanates from a test dipole at distance r from the antenna, the field E_{inc} can be expressed as:

$$\mathbf{E}_{inc} = \hat{\mathbf{a}}_\theta \left[-\frac{j\omega\mu_0}{4\pi r} \exp(-jk_0 r) \right] \exp(j\mathbf{k}_0 \bullet \hat{\mathbf{a}}_r r) \tag{10.16}$$

The term inside the square brackets represents radiation from a point source, and the second term ensures that, of all possible plane waves, the plane wave directed along $\hat{\mathbf{a}}_r$ is mathematically selected. A similar equation can be formed for the TM case with $\mathbf{E}_{inc}$ directed along $\hat{\mathbf{a}}_\phi$. It is also helpful here to define Z_0^{TE} and Z_0^{TM}. Hence:

$$Z_0^{TE} = \frac{E_{inc}}{H_z^{inc}} = \frac{E_{inc}}{H_{inc} \cos\theta} = \frac{\omega\mu_0}{k_{z0}} \tag{10.17}$$

$$Z_0^{TM} = \frac{E_z^{inc}}{H_{inc}} = \frac{E_{inc} \cos\theta}{H_{inc}} = \frac{k_{z0}}{\omega\varepsilon_0} \tag{10.18}$$

For narrow x-directed slots as suggested in Fig. 10.5b the H_x^{inc} field component, generated by the TE illumination of the PRS, can in turn be associated with trial magnetic current densities (unknown at this stage) in the antenna slots. These are denoted by $J_{mx}^s(x_p, y_q)$ on the slot array surfaces at z = 0 in real space, and by $\tilde{J}_{mx}^s(k_{xp}, k_{yq})$ in the plane wave spectral domain. The integers p and q define slot locations. For mathematical convenience, rather as outlined in Chap. 9, Sect. 9.1, p and q are allowed to range in value from $-\infty$ to $+\infty$ since those slots outside the limited number of active examples at the heart of the array contribute nothing to the results.

The adoption of the infinite array concept in Sect. 9.1 yielded the analytical benefit of making it possible to model the reflectarray with minimal complexity and tedious repetition. When the array is of infinite extent its periodicity can be expressed mathematically through the Floquet Theorem. Furthermore, considerable mathematical simplification accrues to the CRA analysis presented below by performing the computation in the spectral domain. When all relevant functions are Fourier transformed into the spectral domain the array can be modelled quite straightforwardly as summations over identically excited elements. The Floquet modes are accommodated automatically within the Fourier transforming process. It is well established in signal theory that an infinite sequence of identical 'pulses' transforms into a frequency spectrum of sinusoidal waves. By analogy the periodic array of identical aperture fields is transformed into Floquet modes, which satisfy the periodic boundary conditions at the substrate surface.

The x-directed magnetic current density $J_{mx}^s(x_p, y_q)$ in the slots of the PRS generate a magnetic field H_{xs}^+ adjacent to the outer surface ($z > 0$) of the PRS, and simultaneously a magnetic field H_{xs}^- adjacent to the inner surface ($z < 0$). If the spectral domain magnetic Green's function ($\tilde{G}_m^\pm(k_{xp}, k_{yq})$) for the periodic surface is known, then we can write:

$$H_{xs}^{\pm} = \pm \frac{1}{p_1 p_2} \sum_{p=-\infty}^{\infty} \sum_{q=-\infty}^{\infty} \tilde{G}_m^{\pm}(k_{xp}, k_{yq}) \tilde{J}_{mx}^{s}(k_{xp}, k_{yq}) \exp -j(k_{xp}x + k_{yq}y) \tag{10.19}$$

where p_1 and p_2 are the element to element pitches in respectively the x and y directions.

In the spectral domain, the above Green's function can be deduced relatively simply by employing a transmission line equivalent network (TEN) [10]. The analogous circuits show downward wave propagation in the negative z-direction in the radiation space $z > 0$, and transmission into the substrate $z < 0$ (Fig. 10.6). The figure presents TEN's for both the TM orientated incident plane wave and the TM alternative. The more general case of oblique incidence is presented in reference [10]. The transmission lines are presumed to extend to infinity in the +z direction where they are terminated in R_{rad} the radiation resistance. In the air region the transmission line characteristic impedances are identified as Z_0^{TM} and Z_0^{TE}, while in the substrate we use Z_1^{TM} and Z_1^{TE}. At z = 0 on the PRS the scatterers, in our case slots, are excited by the H_ξ (TM) and H_η (TE) of the obliquely incident plane wave, where ξ is rotated angle δ from the x axis. Consequently they can induce magnetic current densities $\tilde{J}_{mx}^{TM}$ and $\tilde{J}_{mx}^{TE}$ in the x-aligned slots. Operation in the spectral domain is presumed. For slots $\tilde{J}_{my}$ is usually negligible so we can concentrate on $\tilde{J}_{mx}$. For z-directed propagation, the TEN relates the fields in the interior and the exterior of the substrate and across the interface at z = 0 [10]. In this representation the current density excitations at z = 0 can be replaced by the currents $\tilde{I}_{mx}^{TM}$ and $\tilde{I}_{mx}^{TE}$, where $\tilde{I}_{mx}^{TM} = \tilde{J}_{mx}^{TM} L$ and $\tilde{I}_{mx}^{TE} = \tilde{J}_{mx}^{TE} L$. Hence, on applying network theory to the equivalent circuits in Fig. 10.7, the voltage at the junction at z = 0 is given by:

$$\begin{aligned} \tilde{V}_{in}^{TM} &= \tilde{I}_{mx}^{TM} Z_{in}^{TM} \\ \tilde{V}_{in}^{TE} &= \tilde{I}_{mx}^{TE} Z_{in}^{TE} \end{aligned} \tag{10.20}$$

where $Z_{in}^{TE,TM}$ are the parallel impedances of the lines above and below z = 0. These are respectively $Z_0^{TE,TM}$(see Eqs. (10.17), (10.18)) and

$$Z_{sub}^{TE,TM} = jZ_1^{TE,TM} \tan \beta_1 d \tag{10.21}$$

and for the substrate region

$$Z_1^{TM} = \frac{\beta_1}{\omega \varepsilon_r \varepsilon_0} \text{ and } Z_1^{TE} = \frac{\omega \mu_r \mu_0}{\beta_1} \text{ and } \beta_1 = \omega \sqrt{\varepsilon_r \varepsilon_0 \mu_r \mu_0} \tag{10.22}$$

On combining Eqs. (10.20), (10.21) and (10.22), in the spectral domain, this yields the relation:

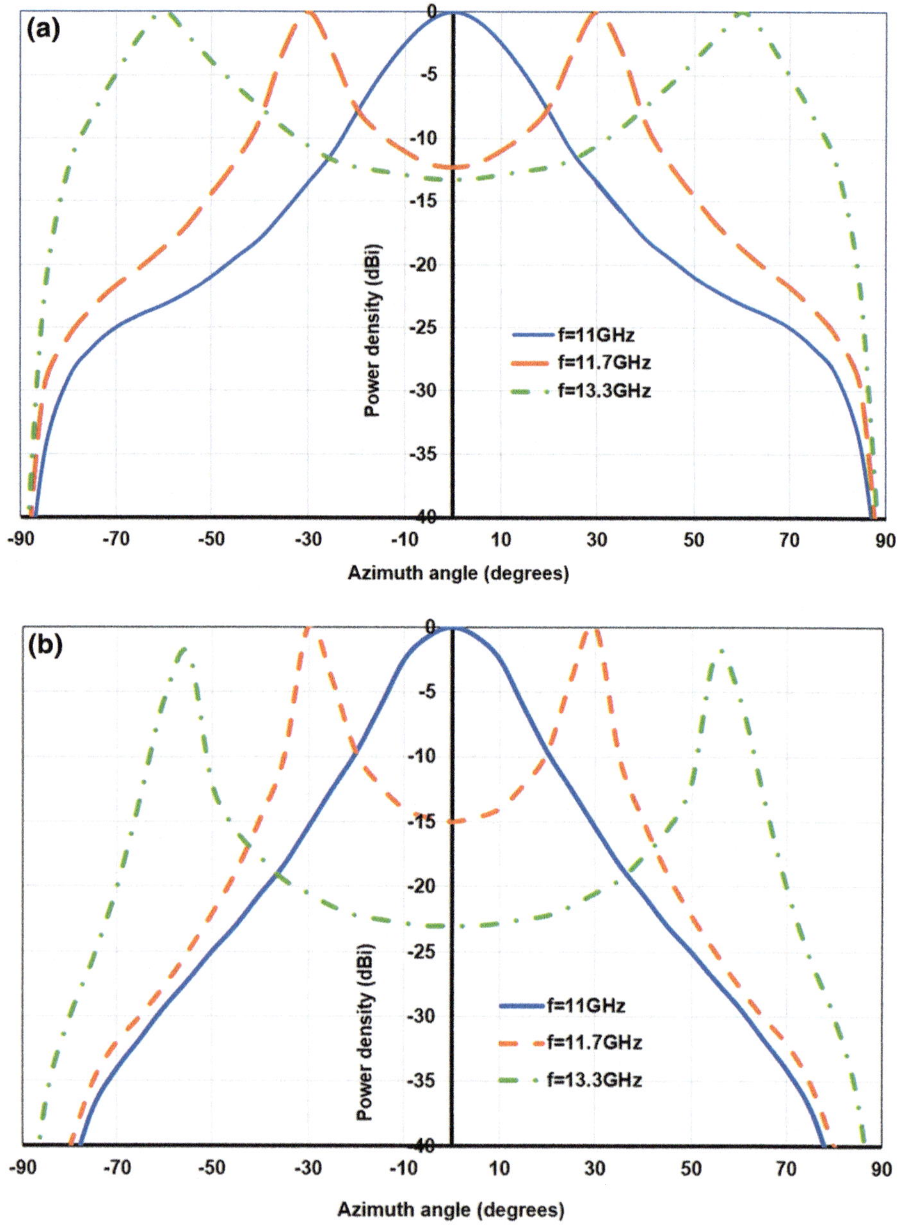

Fig. 10.7 Scanned E-plane (**a**), and H-plane (**b**) radiation patterns for a slot based CRA with a dielectric filled resonator ($\varepsilon_r = 2.2$) for which $L = 0.6$ cm, $w = 0.05$ cm, $p_1 = 1$ cm and $p_2 = 0.3$ cm. The CRA is resonant at 11 GHz for $d = 0.85$ *mm*. Over the frequency range depicted the height (d) varies from $d \approx \lambda/2$ to $d \approx \lambda/3$ (from [6])

$$\tilde{H}_{xs}(k_x, k_y)\big|_0 = -\frac{1}{k_t^2}\left[k_x^2 I_x^{TE}\big|_0 + k_y^2 I_x^{TM}\big|_0\right]\tilde{J}_{ms}(k_x, k_y) \tag{10.23}$$

which is evaluated at z = 0. Referring to the transmission line analogy, at z = 0, we have:

$$I_x|_0 = Y_{in} V_{in}^{TE,TM} \tag{10.24}$$

where V_{in} is the voltage at the surface (Fig. 10.6). To form a Green's function in the spectral domain we choose $V_{in} = 1$ to represent the 'point' source. Hence

$$I_x|_0 = Y_{in} = Y_0^{TE,TM} = \frac{1}{Z_0^{TE,TM}} \tag{10.25}$$

The field due to I_x just above the PRS is therefore

$$\tilde{G}_m^+(k_x, k_y) = -\frac{1}{k_t^2}\left[k_x^2 Y_0^{TE} + k_y^2 Y_0^{TM}\right] \tag{10.26}$$

Z_0^{TE} and Z_0^{TM} and hence Y_0^{TE} and Y_0^{TM} are defined in Eqs. (10.17) and (10.18).

Also

$$\begin{aligned} k_{z0} &= \sqrt{(k_0^2 - k_t^2)} = \sqrt{(k_0^2 - k_x^2 - k_y^2)} \\ k_{z1} &= \sqrt{(k_1^2 - k_t^2)} = \sqrt{(k_1^2 - k_x^2 - k_y^2)} \end{aligned} \tag{10.27}$$

where $k_1 = \beta_1 = \sqrt{\varepsilon_r} k_0$.

Similarly, for the magnetic field just below the PRS and within the substrate (region 1 with relative permittivity = ε_r) the appropriate Green's function is:

$$\tilde{G}_m^-(k_x, k_y) = \frac{1}{k_t^2}\left[jk_x^2 Y_1^{TE}\cot(k_{z1}d) + jk_y^2 Y_1^{TM}\cot(k_{z1}d)\right] \tag{10.28}$$

with

$$Z_0^{TE} = \frac{\omega\mu_r\mu_0}{k_{z0}} \tag{10.29}$$

$$Z_0^{TM} = \frac{k_{z0}}{\omega\varepsilon_r\varepsilon_0} \tag{10.30}$$

The magnetic fields above and below the PRS can now be expressed in terms of the unknown slot magnetic current densities. When the boundary conditions are applied at a reference slot (usually the $p = 0$, $q = 0$ slot), that is:

$$H_{xp}^{+} + H_{xs}^{+} = H_{xs}^{-} \tag{10.31}$$

where H_{xp}^{+} is the tangential magnetic field on the 'closed' metal plate, H_{xs}^{+} is the slot magnetic current induced field just above the slotted surface and H_{xs}^{-} is the equivalent field just below the surface. At this stage we choose to follow the Galerkin approach (see Chap. 5) to the implementation of the moment method and employ entire domain basis functions in the trial form of the 'unknown' slot magnetic current density J_{ms}. Hence

$$J_{ms}(x,y) = \sum_{n=1}^{N} A_n F_n(x,y) \tag{10.32}$$

A suitable choice for F_n is [6]:

$$F_n(x,y) = \frac{\sin\left[\frac{n\pi}{L}\left(x - \frac{L}{2}\right)\right]}{\pi\sqrt{\left[\left(\frac{w}{2}\right)^2 - y^2\right]}} \tag{10.33}$$

In the spectral domain the slot magnetic current density can therefore be written as:

$$\tilde{J}_{mx} = \sum_{n=1}^{N} A_n \tilde{F}_n(k_x,k_y) = \sum_{n=1}^{N} A_n \tilde{f}_n(k_x)\tilde{g}_n(k_y) \tag{10.34}$$

where:

$$\tilde{f}_n(k_x) = \frac{\exp -j\left(\frac{k_x L}{2}\right)\left[-n\pi L + \exp(jk_x L)(n\pi L \cos n\pi)\right]}{(k_x L)^2 - (n\pi)^2} \tag{10.35}$$

$$\tilde{g}_n(k_y) = \mathrm{J}_0\left(\frac{k_y w}{2}\right) \tag{10.36}$$

Hence on applying the Galerkin method the following matrix equation for the unknown A_n's is generated:

$$[Z_{mn}][A_n] = [R_m] \tag{10.37}$$

The entries in the right hand column matrix are:

$$R_m = -\frac{2}{Z_0^{TE,TM}} E_y^{inc} \tilde{F}_m(k_{x0},k_{y0}) \tag{10.38}$$

while the entries in the Z_{mn} matrix are:

$$
\begin{aligned}
Z_{mn} &= \frac{1}{p_1 p_2} \sum_{p=-\infty}^{\infty} \sum_{q=-\infty}^{\infty} \tilde{G}_m^+(k_{xp}, k_{yq}) \tilde{F}_n(k_{xp}, k_{yq}) \tilde{F}_m(-k_{xp}, -k_{yq}) \\
&+ \frac{1}{p_1 p_2} \sum_{p=-\infty}^{\infty} \sum_{q=-\infty}^{\infty} \tilde{G}_m^-(k_{xp}, k_{yq}) \tilde{F}_n(k_{xp}, k_{yq}) \tilde{F}_m(-k_{xp}, -k_{yq})
\end{aligned}
\tag{10.39}
$$

Once the unknown A_n's are determined by solving Eq. (10.37) the field E_y inside the substrate at the position of the source can be calculated, and by reciprocity, which involves sampling the cavity field, the far field becomes calculable.

10.4.2 CRA Radiation Characteristics

Typical E-plane and H-plane far-field computations for slot based CRA's at frequencies near resonance (f = 11 GHz, d = 0.85 mm) have been generated and are depicted in Fig. 10.7. The calculations apply to a substrate with $\varepsilon_r = 2.2$ and for a PRS for which $L = 0.6$ cm, $w = 0.05$ cm, $p_1 = 1$ cm and $p_2 = 0.3$ cm. For an infinite parallel plate waveguide with a dielectric filling ($\varepsilon_r = 2.2$), and a plate separation of 0.85 mm, propagation for the TE_{10} mode cuts-on at 11.89 GHz in the absence of slots. The slot array lowers the resonance to 11.0 GHz, represented by the blue solid line in Fig. 10.7a, b. At this frequency the radiation peak is precisely at broadside and is sidelobe free for both the E-plane and the H-plane, as predicted by Trentini [1]. The symmetry of Eq. (10.36) suggests that the E-plane and H-plane patterns for the CRA should be the same. That they are not is due to the adoption of narrow slots to form the PRS. As we have seen in earlier chapters the E-plane (y-z plane in Fig. 10.5) pattern for a narrow slot in a half-space is semi-circular, while in the H-lane (x-z plane in Fig. 10.5) it is a half 'doughring' with nulls along the major axis of the slot. Consequently, in Fig. 10.7 the E-plane pattern exhibits (Fig. 10.7a) much stronger radiation close to +90° and −90° than occurs in the H-plane (Fig. 10.7b).

More importantly Fig. 10.7 highlights the issue of beam splitting with a CRA, or leaky wave antenna, and how the split beam maxima are influenced by antenna thickness (d) and operating frequency. It is evident that the major influence on beam split angles is the ratio of separation height (d) between the PRS and the ground plane to the wavelength (λ) within the substrate (i.e. d/λ) and that in the H-plane it is possible to locate a maximum beam anywhere between broadside and 75°. In the E-plane it is predicted that the peaks can scan almost to end-fire, as indicated in Fig. 10.7a. What is unexpected is that for 2-D antennas of this type so called 'scanning' is fundamentally associated with the formation of a split beam, particularly at scan angles well away from broadside. The mechanism for this behaviour is suggested in Fig. 10.1. In the H-plane (Fig. 10.1b) a single beam, at broadside, is formed only at resonance since this ensures that the fundamental parallel plate mode is cut-off. This in turn implies that the TE mode in the positive and the negative x-directions in Fig. 10.1b exhibits zero phase shift. The radiating elements of the PRS are thus excited in phase so forming a single broadside beam in both the *x*-*z* plane and the *y*-*z* plane. At higher frequencies the parallel plate TE mode is no longer cut-off in the +*x* and −*x* directions, for mid-point excitation as shown in Fig. 10.1. Leaky-wave beams are thus possible in these directions resulting in the 2-D split beam patterns shown in Fig. 10.7. Note that since d and λ are interlinked, and essentially interchangeable in the basic mode of operation of the Fabry-Perot resonator, beam splitting behaviour similar to Fig. 10.7 can be observed when thickness adjustments (d) replace frequency [5, 6]. The beam-splitting is particularly associated with the mid-point excitation commonly adopted with the CRA. One directional scanning occurs if the array can be end fed.

This split beam characteristic represents an awkward feature of the leaky-wave CRA, which is highlighted in references [5, 6], namely that the desired broadside pencil-shaped radiating beam degrades rapidly with small changes to cavity height from the resonant condition. Given that in theory the CRA periodic surface is presumed to be of infinite extent the PRS could take the form of a circular plate with radiating elements spreading radially outwards from the centre. Propagation from a centrally located feed will also spread radially outwards at above resonance frequencies. In three dimensions, at above cut-off, the split beam maxima of Fig. 10.7 will occur along radii at all possible azimuth angles thus forming a hollow conical beam with a circular rim. While some form of frequency scanning seems possible, the obvious practical implication of conical beam formation, is that the basic centre fed CRA antenna provides a broadside pencil beam radiation pattern only over a relatively narrow bandwidth.

10.5 CRA Thinning

A leaky-wave CRA geometry in which the perfectly conducting ground plane is replaced by a second periodic array, with the characteristics of a high impedance surface (HIS—see Fig. 10.4b), offers some support that a wider band CRA is achievable [8]. The introduction of the HIS primarily makes possible a significant level of antenna thinning. In theory, a leaky-wave CRA can be reduced in thickness to $d \approx \lambda/7$ from the usual $\lambda/2$ by the adoption of an HIS. The mechanism is made clear by considering the simplest possible manifestation of this adjustment, whereby the HIS presents an open-circuit wall to the internal fields the CRA. It is not too difficult to appreciate that in this case the thickness, for no change in resonance, is immediately reduced by 50%. By carefully tailoring the surface impedance it is possible to reduce the thickness even further. Improved bandwidth accrues from the fact that in the thinned CRA the beam scanning becomes much less sensitive to d variations and hence to changes in frequency.

A similar scheme, reported in reference [2], describes a planar Fabry-Perot leaky-wave cavity antenna employing a PRS above an artificial magnetic conductor (AMC—see Fig. 10.8). The AMC is simply an HIS by another name. For an

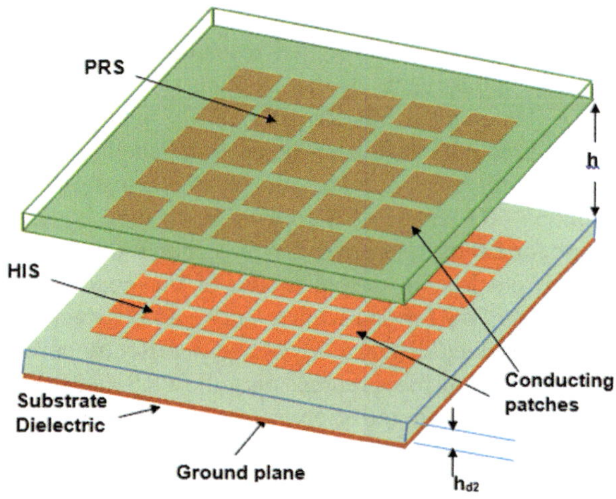

Fig. 10.8 Schematic of a 2-D leaky-wave Fabry-Perot resonator antenna fed by square patch at centre of cavity (from [8])

antenna operating at 14 GHz the AMC is produced by printing a closely packed array of almost square patches (4.1 mm × 4.4 mm) on a 1.13 mm thick grounded substrate ($\varepsilon_r = 2.2$). A highly reflective PRS is adopted to procure high gain. The upper layer PRS is constructed from almost square patches (10 mm × 11 mm) on an ungrounded substrate similar to that used for the AMC. The antenna is fed from an off-square patch embedded at the centre of the HIS. The general arrangement is shown schematically in Fig. 10.8. A full-wave 3-D electromagnetic simulation of the proposed design revealed an antenna thickness of close to λ/4 for resonance at 14 GHz. Modelling and measurements indicate that an optimum gain of 20 dBi at broadside is potentially available for the linearly polarised beam. The bandwidth was estimated at 2%. A CRA with similar performance but exhibiting a much thinner ($d = \lambda/10$) profile has also been reported. The key to thinning is in the nature of the HIS which can in principle provide a surface impedance which lies anywhere between a PEC ($d = \lambda/2$) and a PMC ($d = \lambda/4$).

10.6 Bandwidth Enhancement

The narrowness of the bandwidth of the basic CRA is studied in reference [9] where design guidelines are presented to tailor the CRA's directivity bandwidth. This is predicated on a thorough understanding of the role of higher order modes in the leaky-wave CRA which was achieved by applying reciprocity, and the full-wave spectral domain periodic analysis, as outlined above, to a 2-D geometry. Improved bandwidth is possible, not by replacing the conducting ground plane by an HIS as in Fig. 10.4b, but by introducing a second PRS inserted between the original PRS and the ground plane forming in effect a pair of coupled Fabry-Perot cavities (c.f. Fig. 10.8). The technique is not unlike patch antenna bandwidth enhancement which involves the stacking of two or perhaps three patch radiators in close proximity to form a coupled resonator system. Bandwidth improvement follows separate adjustments of the resonant frequencies of each patch. In the case of the CRA, bandwidth enhancements of six times that of the single layer LWA, at the same directivity level, can be accomplished.

High directivity, high gain, beam formation in a two dimensional leak-wave antenna (2D-LWA) of the Fabry-Perot type is described in references [5–9]. The basic format of the studied LWA is shown schematically in Fig. 10.8 with a PRS of square patches on the underside of the upper substrate, separated by a distance d from an HIS formed from a square array which has twice as many much smaller square patches distance h_{d2} above a perfectly conducting ground plane. Actually, the periodic surfaces representing the PRS and the HIS need not be square patches but could be crossed dipoles, or crossed slots, or round holes, in a conducting screen. In [5] the PRS is formed from printed dipoles, in [6] it is constructed from slots etched into a plane conducting screen, while the antennas in [8, 9] are developments based on square patches. Bandwidth enhancement, as indicated above, is linked to the reduced thickness (d) which the HIS enforces.

High gain with improved bandwidth potential can be procured in theory by omni-directional excitation of a circular plate shaped CRA. In Sect. 10.3, where the basic mechanism of operation of the PRS antenna is examined (c.f. Fig. 10.1), the explanation given there for leaky-wave radiation, presumes that the parallel plate waveguide forming the antenna as a rectangular structure, which is best represented by the Cartesian system. However, given that the Fabry-Perot cavity is theoretically infinite in extent [1] this need not be the case. Cylindrical symmetry could equally be assumed as is done in [5]. In this case the antenna is much more obviously a travelling-wave device with the field pattern of Fig. 10.1b travelling radially outwards from the source. If the source is omni-directional (e.g. a coaxial probe) then the dominant TM mode propagates radially outwards with equal magnitude in all directions exciting the PRS as it does so. This operational model is tentatively hinted at in Fig. 10.3. For an air dielectric the fundamental mode, as in the PRS antenna case, cuts-on at $d \sim \lambda_o/2$, and a pencil beam at broadside is formed.

10.7 Reconfigurable PRS for Fixed Frequency Scanning

CRA modifications, discussed in preceding sections of this chapter, were aimed both at controlling frequency sensitive beam splitting to improve bandwidth, and to take advantage of it to procure practical beam scanning with a CRA. The discussion has, thus far, circumvented the idea of beam scanning at a fixed frequency, or in effect to introduce quasi-phased-array behaviour into the CRA. It is perhaps hardly surprising, nevertheless, that no sooner had the breakthrough to practical Fabry-Perot cavity based antennas been achieved [1–9] largely by employing microwave substrates, that efforts were being made to introduce beam steerability into the mix. The route to this goal has largely been through the introduction of 'reconfigurable' array elements into the PRS [11–16] and/or the HIS [17].

The methodology of electronic control in a CRA for beam scanning is hinted at in the fundamental Eq. (10.11). However, this equation should be treated with caution insofar as it is two dimensional and has been formulated using a simple ray theory and multiple reflections. With $\rho(\theta)$ denoting the internal PRS reflection coefficient at incident angle θ (from x) in Fig. 10.1a, it is clear that control of ρ is key to pattern modification [7]. Note that the $g(\theta)$ term is a magnitude factor only—influencing energy distribution in cavity. It has no influence on phase at the radiating surface. However, in practice control of ρ has been implemented by the introduction of a second periodic surface (HIS) either replacing the ground plane or inserted between the ground plane and the PRS as indicated in Fig. 10.9a. Reconfigurability has generally been advanced by introducing phase agile cells into this additional periodic surface rather than the PRS.

Arguably, the introduction of reconfigurable components into PRS antennas appeared first as a frequency compensation technique [12], whereby PRS directivity could be maintained over an enhanced tunable frequency range. An antenna comprising a PRS formed from an array of square patches is described, and frequency adjustment is achieved by introducing a phase-agile surface on a substrate positioned close to, and just above the ground plane. The geometry is shown schematically in Fig. 10.9, where the lower periodic surface is represented as being 'phase agile'. This is achieved by employing split patches, termed reflection cells, with the gap bridged by one or more varactor diodes (c.f. Fig. 10.9b). The reflectivity of the reflection cells, and hence the phase, is controlled by the suitably located varactor diodes. It is reported [12] that CRA gain can be maintained consistently above 10 dBi over a tuned frequency range of 13.5% from 5.2 to 5.95 GHz. The bandwidth is six time larger than for the standard PRS antenna. By further exploiting the principle enunciated in [7] and encapsulated in Eq. (10.11), it has been suggested that dynamic beamwidth reconfiguration in PRS antennas is feasible and a range of techniques are explored in references [13–15].

While the reconfigurable PRS has presented the primary route toward controlling and optimising beamwidth and directivity, it has also been proposed [15] that PRS beam positioning or scanning can, in principle, be procured by phase control of a two element feed antenna embedded in the Fabry-Perot cavity as shown in Fig. 10.10.

Referring back to Eq. (10.11), it has already been noted that, in the E-plane of the Fabry-Perot resonator antenna the $g(\theta)$ term, representing the feed, only indirectly influences the radiation pattern $P(\theta)$ and consequently, in this plane, phase changes at the two element feed cannot modify $P(\theta)$ and the antenna's broadside radiation characteristic. Consequently beam scanning in this case must be restricted to the H-plane with the PRS excited from a linearly polarised feed. While the simulated beam scanning predictions are presented for the H-plane, that scanning is restricted to this plane is not made clear in references [14, 15]. It is fairly evident from Fig. 10.1b that excitement of the PRS array elements is provided by the axial (z-directed) magnetic field of the close-to-cut-off quasi-TE_{10} mode in the parallel

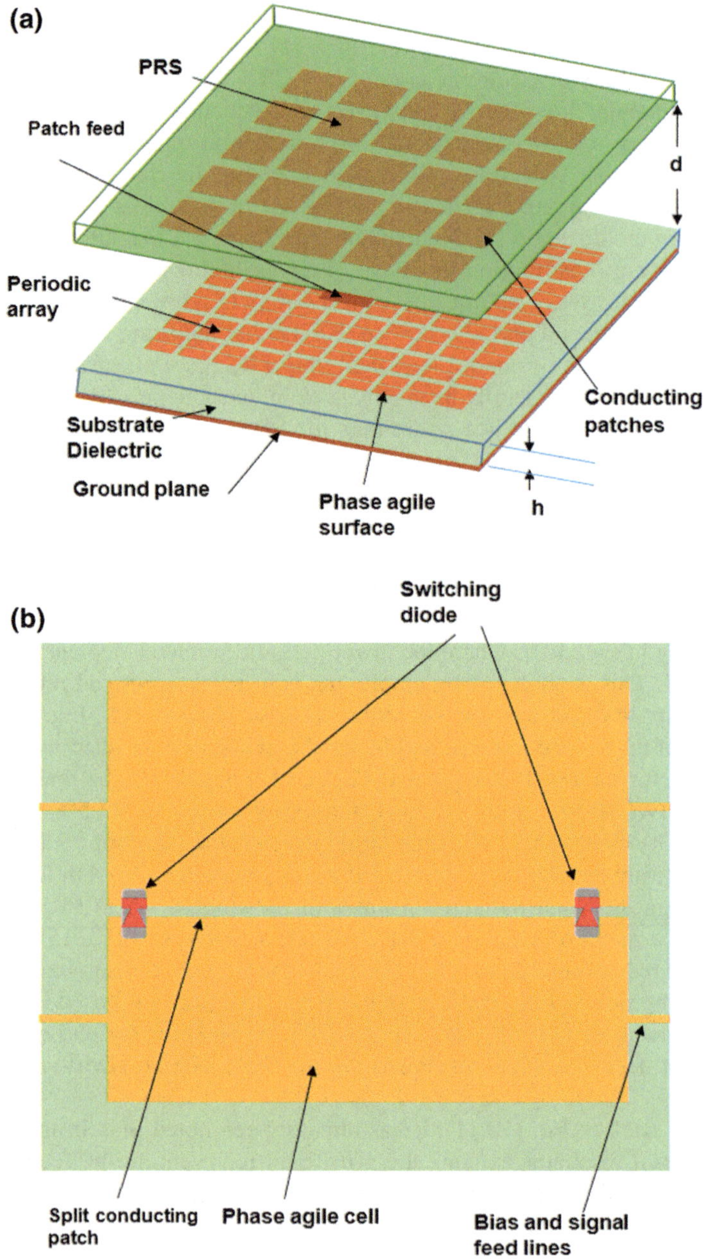

Fig. 10.9 CRA with **a** phase agile HIS and **b** typical phase agile cell incorporating varactor diode switches

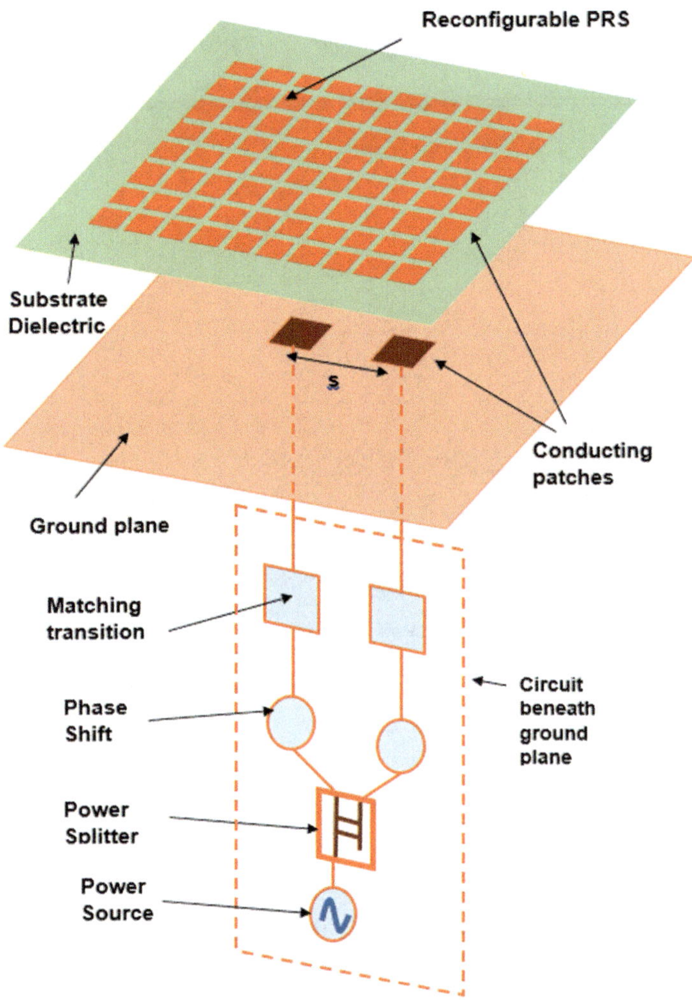

Fig. 10.10 Schematic of a high directivity PRS antenna with additional beam scanning in the H-plane (from [15])

plate waveguide. If the PRS is reconfigurable as suggested in Fig. 10.11, this allows the surface reflection coefficient to be adjusted by controlling the capacitances of the varactor diodes. The introduction of a two-element feed can be made to favour the propagation of this mode in the forward (+z) or backward (−z) directions, thus providing a scanning mechanism which results in three beam positions in the H-plane given by the basic leaky-wave equation.

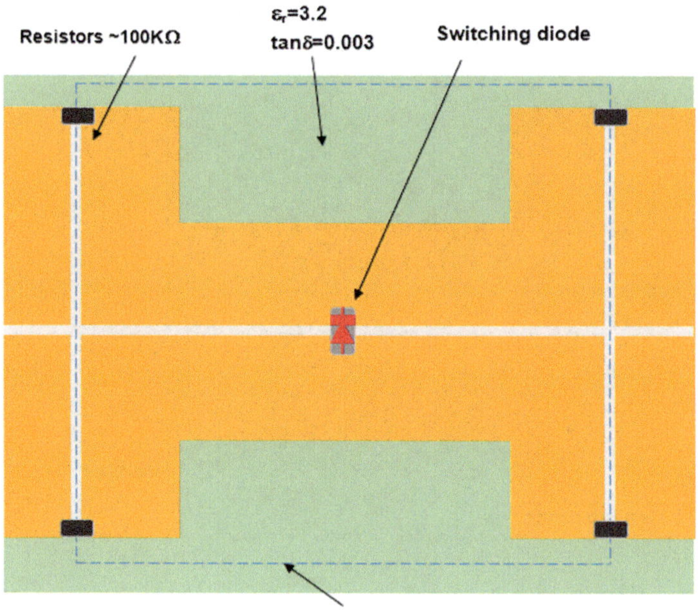

Fig. 10.11 Schematic depicting diode modified patch element for typical reconfigurable patch based PRS antenna (from [15])

$$\Theta = \cos^{-1}\frac{\lambda}{\lambda_g} \tag{10.40}$$

This is confirmed by performing an elementary calculation on the fully dimensioned PRS antenna in reference [15]. The Fabry-Perot cavity can, simplistically, be viewed as an air-filled parallel plate waveguide with d = 78.4 mm. Hence for the quasi-TE_{10} mode we estimate that λ_c = 156.8 mm, while at 2 GHz, the free-space wavelength λ_o = 150 mm. Consequently λ_{g10} = 515.0 mm and from Eq. (10.12) we have *three* possibilities for the mainbeam offset from broadside, namely −17°, 0°, +17°. The paper provides simulated H-plane results showing a −15°, 0°, +15° beam swing for a low directivity example, and −20°, 0°, +20° for a high directivity design. Given the assumptions inherent in the use of Eq. (10.11), this is good agreement. It is perhaps pertinent to note that attempts to achieve significant beam scanning in the H-plane tend to be incompatible with good PRS performance in the E-plane. A not unfamiliar 'trade-off' which is typical of choices regularly encountered in electrical engineering.

10.8 Beam Scanning in a Leaky-Wave CRA

Electronic beam scanning in a CRA represents active research at the time of writing in 2017 but solutions are proving to be elusive. While a range of modifications, which have been introduced into the original Fabry-Perot resonator, such as additional periodic layers variously referred to as an HIS, AMC, FSS or metamaterial, have achieved improved bandwidths, gain and efficiency by controlling beam-splitting, frequency independent scanning has seemingly remained out of reach. Nevertheless, comprehensive modelling of the CRA points to the leaky-wave mechanism as the most promising route to effective non-phased-array beam scanning at a fixed frequency.

Leaky-wave scanning performance of a cavity antenna, based on the application of reconfiguration to a PRS layer, is partly addressed in [16]. Careful inspection of Fig. 10.1b reveals that for a non-directional feed, an in-phase, close-to-cut-off, quasi-TE_{10} mode field propagates in the PRS antenna to the left and right of the source; that is in the −z and +z directions. The PRS thus splits into two 'half-arrays' whose in-phase radiations add constructively into the free-space region contributing in the H-plane to the predicted [1] broadside pencil beam. In reference [16] this split operational mode is emphasised by introducing slightly different array geometries in the left and right regions of the PRS antenna. In summary, the arrangement has arguably produced a limited 1-D leaky-wave scan mechanism in the H-plane of the cavity resonator antenna which was originally presented as a PRS antenna. The investigation outlined in [16] suggests that more significant and more practical levels of scanning in cavity resonator antennas require more emphasis on the leaky-wave mechanism. This is addressed in the next section.

10.8.1 Linear Leaky-Wave Array

The leaky-wave antenna was addressed in several contributions to the literature after 2010. For example, a 1-D Fabry-Pérot resonator, for which the leaky-wave beam scanning requirements are clearly expressed, is reported in [17]. Beam-scanning of up to 40° is demonstrated at a single design frequency. The essence of the method is in the implementation (see Fig. 10.12) of a high impedance surface (HIS) in place of the electrically conducting ground plane, which is normally encountered in the Fabry-Perot cavity antenna (see Sect. 10.4). This leaky-wave structure is analysed using the spectral domain moment method and reciprocity [8] as described in Sect. 10.4.1.

The operational mode of the 1-D leaky-wave antenna can relatively easily be deduced from further consideration of Fig. 10.1a, b. With reference to Fig. 10.1, the conducting side walls in Fig. 10.12 are aligned parallel to the x-z plane, on either side of the feed slot in Fig. 10.1a and above and below the feed in Fig. 10.1b, with no effect on the internal fields. Consequently, if the HIS in Fig. 10.12 were

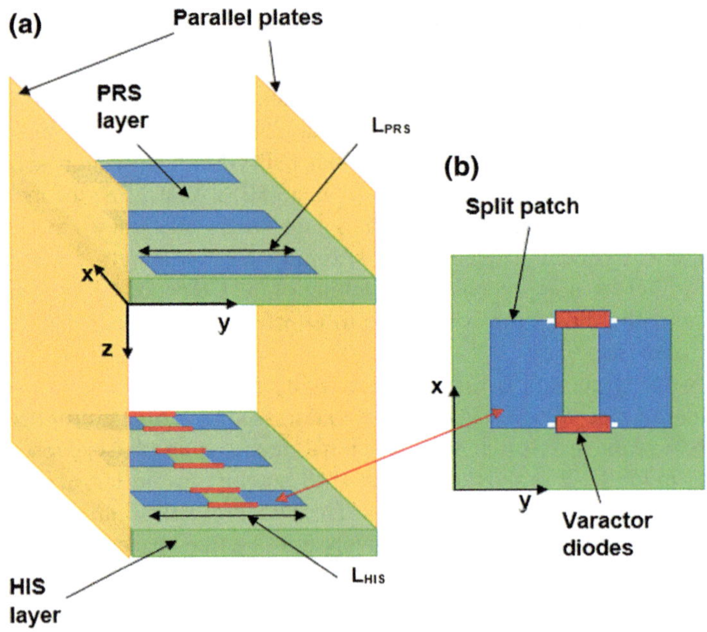

Fig. 10.12 **a** Schematic of 1-D reconfigurable leaky-wave antenna **b** detail of HIS cell (from [17])

replaced by a perfect conductor, then the TE_{10} mode as depicted in Fig. 10.1 would be set up in the 1-D leaky-wave antenna. However, the HIS in its passive state will likely exhibit a high impedance approaching that of an open circuit or a perfect magnetic wall, in which case the TE_{10} mode becomes a $TE_{1/2,0}$ at resonance, with the tangential electric field progressing from a maximum at the HIS to approximately zero at the PRS.

The situation is represented schematically in the accompanying Brillouin diagram (Fig. 10.13). The lowest modes in the parallel plate waveguide, namely the TEM mode and the TE_{10} mode, are required to meet the electromagnetic boundary conditions presented by the periodic inserts which introduces space harmonics and slow-wave modes to the right of the 'light' line (dashed green trace) and leaky-wave modes to the left. The original TEM mode can generally be identified with a slow TM mode (yellow dashed traces in Fig. 10.13), while the empty waveguide TE_{10} mode transmutes into the lowest TE slow mode, retaining the predictable cut-off frequencies associated with the conventional waveguide. So, a leaky wave antenna array supporting a TE mode with a low cut-off frequency at ω_1 will radiate at broadside at this frequency, and at endfire just above ω_4, as dictated by Eq. (10.40). Thus frequency dependent beam scanning proceeds away from broadside toward endfire. Any further increase in frequency moves the operational mode into the slow-wave region where the propagation mechanism becomes a surface wave, and radiation is suppressed. Actually, in a reconfigurable geometry, leaky wave radiation is recoverable at higher frequencies by capacitance changes in the varactor diodes shown in Fig. 10.12. These modify the HIS impedance such that the operational mode is shifted to a higher resonance or cut-off frequency.

But, more importantly, the Brillouin diagram shows how frequency independent beam scanning can be realised with a reconfigurable HIS. At the frequency ω_s for example, if the varactor diode adjustment is capable of shifting the TE mode frequency characteristics to the positions shown, then the propagation coefficient for the leaky wave mode is moved from β_{p1} to β_{p2} to β_{p3}. The beam scans from endfire to broadside. In this 1-D format the antenna is largely omni-directional in the y-z plane (E-plane) above the parallel plates. The main-beam, which is scanned in the H-plane (x-z plane), is therefore 'fan' shaped. In an LWA designed to operate at 5.6 GHz, a beam scan range in the H-plane of 8°–30° from broadside is demonstrated without loss of gain (Fig. 10.14). The influence of HIS changes on the radiation characteristics of similar leaky-wave CRA's are also reported in [11]. To avoid beam splitting (see Fig. 10.9) the 1-D structure is end fed from rectangular waveguide. This represents a complete departure from the normally centre-fed CRA geometry.

That the 1D leaky wave antenna depicted in Fig. 10.12 possesses the range of adjustments required to generate effective H-plane beam scanning is demonstrated in reference [18]. While the evaluation did not include diode switching within the studied device, it is shown that predictable and reliable positioning of the primary beam can be produced at a fixed frequency by simple adjustments in the length of the dipoles forming the PRS (L_{PRS}), and the length of those forming the HIS (L_{HIS}). It is anticipated that these length modifications could be implemented electronically by embedded varactor diodes as illustrated in Fig. 10.13. Moment method

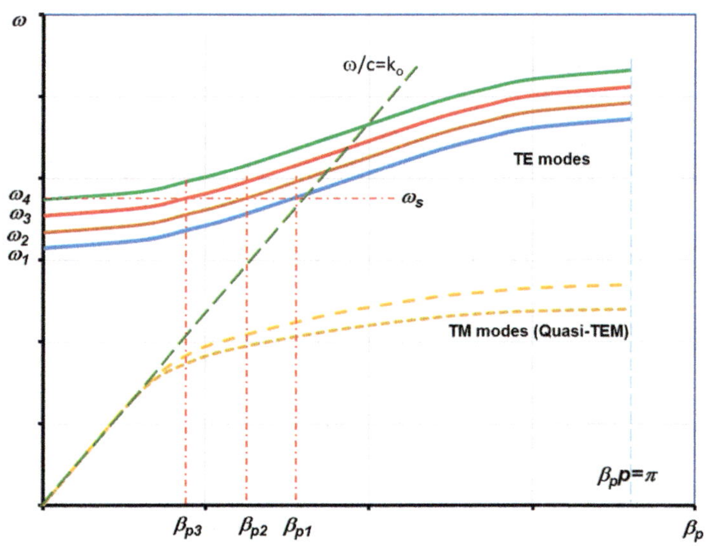

Fig. 10.13 Schematic representation of the Brillouin diagram for a parallel plate waveguide loaded with transverse periodic surfaces

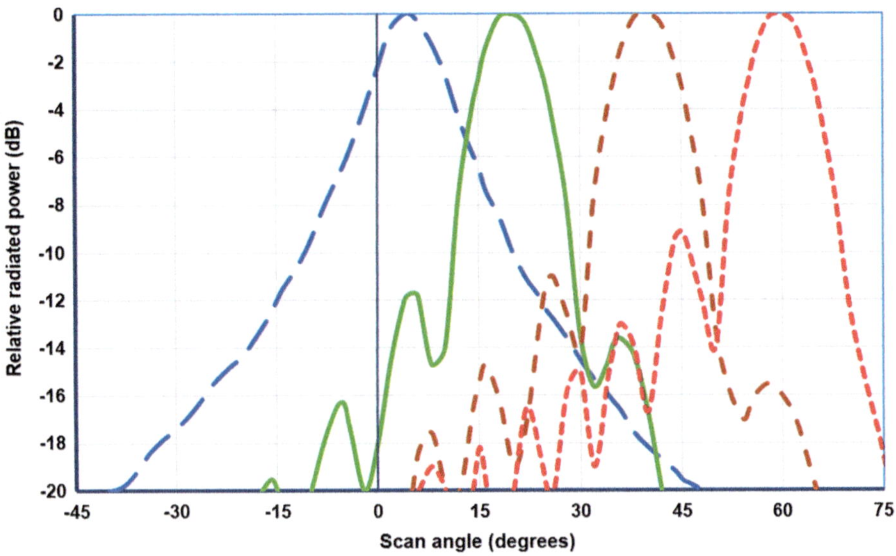

Fig. 10.14 Computed beam scan angles for 15 GHz LWA's design to provide 10° beamwidths at scanned angles of 5° (blue long dashed trace), 20° (green solid trace), 40° (brown dashed trace) and 60° (red short dashed trace)

computations, which have been performed on several versions of the LWA, are summarised in Fig. 10.14. The analysis has been confirmed by measurements [18].

10.8.2 Planar Leaky-Wave Array

Largely in parallel with the linear LWA developments described in the previous section, the introduction of a reconfigurable HIS into a planar antenna geometry has been reported [19–22] with the aim of forming a reflecting surface for use as a reflectarray, or as a compact conformal scanned LWA. However, while competing developments in CRA technology have tended to separate the PRS from the HIS, within the parallel plate structure, and apply reconfigurability by means of the HIS, in the cited articles varactor diode control is incorporated into a periodic structure which notionally combines the two roles. The arrangement is described variously as a tunable textured ground plane, a tunable impedance surface and an artificial magnetic conductor. It is depicted schematically in Fig. 10.15.

The periodic surface comprises a regular array of square patches above a conducting ground plane, not unlike CRA's described earlier in this chapter. However, the inclusion of shorting posts between the patches and the ground plane makes the notion of a parallel plate operating mechanism less sustainable. Nevertheless, references [21, 22] suggest that the Brillouin diagram for this periodic surface exhibits

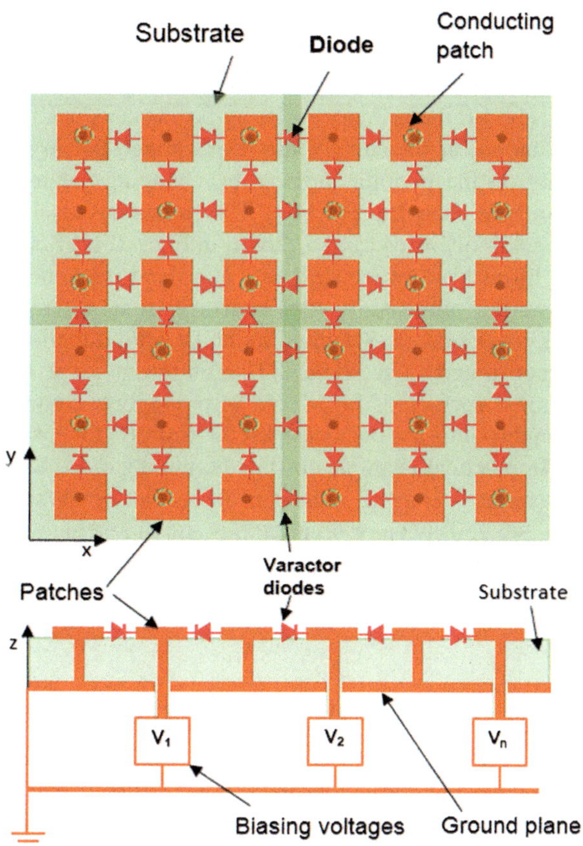

Fig. 10.15 Top and side views of a schematic representation of a planar leaky-wave array antenna formed from a tunable textured surface

similar characteristics to those depicted in Fig. 10.13. The frequency characteristics of this surface (Fig. 10.15) can be constructed by viewing it in LC terms. Capacitance per unit area (C) is furnished by the patch-to-patch gaps while inductance per unit area (L) is created by the metal posts. Consequently, the surface impedance can be equated to that of a matrix of parallel resonant LC circuits. In simplistic terms it can be expressed as [21]:

$$Z_s = \frac{j\omega L}{(1 - \omega^2 LC)} \tag{10.41}$$

Close to resonance, the surface impedance Z_s tends towards infinity and the periodic structure behaves as an HIS, or artificial magnetic conductor, with a zero reflection phase shift. The term textured surface has been introduced to reflect the non-uniformity of the Z_s for this structural format.

As with the linear leaky-wave antenna described in the previous section, scanning can be either frequency dependent, or stepped at fixed frequency by varactor diode adjustment. In the former case, increasing frequency from the dominant TE mode cut-off results in growing β_p (Fig. 10.13) until the chosen curve crosses the 'light' line ($v_p = c$). In this frequency range the TE mode is leaky and according to Eq. (10.40) the primary radiating beam scans from broadside to end-fire. However, the inclusion of varactor diodes as suggested in Fig. 10.15, confers reconfigurability on the periodic surface, by changing patch-to-patch capacitance and hence resonance (see Eq. (10.41)). This manifests itself as repositioning of TE mode cut-off frequencies from ω_1 to ω_4 in Fig. 10.13. So, in much the same manner as for the linear leaky-wave structure, at a fixed frequency ω_s (say) diode adjustment can move the phase shift coefficient β_p from β_{p1} to β_{p2} to β_{p3}, thus scanning the leaky-wave beam. Scanning in the forward direction of the incident mode is clearly demonstrated in reference [21]. Furthermore, it is also suggested there that the textured surface of Fig. 10.15 also supports a backward TE mode which enables scanning into the backward sector from 0° to −50°.

In practical examples of this textured surface leaky-wave antenna, excitation of the surface relies on either an off-set flat notch shaped feed located just above it (by $\sim\lambda_o/50$) [21], or on a flat bow-tie dipole antenna [22]. The scanning is 1-D, in either the x-z or y-z plane depending on the location of the feed. A successful graduation to 2-D scanning had not been reported by the end of 2017.

10.9 The Role of Metamaterials in Compact Slot Arrays

Arguably, without being too disingenuous, there was a period over the past ten to twenty years when the route to virtually guaranteed paper publication on any antenna topic was to include the word 'metamaterial' in the title. Actually many such papers [23–27] did not include metamaterials as strictly defined—namely a material, or medium, displaying both negative permittivity and negative permeability, sometimes referred to as a negative refractive index material (see below). Most of these contributions describe the employment of periodic surfaces in novel antenna forms, using the unique features of such surfaces, in particular frequency selectivity and reconfigurability, as we shall see in the following sections.

10.9.1 Arrays as Passive Periodic Structures

Before we consider the metamaterial role it is important that we fully appreciate the EM characteristics of a regular passive periodic surface from an antenna perspective. As has already been observed, the primary difference, electromagnetically, between an array antenna and passive periodic surface used as a scatterer, such as a frequency selective surface (FSS/metamaterial), is in the nature of the energy source. For the array antenna, as we have seen, each element of the array is an individual source of EM radiation (Chap. 6, Fig. 6.1), whereas the passive FSS is swept by an incoming plane wave (Fig. 10.16) which excites simultaneously the periodically spaced elements. This is, of course, closely analogous to the leaky wave array discussed in the preceding section of this chapter. Hence, as elements of an array, the FSS apertures combine following antenna array rules, thus generating scattered waves in the forward (transmitted) and backward (reflected) directions. In practice the periodic array would normally be planar. However, since the planar array is merely a combination of parallel linear arrays, we can deduce the essence of the behaviour of a passive periodic surface by limiting our attention to a linear geometry, as we have done for the array antenna analysis in Chap. 6.

With reference to Fig. 10.16, it is useful initially to consider the limiting cases of EM wave incidence on the conducting screen containing a periodic array of apertures of arbitrary size, usually ranging from zero to half-wavelength long in their major dimension. For infinitesimal slots the screen is perfectly conducting everywhere, and by Snell's laws leads to a single reflected wave equal in magnitude to the incident wave and reflected in the direction $-\phi$ when the incident angle, to the surface normal, is $+\phi$. The opposite limit, which is rather meaningless practically, involves apertures which are large enough to totally replace the screen. In this case, there is a single transmitted wave at $+\phi$ and no reflected wave.

The more practical large iris limit is the half-wavelength option, in which case a transmitted wave and a reflected wave is formed with essentially equal magnitude. At resonance each iris radiates equally into the forward and backward half-spaces. If we limit our attention to a two dimensional scenario, these re-radiated directions can be deduced from the linear array theory presented in Chap. 6. For the incident

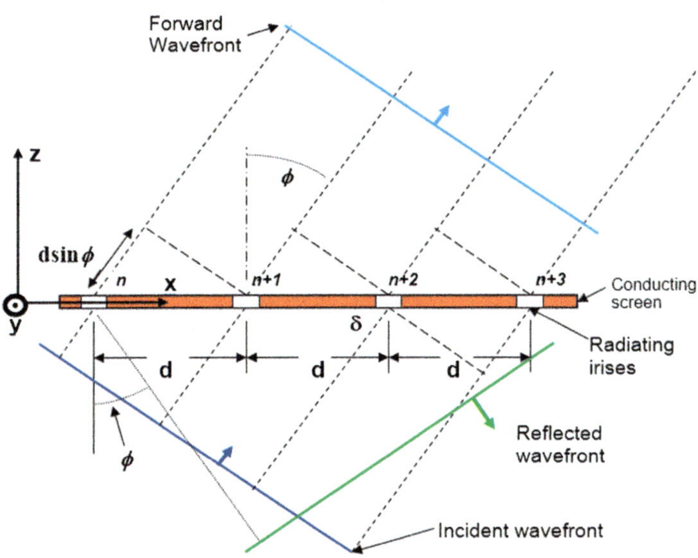

Fig. 10.16 Plane wave scattering at a passive metallic screen supporting a periodic array of apertures separated by distance d

wavefront direction depicted in Fig. 10.16 (dark blue solid trace below the array), the element to element phase shift ψ (Eq. 6.2) is now dictated by the direction of the incoming phase front. For a passive surface there is no independent phase delay or advance and so $\alpha = 0°$. Equation (6.2), therefore, takes the following form for a uniform periodic array excited by an incident plane wave at incident angle ϕ:

$$\psi = k_0 d \sin \phi \tag{10.42}$$

For the n-th element we therefore have:

$$n\psi = nk_0 d \sin \phi \tag{10.43}$$

where n is the element number and d is the element spacing.

For scattering into the forward half-space (z-positive) the periodic surface acts like a uniform array for which (in two dimensions) the forward scattered radiation can be expressed by Eq. (6.8) with ψ given by Eq. (10.42). If $d \leq \lambda_0/2$ the pattern exhibits a single forward scattered beam in the direction ϕ. Perhaps, unsurprisingly, this beam is just the continuation directionally of the incoming plane wave, but attenuated by the apertures as explained above. If $d > \lambda_0/2$ scattering conditions become significantly more complicated due to the onset of grating lobes (see Chap. 6).

Scattering into the backward half-space can be treated in a similar manner, but now scattering is in the $-z$ direction. From a cursory examination of Fig. 10.16, it is clear that the progressive delay $n\psi$ along the array is quite independent of whether or not the scattering is into the forward or backward half-spaces. Consequently, if $d \leq \lambda_0/2$ then a single reflected beam is formed, as dictated by Eq. (6.8), in the backward space, propagating in the $-\phi$ direction (green wavefront). The power density in this reflected beam will be equal to, or less, than that of the incident beam depending of the size of the major dimension of the apertures forming the array. Directionally the system is in accord with Snell's laws.

10.9.2 FSS as a Metamaterial

In preceding sections of this chapter it has generally been assumed that the cavity resonator antennas, which have been chosen for study, have incorporated periodic surfaces formed from scattering elements of relatively regular shape (Fig. 10.17). For passive arrays of such patches or apertures at microwave and millimetre-wave frequencies it is evident, as demonstrated above, that conventional antenna array theory, including Snell's laws comprehensively define the surface characteristics. However the passive surface need not necessarily be composed of regularly shaped scatterers and an indication of the range of possible shapes which could be used in appropriate circumstances are illustrated in Fig. 10.18.

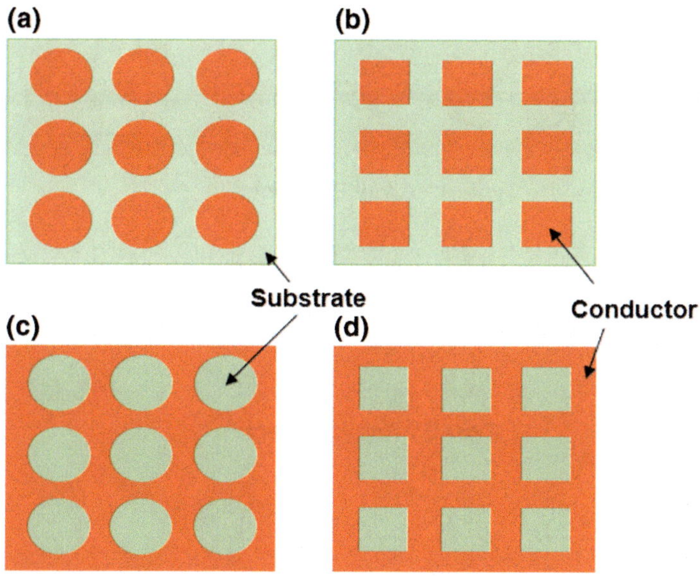

Fig. 10.17 Typical FSS formats for CRA's with (**a**) and (**b**) forming reflecting surfaces while (**c**) and (**d**) present transparent surfaces

Periodic surfaces incorporating these more complex scatterers sometimes get classified as metamaterials. The accepted definition for such a material (usually artificial periodic structures in the form of 2-D surfaces or 3-D volumes) is that macroscopically it displays permittivity and permeability values which are simultaneously negative. The non-simultaneous case is not possible since for any homogeneous isotropic medium we must have from Maxwell's equations:

$$\eta = \sqrt{\frac{\mu_r \mu_0}{\varepsilon_r \varepsilon_0}} \quad \text{and} \quad v = \frac{c}{\sqrt{\mu_r \varepsilon_r}} \tag{10.44}$$

where η is the characteristic impedance and v is the EM velocity in the medium. Such a choice would lead to imaginary values for the characteristic impedance and the EM wave velocity in the medium – a wholly impractical outcome. Interestingly no double negative materials have been discovered in nature. Confusion arises with these materials in the engineering literature because complex periodic structures are often erroneously termed metamaterials even when they do not possess the above property. For example, a relatively simple capacitive/inductive wire-grid combination [23] is classed as a metamaterial, although it is not, while the term is also linked to electromagnetic band gap (EBG) structures, metallodielectric EBG formations [24], metasurfaces and metastructures [25], artificial magnetic conductors (AMC) [2, 26], all of which have roles in compact antenna development [27], but not as metamaterials.

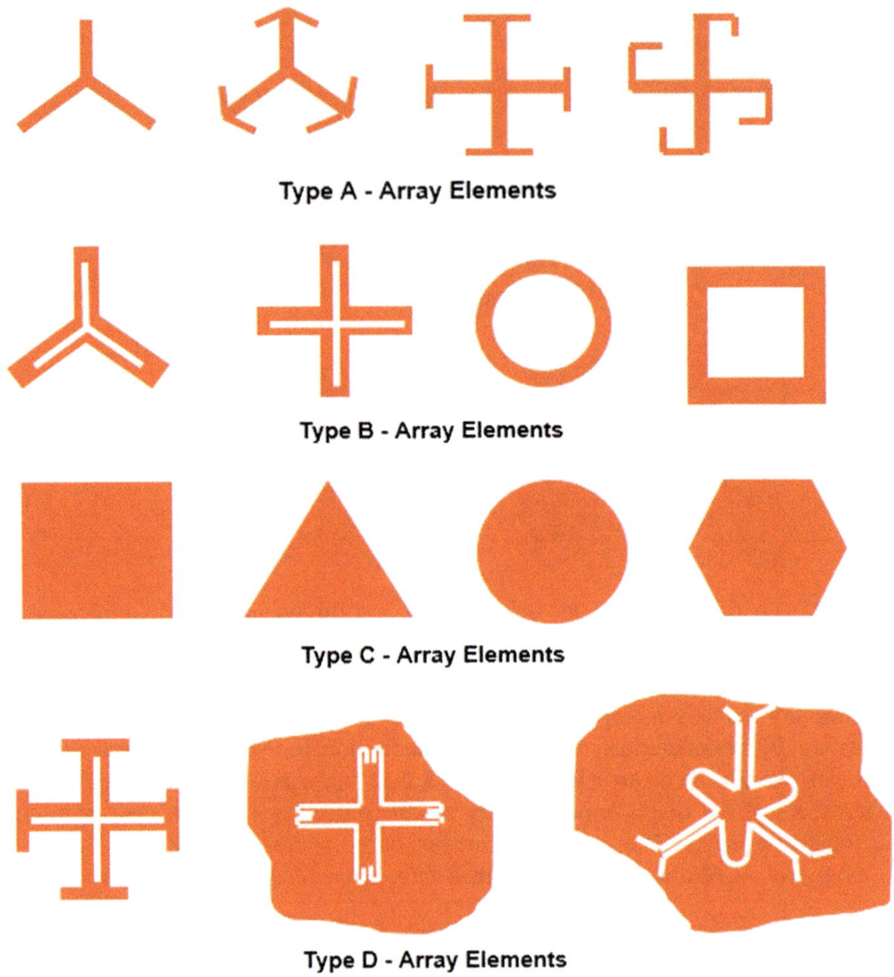

Fig. 10.18 Elemental scatterers for periodic surfaces described as metamaterials fabricated on copper coated planar dielectric substrates. **a** Centre connected, **b** loop types, **c** patch types, **d** combinations

10.9.3 Negative Refractive Index

Unfortunately, considerable scientific 'hype' has been attached to the metamaterial because of its potentially huge engineering significance. From an antenna perspective it has now impacted on the microwave/millimetre-wave scene as unwanted 'baggage' [28–31]. Claims for properties such as

- Negative refractive index n
- Propagation phase of EM wave advances rather than delays

- Evanescent waves can grow rather than attenuate
- For an electromagnetic wave in a metamaterial the electric field vector, the magnetic field vector and the propagation direction vector form a left-handed triplet

all of which seemingly contravene the Snell's Laws and the Maxwell equations, are perplexing. Fortunately, these claims have recently been comprehensively unpicked [32].

The original source of the negative refractive index 'conundrum' is attributed to a 1968 paper [33]. In it the author, V.J. Veselago, explored theoretically the notion of an artificial material possessing both negative permittivity and negative permeability. Referring to Fig. 10.19, we know that if the input medium is air (say), then for a regular dielectric (case (a)):

$$n_{solid} = \frac{v_{air}}{v_{solid}} = \frac{c}{v_{solid}} = \frac{1}{\sqrt{\mu_0 \varepsilon_0}} \cdot \frac{\sqrt{\mu_r \mu_0 \varepsilon_r \varepsilon_0}}{1} = \sqrt{\mu_r \varepsilon_r} \tag{10.45}$$

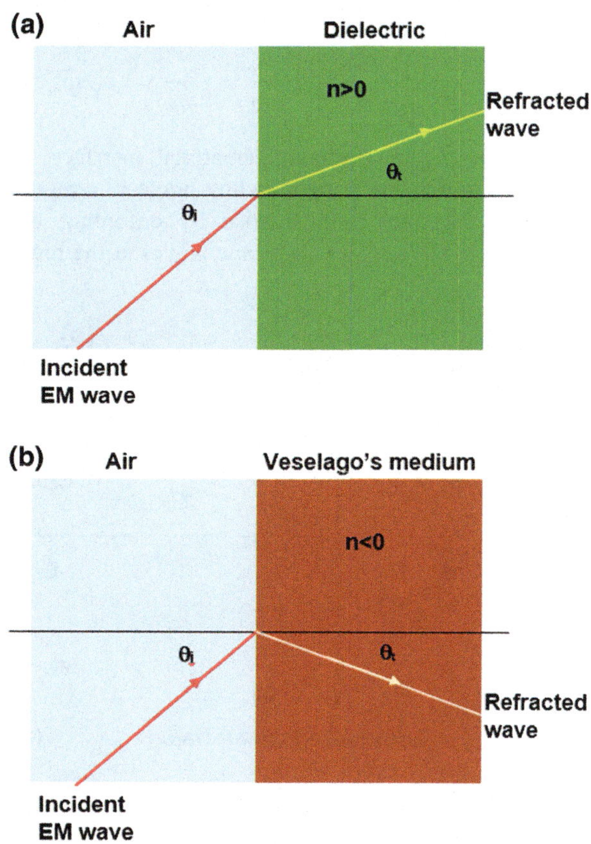

Fig. 10.19 Plane wave refraction at a smooth flat interface between materials with differing refractive index. **a** Regular materials, **b** regular material in contact with metamaterial

For case (b), which replaces the regular solid with a metamaterial we obtain:

$$n_{meta} = \frac{v_{air}}{v_{meta}} = \frac{c}{v_{meta}} = \frac{1}{\sqrt{\mu_0 \varepsilon_0}} \cdot \frac{\sqrt{(-\mu_{rm}\mu_0)(-\varepsilon_{rm}\varepsilon_0)}}{1} = -\sqrt{\mu_{rm}\varepsilon_{rm}} \quad (10.46)$$

if the negative root is chosen as in reference [33]. For the metamaterial the relative permittivity and permeability are respectively $-\varepsilon_{rm}$ and $-\mu_{rm}$. The positive root would imply that Eqs. (10.45) and (10.46) are identical and that EM refraction at the metamaterial surface is no different to the regular material, which is hardly an advance. If substantiated, the electrical ramifications of the negative choice in Eq. (10.46) are not insignificant, and hence the Veselago paper caused quite a 'stir'.

For plane wave refraction at a plane interface between two regular materials, the angle of refraction (θ_t), which is intrinsic to the design of optical devices, is given by Snell's law, namely:

$$\frac{\sin\theta_i}{\sin\theta_t} = \frac{v_1}{v_2} = \frac{c}{v_{solid}} = \sqrt{\mu_r\varepsilon_r} = n_{solid} \quad (10.47)$$

For case (b) we obtain:

$$\frac{\sin\theta_i}{\sin\theta_t} = \frac{v_1}{v_2} = \frac{c}{v_{meta}} = -\sqrt{\mu_{rm}\varepsilon_{rm}} = -n_{meta} \quad (10.48)$$

In other words, for a metamaterial interface, the refraction angle (θ_t) is the opposite sign to the corresponding angle in a regular set up (see Fig. 10.19).

Since the Snell's laws arise by enforcing the electric and magnetic field boundary conditions for the plane waves at the regular interface (Fig. 10.20a) then

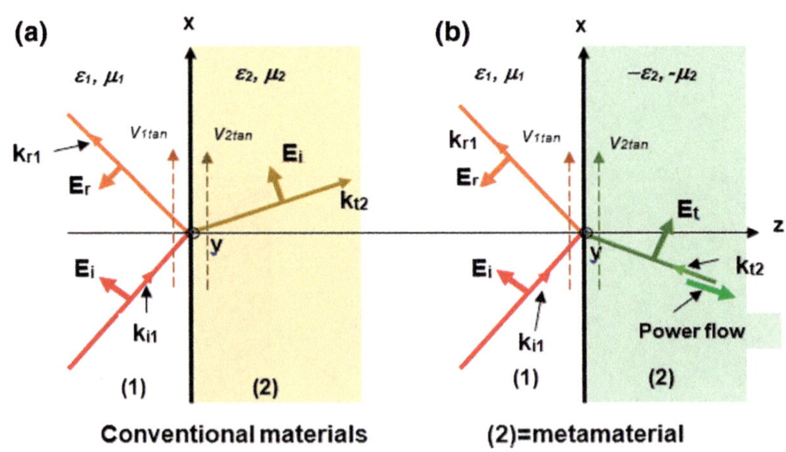

Fig. 10.20 Surface reflection and refraction at (**a**) a conventional interface and (**b**) an interface with a metamaterial. In both cases $n_{1/2} = \pm\sqrt{\frac{\mu_2\varepsilon_2}{\mu_1\varepsilon_1}}$

the same must be true for the metamaterial interface. The situation is illustrated in Fig. 10.20.

Figure 10.20a depicts a mirror flat interface between two conventional materials (1) and (2) for which the permittivities are respectively ε_1 and ε_2, and the permeabilities are respectively μ_1 and μ_2. Satisfying the electromagnetic field boundary conditions at the interface at z = 0 leads to Snell's laws (Appendix A), which are represented in vector form in the figure. The directional vectors $\mathbf{k}_{i1}$, $\mathbf{k}_{r1}$ and $\mathbf{k}_{t2}$ must obey the specified angular relationships while the field vectors $\mathbf{E}_i$, $\mathbf{H}_i$, $\mathbf{E}_r$, $\mathbf{H}_r$, $\mathbf{E}_t$ **and** $\mathbf{H}_t$ automatically satisfy the boundary conditions. Power flow requirements dictate that for the incident, reflected and transmitted plane waves **E**, **H** and **k** obey the clockwise rule (Appendix A) with the Poynting vector, and hence power flow, directed along the relevant **k** vector. Note that the boundary conditions must be met everywhere over the z = 0 interface at any given instant in time. This is only possible if

$$v_{1\,\tan} = v_{2\,\tan} \tag{10.49}$$

It is not difficult to show that adherence to the Snell's laws ensures that this is the case.

So, does case (b) in Fig. 10.20 also meet all of these fundamental physical constraints? The negative refractive index for the metamaterial requires that the refractive wave is directed along $\mathbf{k}_{t2}$ at an angle $-\theta_t$ below the z-axis. However, in addition a negative refraction index implies that, assuming for simplicity that region (1) is air or vacuum:

$$k_{t2} = \frac{\omega}{c}\sqrt{\mu_{mr}\varepsilon_{mr}} = -\frac{\omega n_{1/2}}{c} \tag{10.50}$$

The significance of Eq. (10.50) is that $\mathbf{k}_{t2}$ is directed towards the interface. So the field boundary conditions are in accord with case (a) with the tangentially directed wave components on each side of the interface meeting the dictates of Eq. (10.50). However, while the boundary conditions are seemingly satisfied for the negative index metamaterial, power flow directions are puzzling. The directions of $\mathbf{E}_t$ and $\mathbf{H}_t$ in case (b) dictate that the Poynting vector $\mathbf{S} = \mathbf{E} \times \mathbf{H}^*$ and hence power flow is in the direction opposite to $\mathbf{k}_{t2}$. While for power flow normal to the interface energy conservation is satisfied, for tangential power flow in the $z = 0$ plane the power flow in the regular medium (1) opposes the flow in the metamaterial (2), which suggests that the negative index material is not physically possible. This embodiment of the refracted wave has been compared to a backward wave which is a possible solution to the Maxwell equations in periodic structures. But in backward wave modes, such as those that occur in backward wave microwave oscillators, for example, power flow everywhere within the modal field structure propagates with the same group velocity, which is in opposition to the phase velocity direction – hence the description 'backward wave'. The electromagnetic field solution in case (b) of Fig. 10.20 contravenes this behaviour.

10.9.4 The Flat Lens

The 'coup de gras' for artificial metamaterials displaying negative refractive index is arguably the 'flat lens' proposed by Veselago [33], although many other practical anomalies have been explored and discarded [32]. These intensive studies suggest that most metamaterial claims are identifiable with established properties of periodic structures.

The Veselago flat lens is viewed in conjunction with a conventional lens performing the same focusing role in Fig. 10.21. For the conventional lens the focusing mechanism is easily explained. A source at point (A) in Fig. 10.21a is imaged at point (B) by ensuring the electrical path lengths between (A) and (B) are the same, balancing electrical path lengths in the air with path lengths in the lens of lower phase velocity. In addition the lens surface is parabolically shaped to ensure that the glass/air refractive exit angles match the entry angles. The conventional lens is predominantly used at optical frequencies and consequently it is usually made of hard transparent material such as glass with $\mu_r = 1$. For the flat lens this restriction does not necessarily apply. It is generally imagined to operate at below light frequencies as a 3-D periodic structure for which μ_r and ε_r are negative and greater than unity. The operational principles of the flat lens are not too different to those of the conventional lens. The electrical path lengths from the source at (A) to the image at (B) must be the same for all rays, while the negative refraction angles must be arranged to secure a focused image on the opposite side of the lens as suggested in Fig. 10.21b. If the path between (A) and the secondary focus at (C) is examined for both an edge ray, and a straight through ray, then to arrive at point

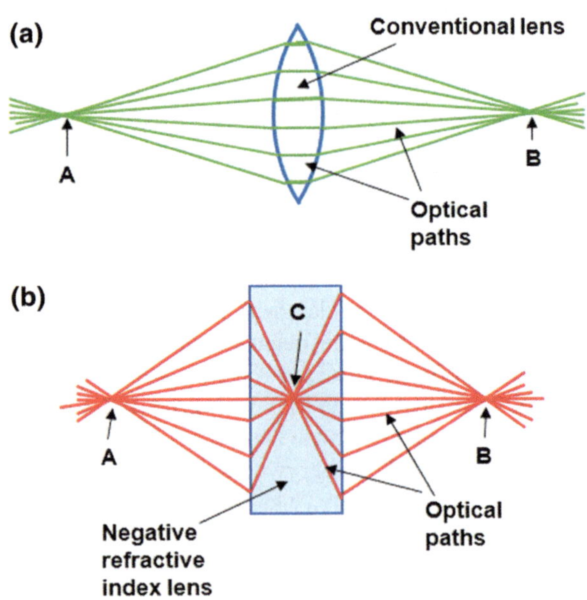

Fig. 10.21 Conventional optical lens compared with 'flat' lens. For **a** glass lens in air ($\mu_r = 1$, $\varepsilon_r = 1$) and for convex lens ($\mu_r = 1$, $\varepsilon_r > 1$) and for **b** the flat lens in air it has $\mu_r < 0$, $\varepsilon_r < 0$

(C) in phase requires that the additional phase delay in the air for the edge ray must be negated by an extra phase *advance* in the metamaterial. A similar requirement applies to all other rays. Equations (10.48) and (10.50) suggest that this is possible for phase. The difficulty with this argument is that it omits the time element. Phase coherence at (C), and hence at (B), must occur at the same instant of time, at all times. This implies that the additional time delays in the air for the longer paths must be balanced by time *advances* in the metamaterial, and this is synonymous with negative time or time reversal! In real physical structures and systems, at least in this corner of the universe, cause and effect is sacrosanct, and hence the flat metamaterial lens is not possible.

10.10 Chapter Summary

The Fabry-Perot resonator antenna or cavity resonator antenna in its most basic form, is shown to be a leaky parallel plate waveguide operated close to the cut-off of the fundamental TE mode. It is examined initially from a plane wave ray perspective and to advance the analysis the periodic waveguide wall is modelled as a uniform and partially reflective surface (PRS). This leads to simple but actually rather useful estimates of directivity and in particular how it relates to the reflectivity of the periodic surface. This type of antenna is usually realised as a copper clad microwave substrate with one surface made periodic by the presence of a suitably designed array of radiating patches or radiating slots.

More generally, the CRA is shown to represent a class of compact planar antenna which includes the partially reflective surface (PRS) antenna, the Fabry-Perot cavity antenna, and the leaky wave CRA. All of these forms can be modelled by the reciprocal periodic moment method which is fully developed in Sect. 10.4. Supported by evidence from the literature, modelling demonstrates that high gain and/or directivity is available from a PRS antenna, but with very restricted bandwidth. Inconvenient beam splitting and scanning occur as the operating frequency is raised above cut-off. Nevertheless bandwidth enhancement is an active pursuit.

Splitting of the primary radiated beam also occurs with centre fed leaky-wave antennas if the applied frequency is above the fundamental mode cut-off value. Consequently, as with the PRS antenna, beam scanning using frequency is problematic. However, beam scanning remains a goal of CRA development and it is shown how this can be achieved, in 1-D embodiments, by using end-fed arrays incorporating one or more additional reconfigurable periodic layers in the form of high impedance surfaces (HIS). The planar leaky-wave antenna incorporating a textured surface has not yet demonstrated 2-D scanning although it possesses the potential to do so. The conundrum then remains of applying leaky-wave methods to secure 2-D beam scanning from a compact planar CRA. To do so will require novel feed arrangements which probably means abandonment of conventional centre fed geometries associated with the CRA. With 1-D scanning the complex diode switching algorithm, which is required to control the multi-diode network, becomes much more multi-layered and difficult where 2-D scanning in arbitrary directions is demanded.

Complex periodic surfaces and structures, for which element sizes and spacings are much less than the wavelength of any penetrating EM waves, are increasingly being introduced into the literature. They are usually correctly referred to as artificial dielectrics. Unfortunately, in some contributions they have simultaneously been described as metamaterials which is also the generic term for so called negative index materials. As is indicated in this chapter the latter materials have never been realised in practice. It is to be hoped that as long as new uses of periodic structures are clearly linked to their periodic behaviour, the metamaterial description will gradually lose its power to confuse.

References

1. G. von Trentini, Partially reflective sheet arrays. IRE Trans. Antennas Propag. **4**(4), 666–671 (1956)
2. A.P. Feresidis, G. Goussetis, S. Wang, J.C. Vardaxoglou, Artificial magnetic conductor surfaces and their application to low-profile high-gain planar antennas. IEEE Trans. Antennas Propag. **53**(1), 209–215 (2005)
3. N. Guerin, S. Enoch, G. Tayeb, P. Sabouroux, P. Vincent, H. Legay, A metallic Fabry-Perot directive antenna. IEEE Trans. Antennas Propag. **54**(1), 220–224 (2006)
4. A. Foroozesh, L. Shafai, Investigation into the effects of the patch-type FSS superstrate on the high-gain cavity resonance antenna design. IEEE Trans. Antennas Propag. **58**(2), 258–270 (2010)
5. T. Zhao, D.R. Jackson, J.T. Williams, H.-Y.D. Yang, A.A. Oliner, 2-D periodic leaky-wave antennas-part I: metal patch design. IEEE Trans. Antennas Propag. **53**(11), 3505–3514 (2005)
6. T. Zhao, D.R. Jackson, J.T. Williams, 2-D periodic leaky-wave antennas-part II: slot design. IEEE Trans. Antennas Propag. **53**(11), 3515–3524 (2005)
7. A.P. Feresidis, J.C. Vardaxoglou, High gain planar antenna using optimised partially reflective surfaces. IEEE Proc. Microwaves Antennas Propag. **148**(6), 345–350 (2001)
8. C. Mateo-Segura, G. Goussetis, A.P. Feresidis, Sub-wavelength profile 2-D leaky-wave antennas with two periodic layers. IEEE Trans. Antennas Propag. **59**(2), 416–424 (2011)
9. C. Mateo-Segura, A.P. Feresidis, G. Goussetis, Bandwidth enhancement of 2-D leaky-wave antennas using double-layer periodic surfaces. IEEE Trans. Antennas Propag. **62**(2), 586–593 (2014)
10. T. Itoh, W. Menzel, A full-wave analysis method for open microstrip structures. IEEE Trans. Antennas Propag. **AP-29**(1), 63–68 (1981)
11. L. Zhou, H. Li, Y. Qin, Z. Wei, C.T. Chan, Directive emissions from subwavelength metamaterial-based cavities. Appl. Phys. Lett. **86**, 101101-1–101101-3 (2005)
12. A.R. Weily, T.S. Bird, Y.J. Guo, A reconfigurable high–gain partially reflecting surface antenna. IEEE Trans. Antennas Propag. **56**(11), 3382–3390 (2008)
13. T. Debogovic, J. Perruisseau-Carrier, J. Bartolic, Partially reflective surface antenna with dynamic beamwidth control. IEEE Antennas Wirel. Propag. Lett. **9**, 1157–1160 (2010)
14. T. Debogovic, J. Bartolic, J. Perruisseau-Carrier, Array-fed partially reflective surface antenna with dynamic beamwidth control and beam-steering. Paper presented at the 6th European Conference on Antennas and Propagation (EuCAP 2012, Prague, 2012)

15. T. Debogovic, J. Perruisseau-Carrier, Array-fed partially reflective surface antenna with independent scanning and beamwidth dynamic control. IEEE Trans. Antennas Propag. **62**(1), 446–449 (2015)
16. L.-Y. Ji, Y. Jay Guo, P.-Y. Qin, S.-X. Gong, R. Mittra, A reconfigurable partially reflective surface (PRS) antenna for beam steering. IEEE Trans. Antennas Propag. **63**(6), 2387–2395 (2015)
17. R. Guzman-Quirós, J.L. Gómez-Tornero, A.R. Weily, Y.J. Guo, Electronically steerable 1-D Fabry-Perot leaky-wave antenna employing a tunable high impedance surface. IEEE Trans. Antennas Propag. **60**(11), 5046–5055 (2012)
18. M. Garcia-Vigueras, J.L. Gomez-Tornero, G. Goussetis, A.R. Weily, Y.J. Guo, 1D-leaky wave antenna employing parallel-plate waveguide loaded with a PRS and HIS. IEEE Trans. Antennas Propag. **59**(10), 3687–3694 (2011)
19. D. Sievenpiper, J. Schaffner, J.J. Lee, S. Livingston, A steerable leaky wave antenna using a tunable impedance ground plane. IEEE Antennas Wirel. Propag. Lett. **1**, 179–182 (2002)
20. D. Sievenpiper, J. Schaffner, H. Jae Song, R.Y. Loo, G. Tangonan, Two-dimensional beam steering using an electrically tunable impedance surface. IEEE Trans. Antennas Propag. **51** (10), 2713–2722 (2003)
21. D. Sievenpiper, Forward and backward leaky wave radiation with large effective aperture from an electronically tunable textured surface. IEEE Trans. Antennas Propag. **53**(1), 236–247 (2005)
22. F. Costa, A. Monorchio, S. Talarico, F.M. Valeri, An active high-impedance surface for low-profile tunable and steerable antennas. IEEE Antennas Wirel. Propag. Lett. **7**, 676–680 (2008)
23. A. Ourir, S.N. Burokur, A. de Lustrac, Phase-varying metamaterial for compact steerable directive antennas. Electron. Lett. **43**(9), 493–494 (2007)
24. Y.R. Lee, A. Charaya, D.S. Lockyer, J.C. Vardaxoglou, Dipole and tripole metallodielectric photonic bandgap (MPGB) structures for microwave filter and antenna applications. Proc. IEEE Optoelectron. **127**(6), 395–400 (2000)
25. H.L. Zhu, S.W. Cheung, X.H. Liu, T.I. Yuk, Design of polarisation reconfigurable antenna using metasurface. IEEE Trans. Antennas Propag. **62**(6), 2891–2898 (2014)
26. Y. Zhang, J. von Hagen, M. Younis, C. Fischer, W. Wiesbeck, Planar artificial magnetic conductors and patch antennas. IEEE Trans. Antennas Propag. **51**(10), 2704–2712 (2003). (Special Issue on Metamaterials)
27. S. Enoch, G. Tayeb, P. Sabouroux, N. Guerin, P. Vincent, A metamaterial for directive emissions. Phys. Rev. Lett. **89**(21), 2139–2142 (2002)
28. J.B. Pendry, Negative refraction makes perfect lens. Phys. Rev. Lett. **85**(18), 3966–3969 (2000)
29. J.B. Pendry, A.J. Holden, W.J. Stewart, I. Young, Extremely low frequency plasmons in metallic mesostructure. Phys. Rev. Lett. **76**(20), 4773–4776 (1996)
30. J.B. Pendry, A.J. Holden, D.J. Robbins, W.J. Stewart, Magnetism from conductors and enhanced non-linear phenomena. IEEE Trans. Microwave Theor. Tech. **47**(11), 2075–2084 (1999)
31. R.A. Shelby, D.R. Smith, S. Schultz, Experimental verification of a negative index of refraction. Science **292**, 77–79 (2001)
32. B.A. Munk, *Metamaterials: Critique and Alternatives* (Wiley, New York, 2009)
33. V.G. Veselago, The electrodynamics of substance with simultaneously negative values of ε and μ. Sov. Phys. Usp. **10**(4), 509–524 (1968)

Chapter 11
Retro-directive Compact Array Antennas

11.1 Retro-directive Tracking—Basic Concept

There are essentially three routes to tracking a moving signal source, such as a satellite, with array technology. Firstly, in a full blown phased array antenna with a transmit/receive unit behind each element of the array, feeding into a high speed digital signal processor. In this case the antenna radiation pattern can be configured almost instantaneously into any form required. This is the expensive option. An undoubtedly cheaper alternative is to use a form of monopulse tracking commonly seen in radar and in parabolic reflector tracking systems [1, 2]. But this is also technically challenging for commercial applications aimed at satellite/mobile/wireless communications.

A third form of tracking, which has been given impetus by recent reconfigurable developments in array antennas, is provided by the 'retro-directive' array [3, 4]. The principle is not new. Retro-direction, whereby an incoming radar signal, on striking a target is returned back in the direction from which it came, is achieved (see Chap. 1) by 'corner reflectors', often seen on the masts of yachts [5]. The device increases the radar cross-section of the target thereby enhancing its detectability against the background of the sea. Arguably the retro-direction in an array offers the most cost effective option for emerging wireless systems.

The retro-directive array topology can be further sub-divided into two basic classifications, depending on whether or not the front-end transmit/receive devices are 'mixer' based, or 'phase-lock-loop (PPL)' based. The essence of the two alternatives is outlined below.

A. J. Sangster, *Compact Slot Array Antennas for Wireless Communications*, Signals and Communication Technology, https://doi.org/10.1007/978-3-030-01753-8_11

11.2 Retrodirective Implementation Through Mixing

The existence of retro-directive antenna action lies in the nature of travelling waves in free space and on how this behaviour is expressed mathematically. An electromagnetic wave travelling away from a source in the direction r varies with time and with distance from the source. At a single frequency ω the variation will be sinusoidal and this can be expressed as:

$$E = A \exp(j\omega t) \exp(-jk_0 r) \tag{1}$$

where A is an arbitrary amplitude parameter representing the source signal magnitude and $k_0 = \omega/c$, where c is the speed of light. The above equation can be re-written in the form:

$$E = A \exp j\omega(t - \frac{r}{c}) \tag{2}$$

which has the functional form of a travelling wave moving in the direction r at velocity c. Now for a wave of the same frequency ω, travelling in the opposite direction $(-r)$ with amplitude A' the electric field can be expressed as:

$$E' = A' \exp(j\omega t) \exp(jk_0 r) = A' \exp j\omega(t + \frac{r}{c}) \tag{3}$$

and E' is the retro-directive version of E. The fundamental difference between E' and E lies in the $exp(jk_0 r)$ factor, which is the complex conjugate of $exp(-jk_0 r)$. In communication terms this implies that in order to transmit an electromagnetic wave back toward the source of the received wave, the phase of the incoming wave must be conjugated before retransmission.

Interestingly it is a property of signal mixing that when a received signal is mixed with the output of a reference local oscillator at exactly twice the frequency of the incoming signal (or the carrier frequency if modulated) the resultant difference signal is the complex conjugate of the original [3–10]. This is illustrated in Fig. 11.1 where a conjugating mixer circuit is attached to one element of a supposed array via a microwave circulator (directional filter). All other elements would be backed by exactly similar circuits all fed from the common local oscillator. It is easy to see that the transmitted (difference) signal is the conjugate of the incoming wave except for the fixed phase ϕ_0 introduced by the local oscillator. This phase is not relevant to the operation of the circuit. The phase ϕ occurs when the incoming wave direction is at an angle θ to the array normal. It is basically the phase delay of the received signal for two adjacent elements separated by a distance d, and is given by:

$$\phi = k_0 d \sin\theta \tag{4}$$

So phase conjugation occurs automatically for any angle of arrival of the received wave at the retrodirective array. A significant disadvantage of this circuit

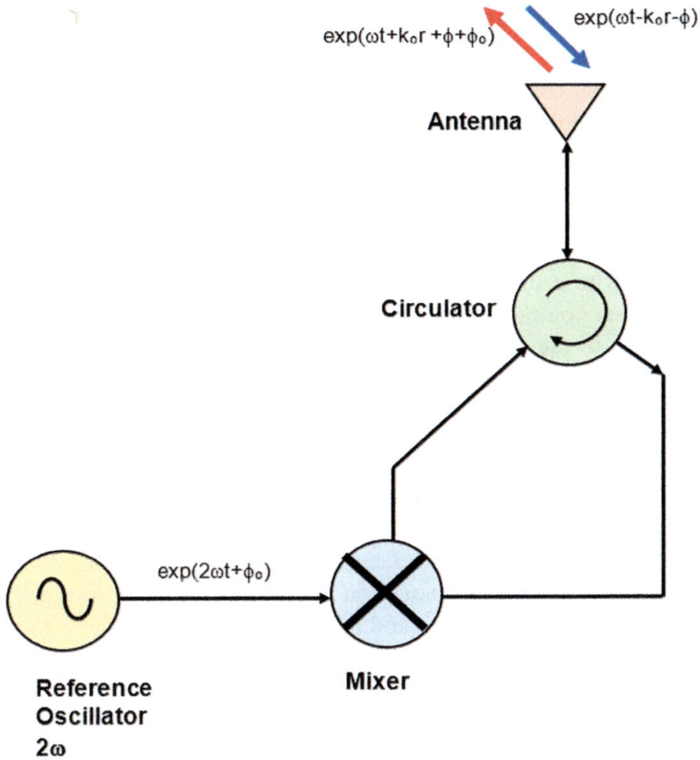

Fig. 11.1 Phase conjugation through mixing

arrangement, apart from the requirement for a local oscillator at double the operating frequency, is the presence of the circulator. This has to be a high quality device, to minimise tracking errors, and at microwave frequencies such a device is very expensive. Furthermore, this cost is multiplied by the number of elements in the array.

One way around the circulator difficulty is to provide separate transmit and receive arrays, as advocated by Myamoto [9, 10] which allows the circulator to be replaced by a ring hybrid coupler, or rat-race. However, the requirement to duplicate the transmitter and receiver array hardware is clearly a disadvantage. The Myamoto solution requires up and down links at different frequencies which, in turn, demands frequency doubling and frequency shifting circuitry in the device being tracked, such as a satellite. This could present unwelcome hurdles in certain applications. Such difficulties have led to the development of phase locked loop (PLL) based conjugating circuits originally proposed by Ghose [4], and later pursued in Refs. [11–16]. These are discussed in the next section.

11.3 Conjugation Employing PLL's

The earliest example of a conjugating circuit incorporating a phase comparison technique is attributable to Chernoff [6], with the basic circuit shown in Fig. 11.2. The voltage controlled oscillator (VCO), which is adjusted by a voltage signal emanating from the phase comparator (at B) and passed through a loop filter and a voltage amplifier. The VCO supplies the transmit antenna with a signal of magnitude F_1 and frequency ω_1, and a presumed phase shift of ϕ. Part of the VCO output is mixed with a signal from the receive antenna (F_2 with phase shift Φ) and the summed output component is split at point A. From there a component (divided by n) is fed back to the phase comparator while the other component is demodulated to extract data. That this circuit produces phase conjugation is easily demonstrated as follows. At position A, the up-converted RF signal can be expresses as

$$V = V_A \exp(\omega_1 t + \phi + \omega_2 t + \Phi) \tag{5}$$

where V_A is an arbitrary amplitude factor, and ϕ and Φ are transmit and receive phases associated with a main beam at an angle relative to the array normal. If we assume (Fig. 11.2) that for transmit and receive magnitudes $F_1 \approx F_2$ then when the loop is phase locked we must have $\Phi = -\phi$, i.e. phase conjugation. Note that this phase condition for phase conjugation requires that phase locking is performed on the ***sum*** signal from the mixer at A. Physically, this means that at phase lock, the signal transmitted by the array is forced to return in the direction of the incoming received signal—or expressed another way the primary beam of the array is forced to be pointing at the source. The second mixer at the lower left of the diagram

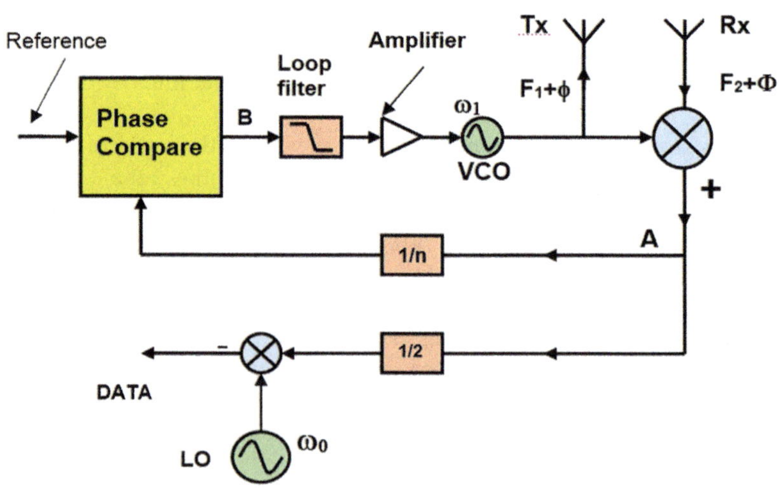

Fig. 11.2 Phase conjugation circuit employing phase comparison

provides heterodyning of the received signal at A with a local oscillator signal at frequency ω_0, to produce an intermediate frequency (IF) signal modulated by the required system data.

A marginally more efficient circuit is presented in Fig. 11.3 [11–14]. In the version described in Ref. [12], the circuit is implemented around a 200 kHz phase detector shown on the left of the circuit. The input from a 10 MHz crystal stabilised oscillator feeds one port of the phase detector once divided by 50, while at the other port a signal emanating from a receive array element is applied. The received signal at 1050 MHz is mixed with the transmit signal from the VCO and the ***sum*** output at 1990 MHz is then divided by 9950. The voltage at the output of the detector at B, essentially at DC, is proportional to the detected phase offset. It is added to an externally applied DC offset voltage to provide a pre-setting voltage for the voltage controlled oscillator (VCO) at $F_1 = 0.94$ GHz. The 'error' voltage from the phase detector provides continual adjustment to the VCO in a direction which ensures that the 200 kHz signal entering the phase detector exhibits zero phase offset. Part of the VCO output passes through a buffer amplifier to the transmit radiator, while part enters a following mixer. As indicated above, there it is mixed with the received signal, and the resultant sum signal is directed through a bandpass filter back to the phase locked loop (PLL). Note that the ***difference*** signal from the mixer provides (see [12]) a constant phase IF signal. These signals from all other receive elements of the array antenna are then added before demodulation. This obviates the need for a second mixer and local oscillator as in the Chernoff circuit.

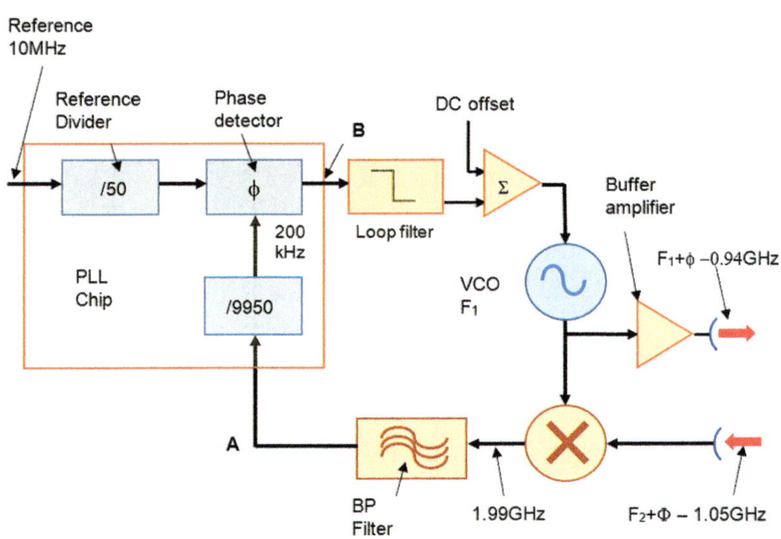

Fig. 11.3 Phase conjugation using PLL chip

11.4 Phase Conjugation Using IQ Processing

The phase conjugation scheme represented by Fig. 11.3 assumes that a double array exists with the receive array separate from the transmit array. This configuration, as already observed, could obviously be a disadvantage in many situations, as could the employment of differing transmit and receive frequencies. Furthermore at microwave frequencies the very high sum frequency fed back to the PLL requires a costly BPF and a high degree of division. An 'in-phase/quadrature phase' (IQ) modification has been proposed [15, 16] to circumvent these difficulties and is shown diagrammatically in Fig. 11.4. The IQ modulator on the right of the circuit is pivotal to the approach. The circuit as a whole (one to each array element) conjugates the signal (ω_{RF}) received at the left port, from the appropriate array element, before dispatching it at the right (ω_{out}) to the transmit radiating element. As will be demonstrated below, the transmit and receive elements need not be separate radiators, as Fig. 11.4 seems to suggest. The IQ modulator, in analogue form, primarily comprises a pair of up-converting mixers, a 90° phase shift and a summing junction.

Phase conjugation is procured as follows. A received signal (ω_{RF}) at frequency 2400.5 MHz (say), and phase delay ϕ, is mixed with a 2400 MHz signal from the frequency controlled local oscillator (VCO1) to form a 0.5 MHz IF signal at A, i.e.

$$V = V_A \cos(\omega_{RF}t - \omega_{LO1}t + \phi) \tag{11.6}$$

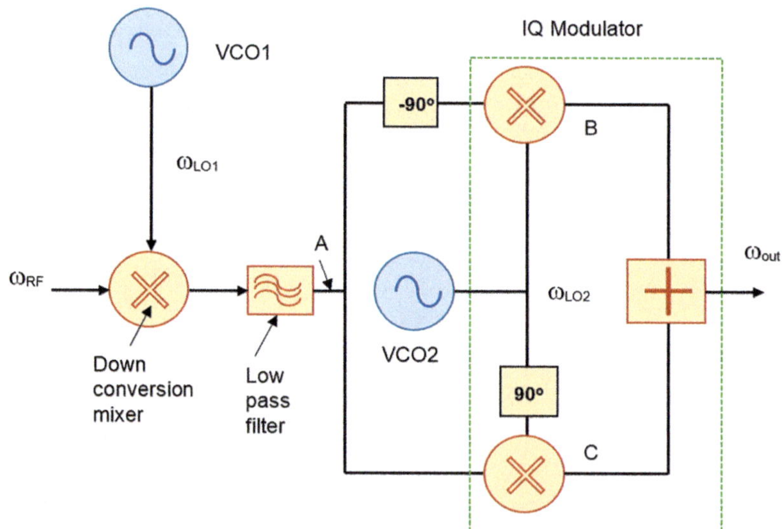

Fig. 11.4 Phase conjugation using I and Q processing

where V_A is an arbitrary complex magnitude. At A this signal is split equally into two separate paths. In one of these paths a −90° phase shift is applied before the signals enter the IQ modulator. There they are mixed with a second local oscillator (ω_{LO2}) signal at 2410 MHz say so that at point B we have:

$$\begin{aligned} V_B = {} & V_1 \cos(\omega_{RF}t - \omega_{LO1}t + \omega_{LO2}t + \phi - 90^\circ) \\ & + V_1 \cos(-\omega_{RF}t + \omega_{LO1}t + \omega_{LO2}t - \phi + 90^\circ) \end{aligned} \tag{7}$$

and at C:

$$\begin{aligned} V_C = {} & V_1 \cos(\omega_{RF}t - \omega_{LO1}t + \omega_{LO2}t + \phi + 90^\circ) \\ & + V_1 \cos(-\omega_{RF}t + \omega_{LO1}t + \omega_{LO2}t - \phi + 90^\circ) \end{aligned} \tag{8}$$

When summed, the resultant signal at ω_{out} has the form:

$$V_{out} = V_2 \cos(\omega_{LO1}t + \omega_{LO2}t - \omega_{RF}t - \phi + 90^\circ) \tag{11.9}$$

which is phase conjugated relative to the input as required. The output frequency is equal to the input frequency if $\omega_{LO1} + \omega_{LO2} = 2\omega_{RF}$.

11.5 Retro-directive Array Implementation

It is clear from the literature that the performance of retro-directive array antennas is measurably susceptible to direct leakage between the transmission and receiver sections of the system, particularly if transmit and receive operations are performed at the same frequency. The main source of this leakage is mutual coupling between closely spaced array elements, and to avoid this, the transmit and receive functions are often totally separated, by adopting two distinct arrays [6, 12, 16]. Leakage between the transmission and reception circuits both reduces the sensitivity of the system and diminishes tracking accuracy.

It is not difficult to appreciate that doubling up on array hardware to counteract leakage seems like technological ‘over kill’. It certainly adds significantly to build costs. The usual way around this is to employ a duplexer behind each array element which allows the transmit and receive channels to use the same antenna element. This is suggested in Fig. 11.1 where the duplexer takes the form of a circulator. Unfortunately, as indicated previously, duplexers, which display high enough isolation between transmit and receive ports to permit reliable and accurate retro-directive behaviour in an array, tend to be very expensive.

Given advances in planar microwave circuit fabrication techniques, a potentially much more cost effective solution is to employ array radiator elements which have dual-mode capability [17]. An example is depicted in Fig. 11.5, where the retrodirective array is formed from square patches capable of receiving/transmitting vertically or horizontally polarised electromagnetic waves. For this particular dual-mode antenna array the transmit circuit port in each phase conjugating circuit (PCC) is linked to the vertical polarisation connector at the patch, while the receive port is linked to the horizontal polarisation connector. Isolation between the vertical polarisation and horizontal polarisation ports is required to be of the order of 30 dB [16, 17]. Radiators which can potentially provide high isolation between orthogonal modes of operation are discussed below.

For some communications applications, for example between a ground station, or a moving vehicle on the surface of the earth, and a potentially gyrating satellite, it is almost inevitable that severe signal fading would occur if this dual-mode linearly polarised technique were adopted in a retrodirective tracking array. Generally the answer to the problem of sustaining a channel of communication between platforms, which could be rotating and moving, involves the introduction of circularly polarised transmission and reception channels. However, circularly polarised patch antennas, which are also dual mode, are less well established than linear alternatives, as the patent situation [18–21] and the literature [22–30] attests. Thus, the adoption of retrodirective technology, if it is to be commercially viable, calls for the use of dual-polarised array elements to eliminate the need for separate transmit and receive arrays while maintaining signal fidelity in a mobile communications system. Unfortunately, in microstrip technology, this in turn can lead to rather complex feed arrangements, particularly when the radiators are incorporated into arrays of four or more elements, and particularly when beam-switching is involved [3, 4]. As has been noted earlier, it is clear from the literature that circularly polarised slot or patch antennas, which are also dual-mode, are less well established than their linear alternatives. Nevertheless, microstrip transmission line technology is the preferred platform for mobile communications. This limits possible antenna array options to three distinct alternatives, as is outlined below.

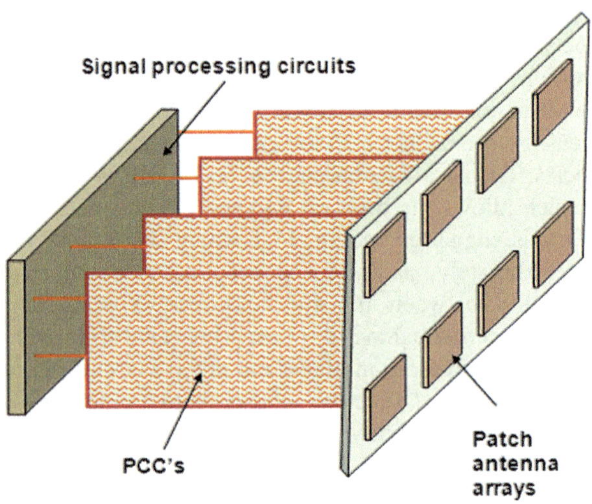

Fig. 11.5 Retro-directive transmit/receive array with orthogonally polarised patch radiators

11.6 Cross-slot Fed Microstrip Patch Antenna

Arguably, the most commonly employed mechanism for securing a compact dual-polarised array element based on microstrip technology is the square conducting patch excited parasitically from a circularly symmetric cross-slot [25, 28]. Edimo et al. describe arguably the simplest arrangement, which is illustrated schematically in Fig. 11.6. The non-resonant X-slot is located centrally below the square patch radiator which means, referencing Fig. 1.4, that the orthogonally directed slots of the X-slot separately excite the degenerate orthogonal TE_{101} modes of the patch, thus securing dual-polarised radiation. Each of the two microstrip feed lines, in principle, excites only the slot element whose major dimension is transverse to it. However, the geometry enforces the feed lines to be on opposite sides of the slot plane and thus occupy separate substrates, which adds undesirable complexity in fabrication terms.

The patch excitation arrangement depicted in Fig. 11.6 displays a degree of imbalance between the patch and the feed lines—the fields of one line are shielded from the patch, while for the other they are not. This leads to issues of differential gain between the desired polarisations, cross-polarisation differences and poor axial ratio outcomes. Nevertheless the simple arrangement depicted in Fig. 11.6 has been demonstrated [25] to operate over an $\sim$5% bandwidth, for a 5 GHz design example. The cross-coupling between the orthogonal ports is modest, and predicted to be of the order of −20 dB.

The troublesome cross-coupling generated by the slot excitation scheme depicted in Fig. 11.6 can be significantly reduced [28] by simply incorporating microstrip T-junctions into each feed line (see Fig. 11.7) so that it is no longer in close proximity to the orthogonal uncoupled slot. The arrangement introduces a modicum of additional complexity to the layout and to the fabrication of the antenna, but not without considerable performance gains. For an antenna designed to operate at 1.5 GHz, Sanford and Tengs report a 20% bandwidth capability (defined by $S_{11} < -10$ dB) with port isolation (S_{21}) approaching −35 dB over the band.

11.7 Dual-Polarised Radiation from Apertures

While retro-directive array antennas for mobile applications are, on the basis of current technology, likely to be optimally realised using patch radiating elements, needless to say, non-patch alternatives, providing circularly polarised and/or dual-mode capability, are also scattered through the literature [29]. Generally, these competing possibilities comprise either microstrip fed cross-slot radiators, or microstrip fed circular apertures. An example which gives a flavour of non-patch possibilities is depicted in Fig. 11.8. The four holes etched into the metallic ground plane of the middle substrate (b) are sandwiched between substrates (a) and

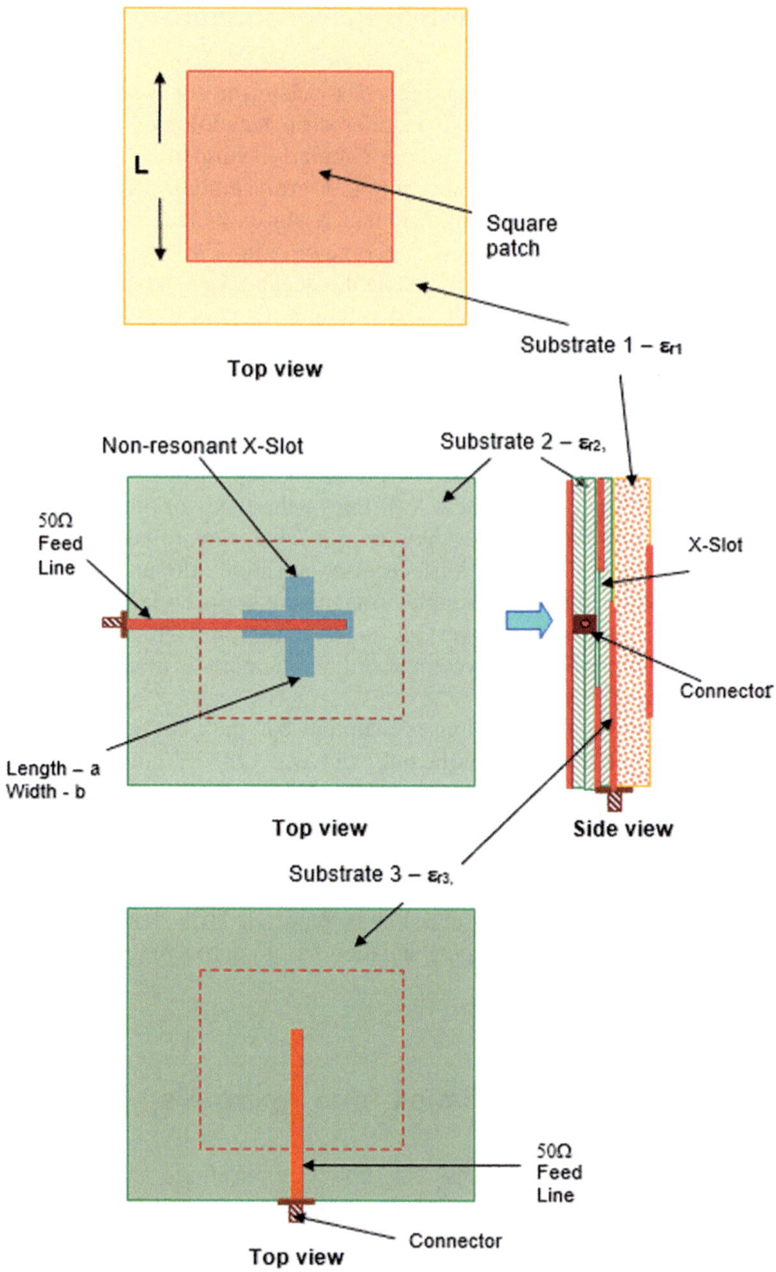

Fig. 11.6 Schematic of dual-mode dual-polarised patch antenna parasitically fed from a non-resonant X-slot by means of microstrip lines above and below the slot

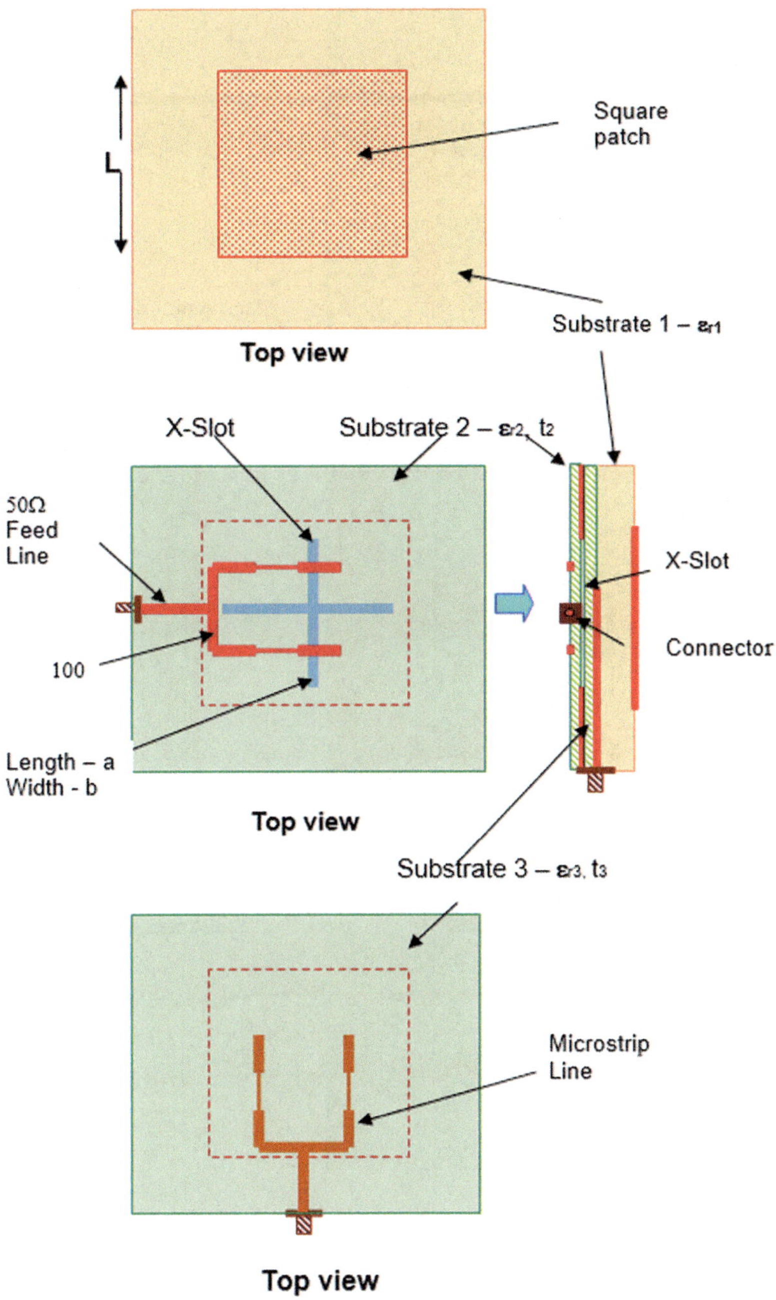

Fig. 11.7 Schematic of a dual-mode dual-polarised patch antenna parasitically fed from a non-resonant X-slot by means of balanced microstrip lines above and below the slot to improve bandwidth and axial ratio

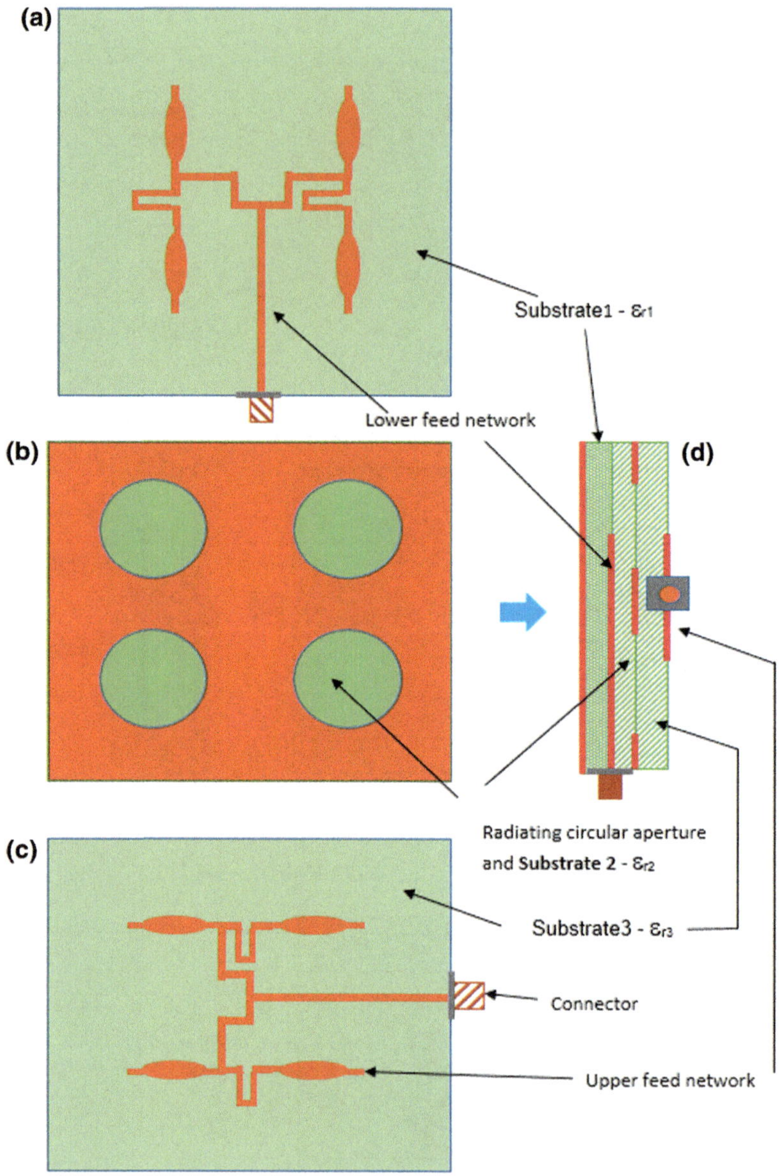

Fig. 11.8 Schematic of dual-mode dual-polarised circular aperture array antenna parasitically fed from balanced microstrip lines above and below the four elements

(c) which carry branching microstrip lines. The elliptic shapes, incorporated into the branch-lines, should register with the circular apertures, and are designed to improve the matching process. The upper substrate ε_{r1} excites vertically polarised radiations from the four apertures while the lower substrate ε_{r3} generates transversely polarised waves. If the combination is fed from a quadrature hybrid then this four hole antenna element provides circular polarised radiation. The metallic layer on the back surface of substrate ε_{r1} ensures one-directional radiation —to the right on the end view in Fig. 11.8.

While this concept of circular apertures, fed from back-to-back microstrip feed structures, offers some advantages in fabrication terms, cited performance figures tend to be modest. The -10 dB bandwidth is of the order of 5% while cross-polarisation levels tend to be high at approximately -20 dB. In addition, it should be noted that, like slot radiators, holes are prone to high mutual coupling, which is best avoided in retro-directive array applications.

11.8 Parasitically Fed Microstrip Patch Antenna

In order to eliminate the disadvantage of requiring two separate feed substrates to procure dual polarisation in a single parasitically coupled radiating element, it is necessary to feed the patch through two orthogonally directed non-contacting apertures or slots. The general arrangement takes the basic form shown in Fig. 11.9, where the square patch (or patches) is excited by a pair of orthogonally directed slots which in turn set up spatially orthogonal TE_{101} modes under the patch. This geometry permits slot excitation by a single microstrip substrate with feed lines on one face. Such dual-mode aperture coupled patch arrangements are described in papers by Adrian and Schaubert [22], Gosalia and Lazz [23] and Gao et al. [26]. In [22], the antenna is designed to operate at ~2.215 GHz, and is fabricated on a substrate with $\varepsilon_r = 2.55$. It displays a modest 3 dB gain bandwidth of 3.5% with an axial ratio of 2 dB across this band. This equates to a bandwidth of approximately 8 MHz. However this can be improved by introducing a second patch as described by in reference [26]. Designed for 2.6 GHz their dual patch design with dumbbell coupling slots, as shown in Fig. 11.9, is reported to provide an $S_{11} = -10$ dB over a bandwidth of ~21% with cross-polarisation levels of better than minus 22 dB. To secure dual polarisation and circular polarisation simultaneously with the dual mode patch antennas described above, it is necessary to feed the antenna through a 3 dB hybrid coupler [30] (see Fig. 11.10). This scheme generates the required quadrature phase relationship at the antenna ports. However, the combination can be space demanding, which can be an impediment to the formation of a planar array with grating-lobe free element spacing (i.e. $\leq \lambda_0/2$). The problem is partially solved in a paper by Sangford and Tengs [28] where the hybrid is merged, with the slot feed lines and the coupling slots, below the patch. However, the outcome is very complex in fabrication terms, making it potentially unattractive in any cost conscious development.

While good impedance bandwidth of the order of ~10% is most often quoted in assessing antenna suitability to communication systems, some applications also demand comparable circular polarisation ellipticity bandwidths [31–34]. Typically ellipticity bandwidths can be 30% less than the impedance bandwidth of dual

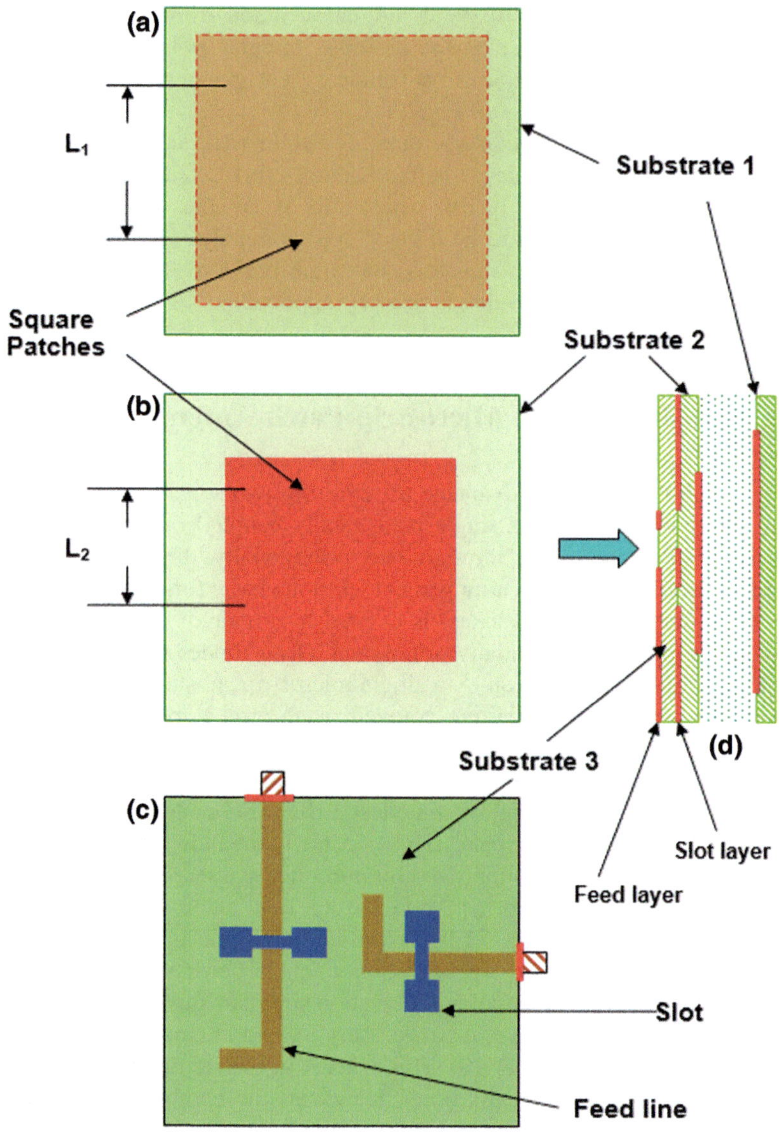

Fig. 11.9 Schematic of dual-mode dual-polarised patch antenna. **a** Substrate 1—copper patch $L_1 \times L_1$, thickness t_1, relative permittivity ε_{r1}, **b** substrate 2—copper patch $L_2 \times L_2$, thickness t_2, relative permittivity ε_{r2}, **c** two dumbbell coupling slots in top face, 50Ω feed lines in lower face (see Gao [26] for detail), substrate thickness t_3, relative permittivity ε_{r3}, **d** edge view showing air/polystyrene layer between substrate 1 and substrate 2, thickness t_4, relative permittivity ε_{r4}

polarised patch antennas. Dual-mode dual polarised microstrip antennas have generally employed square patches, but a desire for wider ellipticity bandwidth has generated interest in circular patches, as is shown by Karmakar and Bialkowski

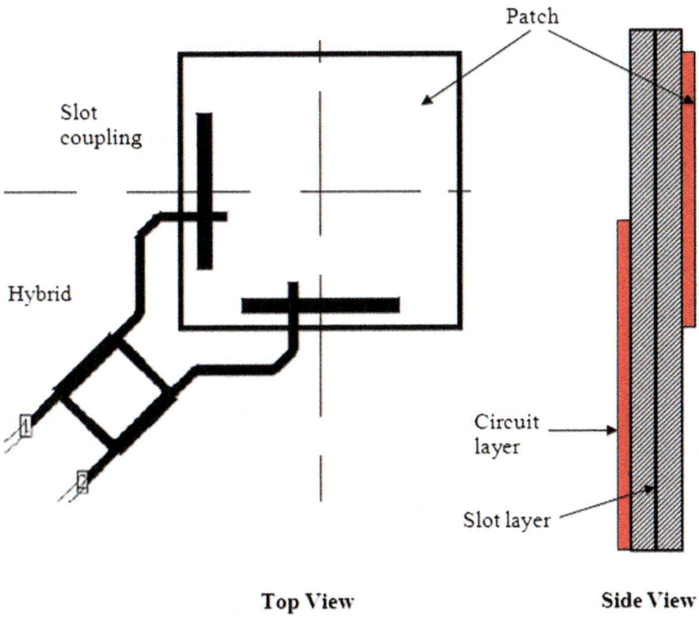

Fig. 11.10 Circularly polarised dual-mode square patch antenna with slot coupling and microstrip hybrid feed arrangement

[31]. A typical manifestation of this approach is depicted in Fig. 11.11. The circular symmetry of the round patch confers sufficient marginal improvement in polarisation circularity, at wide angles in the radiation pattern, to acquire enhanced ellipticity bandwidth for the circular radiator, when compared with square patch alternatives.

11.9 Phase Conjugated Active Antenna Element

Currently, the most attractive method of securing phase conjugation in a retro-directive array antenna is furnished by heterodyne mixing. However, as we know from the literature this essentially requires that the RF (receive) and the IF (transmit) frequency be the same, or very nearly the same. This means that RF leakage reflected or radiated back toward the source can significantly interfere with the retro-directive signal. Getting rid of the unwanted RF leakage is not possible using conventional filtering techniques because the IF is at the same frequency. The active antenna element represents a possible solution to this problem [34, 35].

The signal (RF at 6 GHz) received by the active phase conjugating antenna (the red patch—Fig. 11.12), is split into two components, to the right of the figure. These are then forced into an antiphase relationship by using line length. The local

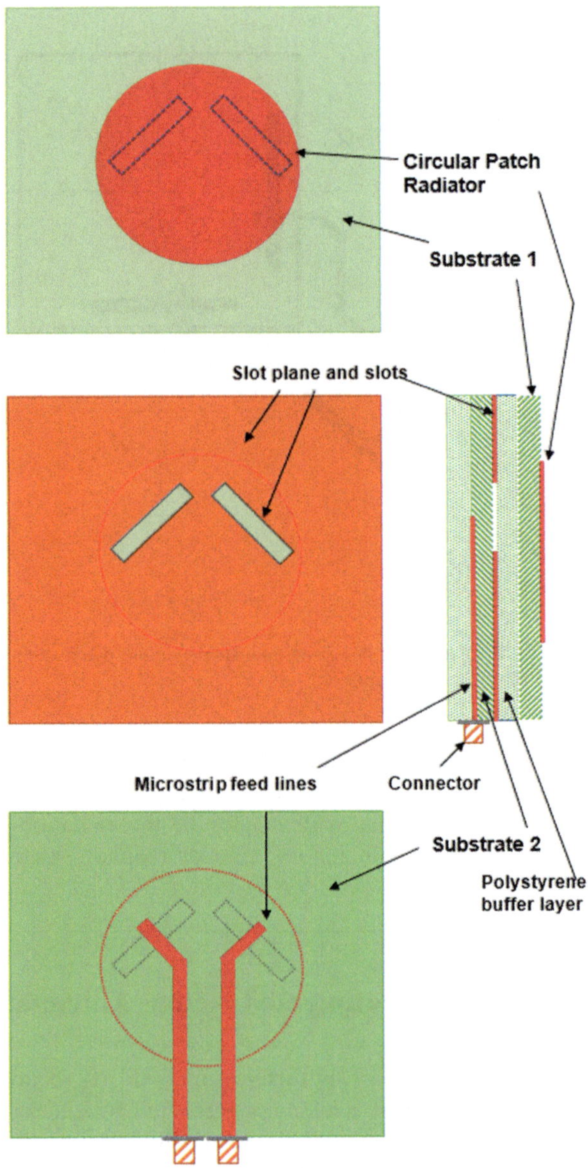

Fig. 11.11 Dual-mode dual-polarised microstrip patch antenna implemented using circular patches for enhanced axial ratio bandwidth

oscillator signal at twice the RF frequency is introduced at the left of the circuit. It is also split into two equal components with a differential 180° phase shift added by line length. The advantage of the RF frequency signals on the microstrip lines, in the upper and lower portions of the circuit, being in antiphase is that leakage or cross-coupling across the channels, which can degrade the retro-directive operation, is minimised. The FET gate mixers provide both phase conjugation and gain, with

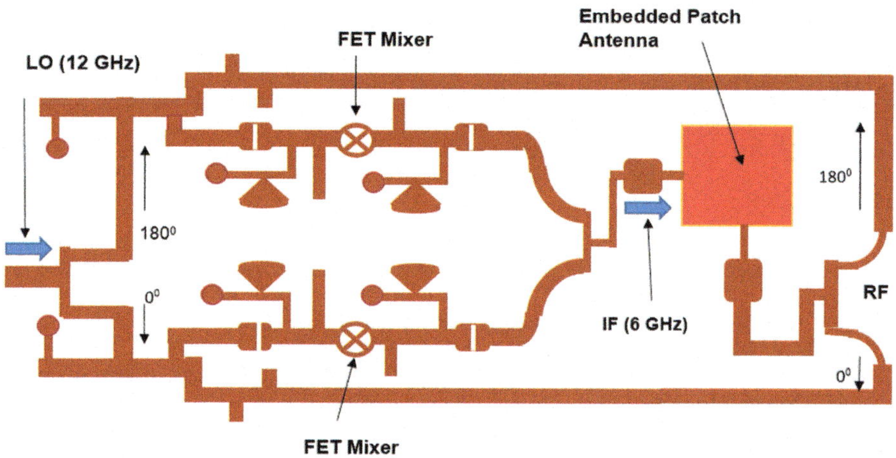

Fig. 11.12 Phase conjugating active antenna element incorporating a balanced quasi-optical mixer

the difference frequency IF signals adding at the T-junction before passing to the antenna for transmission. Note that at the patch the transmit and receive signals are isolated from each other by employing the orthogonal linear polarisations of the square patch. For retro-directive communication between a mobile platform and a satellite, which depends on circularly polarised radiations, the Fig. 11.12 circuit is inappropriate. While the concept is promising, clearly further research is needed to achieve both a dual-polarised and a circularly polarised active antenna element for a mobile communications system with fully operational retro-directive capability.

11.10 Chapter Summary

A review is furnished here of state-of-the-art technology (relevant up to 2017), which aims to provide compact antenna designs for wireless applications where relative motion between the receive antenna and the source of a desired communications signal exists. The development involves the introduction of retrodirective capability into compact transmit/receive arrays on, for example, mobile platforms and similar applications. It is shown that the technology follows three distinct paths which are presented in detail.

Several antenna formats, exhibiting potential compatibility with the retrodirective mode of operation, are studied, and it is concluded that compactness and low cost are best served with antennas fabricated in microstrip technology. Furthermore, current developments indicate that the antenna forms which are generally preferred are those using one or more square patches excited through separated parasitic coupling slots, in order to secure highly desirable circularly polarised radiation.

It is tentatively suggested that the future may reside with active antenna elements, but this remains some distance away, until dual-polarised, circularly polarised capability can be confidently demonstrated in commercial products.

References

1. R.C. Johnson, *Antenna Engineering Handbook*, 3rd edn. (McGraw-Hill Inc., New York, 1993)
2. A.W. Rudge, K. Milne, A.D. Olver, P. Knight (eds.), *The Handbook of Antenna Design: volumes I and II* (Peter Peregrinus Ltd., London, 1986)
3. C.Y. Pon, Retrodirective array using heterodyne technique. IEEE Trans. Antennas Propag. **AP-12**, 176–180 (1966)
4. R.N. Ghose, Electronically adaptive antenna systems. IEEE Trans. Antennas Propag. **AP-12** (2), 161–169 (1964)
5. L.G. Van Atta, Electromagnetic Reflector. US Patent 2,908,002, 6 Oct 1959
6. R.C. Chernoff, Large active retrodirective arrays for space applications. IEEE Trans. Antennas Propag. **AP-27**(4), 489–496 (1979)
7. K. Leong, Y. Wang, T. Itoh, A full duplex capable retrodirective array system for high speed beam tracking and pointing applications. IEEE Trans. Microwave Theory Tech. **MTT-52**(5), 1479–1489 (2004)
8. G.S. Shiroma, R.Y. Miyamoto, W.A. Shiroma, A full-duplex dual-frequency self-steering array using phase detection and phase shifting. IEEE Trans. Microwave Theory Tech. **MTT-54**(1), 128–134 (2006)
9. R.Y. Myamoto, Y. Qian, T. Itoh, An active integrated retrodirective transponder for remote information retrieval. IEEE MTT-S, Int. Microwave Symp. **3**, 1431–1434 (2000, June)
10. R.Y. Myamoto et al., Digital wireless sensor server using an adaptive smart-antenna/ retrodirective array. IEEE Trans. Vehicular Technol. **52**(5), 1181–1189 (2003)
11. S.L. Karode, V.F. Fusco, Frequency offset retrodirective antenna array. Electron. Lett. **33**(16), 1350–1351 (1997)
12. S.L. Karode, V.F. Fusco, Multiple target tracking using retrodirective antenna arrays. Nat. Conf. Antennas Propag. Publication No. 461, 178–181 (1999)
13. V.F. Fusco, C.B. Soo, N. Buchanan, Constant phase IF self-tracking receiver analysis and characterisation. IEEE Proc. (MAP) **151**(6), 507–513 (2004)
14. N.B. Buchanan, V.F. Fusco, Triple mode PLL antenna array. IEEE MTT-S Int. Microwave Symp. Digest, 1691–1694 (2004)
15. V.F. Fusco, C.B. Soo, N. Buchanan, Analysis and characterisation of PLL-based retrodirective array. IEEE Trans. Microwave Theory Tech. **MTT-53**(2), 730–738 (2005)
16. V.F. Fusco, N. Buchanan, High performance IQ modulator based phase conjugator for modular retrodirective antenna array implementation. IEEE Trans. Microwave Theory Tech. **MTT-57**(10), 2301–2306 (2009)
17. V.F. Fusco, N. Buchanan, Dual mode retrodirective phased array. Electron. Lett. **45**(3), 139–141 (2009)
18. T.D. Huynh, Dual polarised aperture coupled microstrip patch antenna system. US Patent No. 5,896,107, 1999
19. L. Desclos et al., Single and dual mode patch antennas. US Patent No. 6181281, 30 Jan 2001
20. N. McDonald, Dual polarised patch antenna. European Patent No. EP 0901185, 14 Nov 2001

21. S.M. Sanzgiri et al., Dual polarised microstrip patch antenna array for PCS base stations. US Patent No. 5,923,296, 13 July 1999
22. A. Adrian, D.H. Schaubert, Dual aperture-coupled microstrip antenna for dual or circular polarisation. Electron. Lett. **23**(23), 1226–1228 (1987)
23. K. Gosalia, G. Lazz, Reduced size dual-polarised microstrip patch antenna for wireless communications. IEEE Trans. **AP-51**(9), 2182–2186 (2003)
24. S.L. Karode, V.F. Fusco, Dual-polarisation microstrip patch antenna using feedforward isolation enhancement for simultaneous transmit/receive applications. Nat. Conf. Antennas Propag. Conference Publication No. 461, 49–50 (1999)
25. M. Edimo, A. Sharaiha, C. Terret, Optimised feeding of a dual-polarised broadband aperture-coupled printed antenna. Electron. Lett. **28**(19), 1785–1787 (1992)
26. S.C. Gao, L.W. Li, P. Gardner, P.S. Hall, Wideband dual-polarised microstrip patch antenna. Electron. Lett. **37**(20), 1213–1214 (2001)
27. A. Uz Zaman, L. Manholm, D. Derneryd, Dual-polarised microstrip patch antenna with high port isolation. Electron. Lett. **43**(10), 551–552 (2007)
28. J.R. Sandford, A. Tengs, A two substrate dual-polarised aperture-coupled patch. IEEE Antennas Propag. Soc. Int. Symp. AP-S Digest **3**, 1544–1547 (1996)
29. E. Rammos, A new wideband high-gain stripline planar array antenna for 12 GHz TV. Electron. Lett. **18**(6), 252–253 (1982)
30. A.J. Sangster et al., A dual-polarised stripline fed slot antenna incorporating signal cancellation. IEE Proc. (MAP) **148**(6), 357–362 (2001)
31. N.C. Karmakar, M.E. Bialkowski, Circularly polarised aperture coupled circular microstrip patch antennas for L-band applications. IEEE Trans. **AP-47**(5), 933–940 (1999)
32. N.C. Karmakar, M.E. Bialkowski, A beam forming network for a circular switched-beam phased array antenna. IEEE Microwave Wireless Lett. **11**(1), 1–9 (2001)
33. M.E. Bialkowski, N.C. Karmakar, A two ring circular phased array antenna for satellite communications. IEEE Antennas Propag. Mag. **41**(3), 14–23 (1999)
34. K. Chang, R.A. York, P.S. Hall, T. Itoh, Active integrated antennas. IEEE Trans. **MTT-50**(3), 937–944 (2002)
35. R.Y. Miyamoto, Y. Qian, T. Itoh, Active retrodirective array for remote tagging and wireless sensor applications. IEEE MTT-S Int. Microwave Symp. Dig., Boston, MA, 1431–1434 (2000, June)

Appendix A
Free-Space Electromagnetic Waves

A.1 Plane Wave Solution to the Maxwell Equations

All materials contain electric charges bound loosely or otherwise within atoms and molecules. If these materials exist in an environment which naturally or artificially causes agitation of the charge, and hence changes in the associated electric and magnetic fields, then electromagnetic waves are unavoidable. These waves can appear in quite complex forms such as trapped wave, fast waves, slow waves, surface waves, leaky waves, evanescent fields and radiating waves. To describe these various and quite diverse formats a set of equations is required which represent the field components of the wave at a point in space. The point forms (differential forms) of Maxwell's equations are developed in a host of text books on the topic (see Bibliography) and essentially entails the recruitment of well known vector-differential theorems such as the divergence theorem and Stokes' law to accomplish the transitions.

Many electromagnetic wave guiding and manipulating problems are of the source-free variety, which implies that the source is so far distant that the waves of interest are freely propagating and in an unbounded region (free-space) are plane waves. These waves, also termed transverse electromagnetic (TEM) waves, are described as 'plane' because the radius of curvature of the wave-front is very large, and thus the natural rate of curvature of the front, can be deemed mathematically insignificant allowing it to be fully described by means of Cartesian coordinates. In this scenario, the EM problem reduces to a boundary value problem, for which Maxwell's equations, in differential form, become:

$$\nabla \cdot \mathbf{D} = 0 \tag{A.1}$$

$$\nabla \cdot \mathbf{B} = 0 \tag{A.2}$$

A. J. Sangster, *Compact Slot Array Antennas for Wireless Communications*, Signals and Communication Technology, https://doi.org/10.1007/978-3-030-01753-8

$$\nabla \times \mathbf{E} = -\frac{\partial \mathbf{B}}{\partial \mathbf{t}} \tag{A.3}$$

$$\nabla \times \mathbf{H} = \frac{\partial \mathbf{D}}{\partial \mathbf{t}} \tag{A.4}$$

where **E** and **H** represent the electric and magnetic field intensities in the region of interest. As always, $\mathbf{D} = \varepsilon\mathbf{E}$ is the electric flux density while $\mathbf{B} = \mu\mathbf{H}$ is the magnetic flux density. The 'del' operator (∇) expresses directional derivatives in the three space directions. It is a vector, which in the Cartesian system (for example) has the form:

$$\nabla = \hat{\mathbf{a}}_\mathbf{x}\frac{\partial}{\partial \mathbf{x}} + \hat{\mathbf{a}}_\mathbf{y}\frac{\partial}{\partial \mathbf{y}} + \hat{\mathbf{a}}_\mathbf{z}\frac{\partial}{\partial \mathbf{z}} \tag{A.5}$$

where $\hat{\mathbf{a}}_\mathbf{x}, \hat{\mathbf{a}}_\mathbf{y}$ and $\hat{\mathbf{a}}_\mathbf{z}$ are unit vectors directed along x, y, and z respectively. When the del operator is multiplied to a scalar (ϕ(x, y, z) say) the result is a vector which expresses the gradient or slope of the function ϕ in all three space directions. i.e.:

$$\nabla\phi = \hat{\mathbf{a}}_\mathbf{x}\frac{\partial\phi}{\partial \mathbf{x}} + \hat{\mathbf{a}}_\mathbf{y}\frac{\partial\phi}{\partial \mathbf{y}} + \hat{\mathbf{a}}_\mathbf{z}\frac{\partial\phi}{\partial \mathbf{z}} \tag{A.6}$$

Cross multiplication of del with a vector produces the operation of 'curl', while dot multiplication produces the operation of 'divergence' ('div'). Crudely, curl is circulation at a point, while divergence is flux at a point.

A.2 Second Order Differential Equation

To solve the Maxwell equations for E-field or H-field behaviour in a bounded region it is first necessary to form an equation either E or H alone. The standard procedure for achieving this conversion is to perform a curl operation on either the curl equation for E or the corresponding equation for H. This gives, for example, using Eq. (A.3)

$$\begin{aligned} \nabla \times \nabla \times \mathbf{E} &= -\frac{\partial}{\partial t}\mu\nabla \times \mathbf{H} \\ &= -\frac{\partial}{\partial t}[\frac{\partial}{\partial \mathrm{t}}\mu\varepsilon\,\mathbf{E}] \\ &= -\mu\varepsilon\frac{\partial^2\mathbf{E}}{\partial \mathbf{t}^2} \end{aligned} \tag{A.7}$$

Hence, on using a convenient vector identity, which states that for any vector **F**

$$\nabla \times \nabla \times \mathbf{F} = \nabla\nabla \cdot \mathbf{F} - \nabla^2 \mathbf{F} \tag{A.8}$$

Equation (2.39) can be re-expressed as follows:

$$\nabla\nabla \cdot \mathbf{E} - \nabla^2 \mathbf{E} = -\mu\varepsilon \frac{\partial^2 \mathbf{E}}{\partial \mathbf{t}^2} \tag{A.9}$$

But, from Eq. (A.1), $\nabla \cdot \mathbf{E} = 0$, for a linear, homogeneous medium for which permeability μ and permittivity are constants. Therefore

$$\nabla^2 \mathbf{E} = \mu\varepsilon \frac{\partial^2 \mathbf{E}}{\partial \mathbf{t}^2} \tag{A.10}$$

and by analogy:

$$\nabla^2 \mathbf{H} = \mu\varepsilon \frac{\partial^2 \mathbf{H}}{\partial \mathbf{t}^2} \tag{A.11}$$

Equations (A.10) and (A.11) are wave equations. Equations of this nature, with appropriate variables, appear in most branches of science and engineering and their solutions have been studied widely. Solutions depend very much on the boundary conditions—namely the conditions imposed on the variables at the periphery or containing surface of the solution region. They can fix the magnitude of the variable (Dirichlet condition) or the rate of change of the variable (Newman condition) or a mixture of both. A unique solution depends on the conditions being neither under specified or over specified.

For example, let us consider formulating a solution to Eq. (A.10), and inevitably Eq. (A.11) because of the Maxwell linkages, for a region of free space ($\mu = \mu_O$: $\varepsilon = \varepsilon_O$) which is large enough to presume that all boundaries are effectively at infinity. In this case we can choose to represent the region mathematically using Cartesian coordinates, and furthermore since we anticipate that the solution is a waveform we can arbitrarily determine that the waves travel in the z-direction. This implies that the rates of change of the E-field in x and y are zero, and using (A.1) it follows that $E_z = 0$. The equation to be solved, therefore, is:

$$\frac{\partial^2 \mathbf{E}}{\partial \mathbf{z}^2} = \mu_0 \varepsilon_0 \frac{\partial^2 \mathbf{E}}{\partial \mathbf{t}^2} \tag{A.12}$$

where, in general, $\mathbf{E} = \hat{\mathbf{a}}_\mathbf{x} E_x + \hat{\mathbf{a}}_\mathbf{y} E_y$. However, if we choose to align the coordinate system so that **E** lies along the x-axis (x polarised solution), then $E_y = 0$ and the wave equation reduces to the scalar form:

$$\frac{\partial^2 E_x}{\partial z^2} = \frac{1}{c^2}\frac{\partial^2 E_x}{\partial t^2} \tag{A.13}$$

With 'speed of light' $c = \frac{1}{\sqrt{\mu_0 \varepsilon_0}}$.

A.3 General Solution

Equation (A.13) has a wave solution of the general form:

$$E_x = Af(z - ct) + Bf(z + ct) \tag{A.14}$$

This is easily demonstrated by substitution back into the equation. The first term represents a wave travelling in the +z direction while the second allows for a reflected wave, if such exists. Given that velocity is the rate of change of z with respect to time, it is evident that c represents velocity (actually phase velocity) of the electromagnetic wave in 'free space'. For vacuum it is equal to 3×10^8 m/sec. The application of Maxwell's equations also gives $H_z = 0$ and:

$$H_y = \frac{A}{\eta} f(z - ct) + \frac{B}{\eta} f(z + ct) \tag{A.15}$$

Also

$$\frac{E_x}{H_y} = \pm\sqrt{\frac{\mu_0}{\varepsilon_0}} = \pm\eta \tag{A.16}$$

η is termed the free space wave impedance which for air or vacuum has the value $120\pi\ \Omega$. The resultant solution is a plane electromagnetic wave, also termed a transverse electromagnetic (TEM) wave, for which E and H are transverse to the direction of propagation and orthogonal to each other. E and H are also in time phase, as Eq. (A.16) attests (see Fig. A.1).

Electrical engineers are generally very familiar with the relationship between power (P), voltage (V) and current (I) in the form:

$$P = \frac{1}{2} VI \text{ Watts} \tag{A.17}$$

where V and I are defined in peak, rather than in the more common r.m.s., format. But, voltage is simply integrated electric field E (V/m), and from Ampere, current is integrated magnetic field intensity H (A/m), so by analogy we can suggest that for the plane wave:

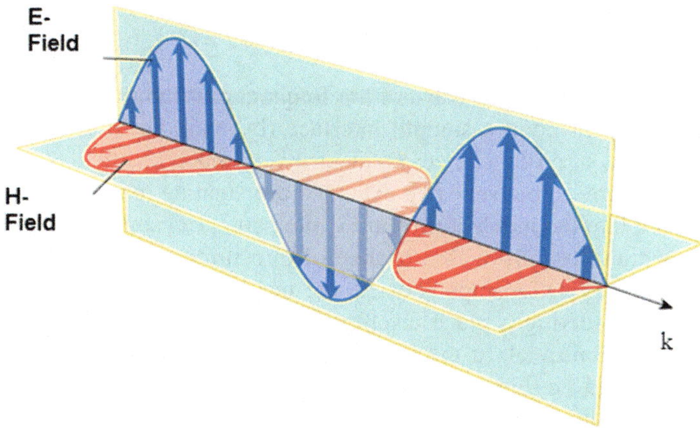

Fig. A.1 TEM wave field and directional relationships

$$p = \frac{1}{2}EH = \frac{1}{2}\mathrm{c}\varepsilon_0\mathrm{E}^2 \text{ Watts/m}^2 \tag{A.18}$$

This means that p is the real power flow density in the TEM wave. In general, complex power flow density in an electromagnetic wave is given by the Poynting vector **S**, where:

$$\mathbf{S} = \frac{\mathbf{1}}{\mathbf{2}}\mathbf{E} \times \mathbf{H} \text{ Watts/m}^2 \tag{A.19}$$

In electrical engineering it is much more usual to examine wave solutions at a single frequency (ω rad/sec)—namely sinusoidal solutions. This actually incurs little loss of generality, since any arbitrary time variation carried on a radio-wave can be resolved into a spectrum of single frequency components. The adoption of a single frequency, or a spectral frequency, in carrying through time varying computations has the distinct advantage that the time variable can be omitted. The calculations are then progressed in phasor notation. In trigometric form Eq. (A.14) becomes:

$$E_x = A \exp j(\omega t - \beta z) + B \exp j(\omega t + \beta z) \tag{A.20}$$

where A and B are complex constants. The phasor form is:

$$E_x = |A| \exp(-j\beta z + \varphi) + |B| \exp(j\beta z + \theta) \tag{A.21}$$

with φ and θ representing the phases respectively of A and B.

A.4 Snell's Laws

When a plane electromagnetic wave at the frequency of light, or in fact any radio frequency, is incident upon a smooth interface (by 'smooth' it is meant that any surface undulations or protuberances are in size very much less than the wavelength of the impinging waves) between two extended propagating media, part of the wave is reflected back into the incident medium while part is transmitted or refracted into the second medium, usually with a change of direction.

Analytically, the relationships between the incident and reflected waves can be developed by considering a plane electromagnetic wave, incident at a physically real angle θ_i to the normal, at the interface between two semi-infinite regions of space, as suggested in Fig. A.2. Each region is presumed to comprise linear homogeneous medium with a different index of refraction (n). The index of refraction is defined as:

$$n = \frac{c}{v} \tag{A.22}$$

where c is the speed of light in vacuum, or free space, while v is its speed within the specified medium. Also, with reference to Fig. A.2, the following definitions apply:

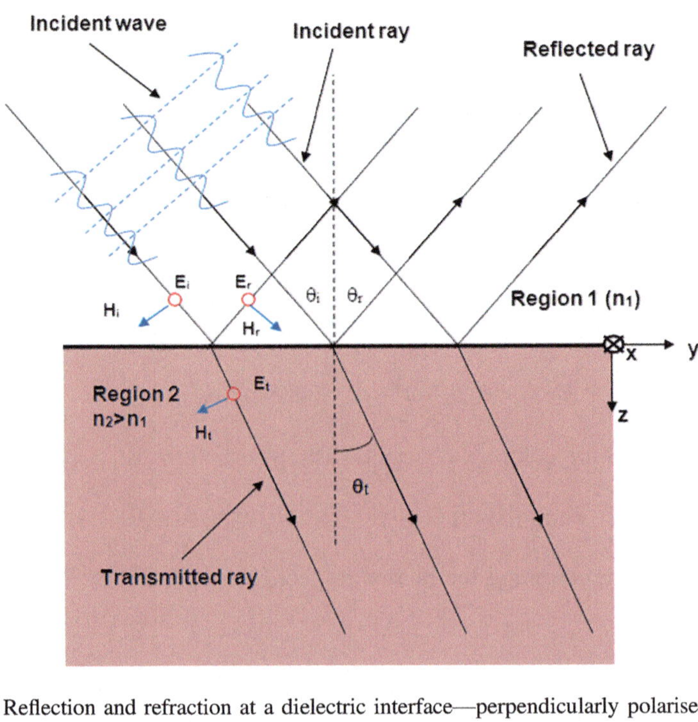

Fig. A.2 Reflection and refraction at a dielectric interface—perpendicularly polarised case

$$c = \frac{1}{\sqrt{\mu_0 \varepsilon_0}} \tag{A.23}$$

and

$$v_1 = \frac{1}{\sqrt{\mu_0 \varepsilon_0 \varepsilon_{r1}}} \tag{A.24}$$

$$v_2 = \frac{1}{\sqrt{\mu_0 \varepsilon_0 \varepsilon_{r2}}} \tag{A.25}$$

Here, ε_O and μ_O are the free space permittivity and permeability respectively, while ε_{r1} is the relative permittivity of medium 1 and ε_{r2} is the relative permittivity of medium 2. Both media are assumed to be loss-less and non-magnetic in which case $\mu_1 = \mu_2 = \mu_O$. The indices of refraction for the two media then become:

$$n_1 = \sqrt{\varepsilon_{r1}}; \quad n_2 = \sqrt{\varepsilon_{r2}} \tag{A.26}$$

Maxwell's equations in the semi-infinite regions remote from the interface are, as we have seen above, fully satisfied by TEM plane waves. It remains then to satisfy the Maxwell boundary conditions at the interface. If this can be done the resultant solutions represent complete EM solutions for the specified boundary value problem. For an incident TEM electromagnetic wave, as depicted in Fig. A.2, the E-field vector and the H-field vector must be mutually orthogonal to each other and to the direction of propagation, usually defined by a unit vector $\hat{\mathbf{k}}$, directed in the direction of the relevant ray. In this case we can write:

$$\mathbf{H} = \frac{1}{\eta}(\hat{\mathbf{k}} \times \mathbf{E}) \tag{A.27}$$

where η is the wave impedance for the medium containing the wave. Hence for regions 1 and 2 respectively:

$$\eta_1 = \sqrt{\frac{\mu_0}{\varepsilon_{r1}\varepsilon_0}} \tag{A.28}$$

$$\eta_2 = \sqrt{\frac{\mu_0}{\varepsilon_{r2}\varepsilon_0}} \tag{A.29}$$

However, this condition does not fully establish the polarisation direction, which must also be specified. There are two basic choices from which any other polarisation possibilities can be deduced. We can choose the E-field vector of the incident wave to be either normal to the *yz*-plane, or parallel to it. The yz-plane in Fig. A.2 is generally termed the plane of incidence for the incoming wave, being the plane that contains both the direction vector $\hat{\mathbf{k}}$ and the unit normal to the interface ($\hat{\mathbf{n}}$). When the electric field in the incident TEM wave is normal to the plane of incidence, the wave is said to be perpendicularly polarised, and when it is parallel to this plane it is described as parallel polarised. Note that in relation to the surface of the earth, while parallel polarisation equates to horizontal polarisation, perpendicular polarisation can be termed vertical polarisation only if θ_i approaches 90°. Perpendicular polarisation is often termed transverse electric (TE) propagation while parallel polarised waves get the complementary description of transverse magnetic (TM) waves.

Now that we know the electromagnetic field forms (TEM waves) remote from the interface between regions 1 and 2 in Fig. A.2, we can examine the field conditions (boundary conditions) precisely at the interface. For the diffraction set up depicted in Fig. A.2 with a perpendicularly polarised TEM wave incident at θ_i the field directions at a given instant in time can be represented vectorially as shown. Just at the interface a typical ray of the incident TEM wave is both reflected off the surface and transmitted through it. Also, for a 'smooth' surface 'common sense' suggests that it is safe to presume that the reflected and transmitted waves retain the polarisation of the incident wave. Furthermore, there will be a single reflected ray and a single transmitted ray. Actually, this latter assumption is not strictly necessary as we will show presently.

When the TEM wave direction (or ray) lies in paths other than along the coordinate axes, it is usual to define the ray direction by the vector $\mathbf{k}$ which is chosen to be equal in magnitude to the wave coefficient k. That is $\mathbf{k} = \hat{\mathbf{k}}k$. Hence we can express mathematically the wave component in any other direction ($\mathbf{r}$ say). For the case shown in Fig. A.3, where the electric field is x-directed, the expression has the form:

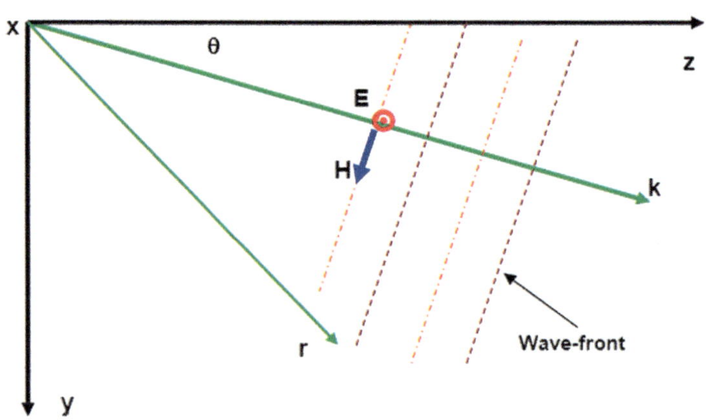

Fig. A.3 Representation of TEM wave with **E**, **H**, and **k** in mutually orthogonal directions

$$\mathbf{E_i} = \hat{\mathbf{a}}_\mathbf{x} E_i \exp(-j\mathbf{k} \bullet \mathbf{r}) \tag{A.30}$$

Consequently, if **r** and **k** lie in the yz-plane as suggested in Fig. A.3, then clearly:

$$\mathbf{k} = \hat{\mathbf{a}}_\mathbf{y} k_y + \hat{\mathbf{a}}_\mathbf{z} k_z \tag{A.31}$$

$$\mathbf{r} = \hat{\mathbf{a}}_\mathbf{y} y + \hat{\mathbf{a}}_\mathbf{z} z \tag{A.32}$$

Also

$$k^2 = k_y^2 + k_z^2 \tag{A.33}$$

so we can conveniently write:

$$k_y = k \sin\theta \tag{A.34}$$

and

$$k_z = k \cos\theta \tag{A.35}$$

Hence, employing these relationships, Eq. (A.30) can be expanded into the non-vectorial form:

$$E_{xi} = -E_i \exp(j(\omega t - k_1 z \cos\theta_i - k_1 y \sin\theta_i) \tag{A.36}$$

where

$$k_1 = \frac{\omega}{v_1} = \frac{\omega}{c} n_1 \tag{A.37}$$

For a transverse electromagnetic wave the electric and magnetic fields are related through Eq. (A.27). Hence on combining Eqs. (A.36) and (A.27), and observing the field directions in Fig. A.2 we obtain for magnetic fields:

$$H_{yi} = -H_i \cos\theta_i \exp[j(\omega t - k_1 z \cos\theta_i - k_1 y \sin\theta_i)] \tag{A.38}$$

$$H_{zi} = H_i \sin\theta_i \exp[j(\omega t - k_1 z \cos\theta_i - k_1 y \sin\theta_i)] \tag{A.39}$$

Also, we note that if these field components represent a TEM wave, then we must have:

$$\frac{E_i}{H_i} = \eta_1 = \frac{\eta_0}{n_1} \tag{A.40}$$

Similar constructions lead to the following equations for the reflected and transmitted field components:

$$E_{xr} = -E_r \exp(j(\omega t + k_1 z \cos\theta_i - k_1 y \sin\theta_i) \tag{A.41}$$

$$H_{yr} = H_r \cos\theta_i \exp[j(\omega t + k_1 z \cos\theta_i - k_1 y \sin\theta_i)] \tag{A.42}$$

$$H_{zr} = H_r \sin\theta_i \exp[j(\omega t + k_1 z \cos\theta_i - k_1 y \sin\theta_i)] \tag{A.43}$$

with

$$\frac{E_r}{H_r} = \eta_1 \tag{A.44}$$

and

$$E_{xt} = -E_t \exp(j(\omega t - k_2 z \cos\theta_t - k_2 y \sin\theta_t) \tag{A.45}$$

$$H_{yt} = -H_t \cos\theta_t \exp[j(\omega t - k_2 z \cos\theta_t - k_2 y \sin\theta_t)] \tag{A.46}$$

$$H_{zt} = H_t \sin\theta_t \exp[j(\omega t - k_2 z \cos\theta_t - k_2 y \sin\theta_t)] \tag{A.47}$$

where

$$\frac{E_t}{H_t} = \eta_2 \tag{A.48}$$

and

$$k_2 = \frac{\omega}{c} n_2 \tag{A.49}$$

The above field expressions for the incident and reflected waves in region 1 and the transmitted waves in region 2 each separately satisfy Maxwell's equations in these regions. A solution that satisfies Maxwell's equations for the entire volume including the interface is achieved by enforcing the electromagnetic field boundary conditions, given in equations at the interface. That is, at z = 0, we require that across the divide between regions 1 and 2:

$$E_x \text{ is continuous} \tag{A.50}$$

$$H_y \text{ is continuous} \tag{A.51}$$

$$B_z \text{ is continuous} \tag{A.52}$$

On combining Eq. (A.50) with the field expressions (A.36), (A.41) and (A.45) we obtain with little difficulty:

$$E_{xi} + E_{xr} = E_{xt} \tag{A.53}$$

on the z = 0 plane. The implication is that:

$$E_i \exp(-jk_1 y \sin\theta_i) + E_r \exp(-jk_1 y \sin\theta_r) = E_t \exp(-jk_2 y \sin\theta_t) \tag{A.54}$$

This equation must remain true over the entire z = 0 boundary, from $-\infty \leq y \leq +\infty$. This is only possible if:

$$k_1 \sin\theta_i = k_1 \sin\theta_r = k_2 \sin\theta_t \tag{A.55}$$

It is pertinent to note here that if at the commencement of this derivation we had, without pre-knowledge of refraction rules, chosen to presume that several reflected waves at angles θ_{r1}, θ_{r2}, θ_{r3} …, and several transmitted waves at angles θ_{t1}, θ_{t2}, θ_{t3} …, were possible, then the equivalent form of Eq. (A.54) would lead to:

$$k_1 \sin\theta_{r1} = k_1 \sin\theta_{r2} = k_1 \sin\theta_{r3} = \ldots\ldots \tag{A.56}$$

and

$$k_2 \sin\theta_{t1} = k_2 \sin\theta_{t2} = k_2 \sin\theta_{t3} = \ldots\ldots \tag{A.57}$$

These equations clearly dictate that $\theta_{r1} = \theta_{r2} = \theta_{r3} = \ldots$, and $\theta_{t1} = \theta_{t2} = \theta_{t3} = \ldots$ In other words, an 'optically smooth' surface produces only one reflected wave and one transmitted wave.

Equation (A.55) is the source of Snell's Laws which state that at an optically smooth interface between two lossless media:

$$\theta_r = \theta_i \tag{A.58}$$

$$\frac{\sin\theta_t}{\sin\theta_i} = \frac{k_1}{k_2} = \frac{n_1}{n_2} \tag{A.59}$$

However, these laws govern only the reflection and refraction angles. We also need to have knowledge of the relative magnitudes of the reflected and transmitted wave, and how these are influenced by material properties.

When Eqs. (A.50), (A.51) and (A.52) are applied to the TEM field components at the boundary, while also applying Snell's laws, the following relations are generated:

$$E_i + E_r = E_t \tag{A.60}$$

$$(H_i - H_r)\cos\theta_i = H_t \cos\theta_t \tag{A.61}$$

$$(B_i - B_r)\sin\theta_i = B_t \sin\theta_t \tag{A.62}$$

Equation (A.60) can be converted to magnetic field form by employing Eqs. (A.44) and (A.48) leading to:

$$\eta_1(H_i + H_r) = \eta_2 H_t \tag{A.63}$$

Consequently, if we choose to define reflection coefficient for this perpendicularly polarised example (TE case) as:

$$\rho_{TE} = \frac{H_r}{H_i} \tag{A.64}$$

then making use of Eqs. (A.61) and (A.63) the following useful relationship is deduced:

$$\rho_{TE} = \frac{\eta_2 \cos\theta_i - \eta_1 \cos\theta_t}{\eta_2 \cos\theta_i + \eta_1 \cos\theta_t} \tag{A.65}$$

This can also be expressed in a slightly more familiar form, which explicitly incorporates the indices of refraction, namely:

$$\rho_{TE} = \frac{n_1 \cos\theta_i - n_2 \cos\theta_t}{n_1 \cos\theta_i + n_2 \cos\theta_t} \tag{A.66}$$

Similarly, if we choose to define the transmission coefficient as:

$$\tau_{TE} = \frac{H_t}{H_i} \tag{A.67}$$

then:

$$\tau_{TE} = \frac{2n_2 \cos\theta_i}{n_1 \cos\theta_i + n_2 \cos\theta_t} \tag{A.68}$$

It is not difficult to demonstrate that:

$$\rho_{TE} = \frac{H_r}{H_i} = \frac{E_r}{E_i} \tag{A.69}$$

and

$$\frac{E_t}{E_i} = \frac{n_1}{n_2}\tau_{TE} \tag{A.70}$$

An analogous derivation can also be followed through for the parallel polarisation case (TM case). If this is done we obtain:

$$|\rho_{TM}| = \frac{E_r}{E_i} = \frac{H_r}{H_i} = \left|\frac{n_2\cos\theta_i - n_1\cos\theta_t}{n_2\cos\theta_i + n_1\cos\theta_t}\right| \tag{A.71}$$

and

$$\tau_{TM} = \frac{2n_1\cos\theta_i}{n_2\cos\theta_i + n_1\cos\theta_t} \tag{A.72}$$

The reflection coefficient, as a function of incident angle for both the TE and TM cases, is plotted in Fig. A.4. Clearly, for an interface between lossless dielectrics of differing refractive indices the reflection behaviours are distinct. While for the TE case it increases monotonically from a magnitude of 0.33 ($n_1 = 1$, $n_2 = 2$) at $\theta_i = 0$, to unity at $\theta_i = 90°$, it drops to zero close to 60° in the TM case. At the zero

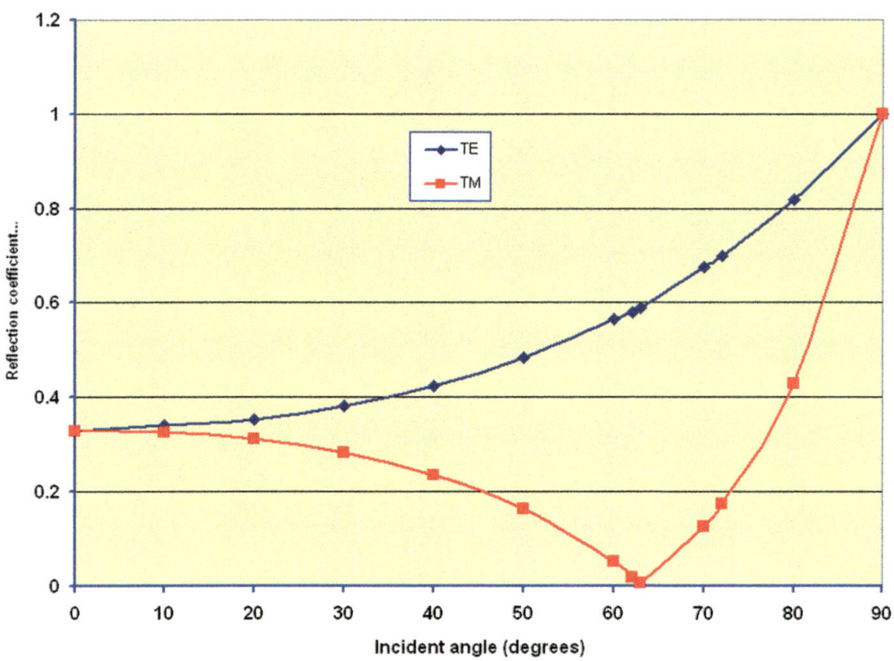

Fig. A.4 Reflection coefficient at an air/dielectric interface for TE and TM incidence

reflection angle the two surfaces are said to be 'matched' for surface normal wave components. It is termed the Brewster angle, a physical property which underpins the design of light polarizers.

Bibliography

1. R. Ferrari, *An Introduction to Electromagnetic Fields* (Van Nostrand Reinhold Co. Ltd., New York, 1975)
2. C.T.A. Johnk, *Engineering Electromagnetic Fields and Waves* (Wiley, New York, 1988)
3. R.P. Feynman, R.B. Leighton, M. Sands, *Lectures in Physics, Vol. II* (Addison-Wesley Publishing Co., London, 1972)
4. P.M. Morse, H. Feshbach, *Methods of Theoretical Physics*, (McGraw-Hill Book Co., New York, 1953)
5. J.D. Kraus, *Electromagnetics* (McGraw-Hill Book Co., London, 1984)
6. P. Lorrain, D. Corson, *Electromagnetic Field and Waves* (W.H. Freeman & Co., San Francisco, 1962)
7. A.J. Baden-Fuller, *Engineering Electromagnetism*, (Wiley, New York, 1993)
8. P. Hammond, *Applied Electromagnetism* (Pergamon Press Ltd., Oxford, 1971)
9. R.M. Bevensee, *Slow-wave Structures* (Wiley, New York, 1964)
10. J.A. Stratton, *Electromagnetic Theory* (Wiley, New Jersey, 2007)

Appendix B
Basic Waveguide Theory

B.1 Introduction

In Chap. 1 it is demonstrated, in an empirical manner, that the hollow rectangular waveguide provides an effective transmission medium for electromagnetic waves by trapping a free-space TEM wave within its confines. Actually it is not only hollow pipes of rectangular cross-section which do this. Circular and elliptical cross-sections are also possible. Given that the hollow pipe exhibits an unbroken metallic structure to a source of electromagnetic waves, it is clear that at DC, and at low enough frequencies, the waveguide will present a short circuit to the source. At what frequency does the waveguide change from a non-propagating to a propagating structure? This question, together with the multi-mode nature of waveguide, will be addressed in this Appendix.

It is pertinent to note, at the outset, that the solutions to Maxwell's equations can be of four distinct types. In the Cartesian frame the categories are related to the existence or otherwise of field components pointing in the propagation direction of the wave. If we take this to be the z-direction, as in Fig. B.1, then we can make the following statements:

$E_z = H_z = 0$	defines a TEM wave
$E_z \neq 0;\ H_z = 0$	transverse magnetic (TM)
$\mathrm{E_z} = 0; H_z \neq 0$	transverse electric (TE)
$\mathrm{E_z} \neq 0; H_z \neq 0$	hybrid mode (EH or HE)

As we have seen the TEM solution occurs in unbounded free space. It also forms in two wire transmission systems such as coaxial line. In hollow pipes, solutions are of the TE or TM type. A preliminary insight into the nature of propagation in a hollow rectangular waveguide is provided in Chap. 1. It is evident that the solutions to Maxwell's equations in this type of structure divide into two classes, namely transverse electric (TE) modes and transverse magnetic (TM) modes. The modes

A. J. Sangster, *Compact Slot Array Antennas for Wireless Communications*, Signals and Communication Technology, https://doi.org/10.1007/978-3-030-01753-8

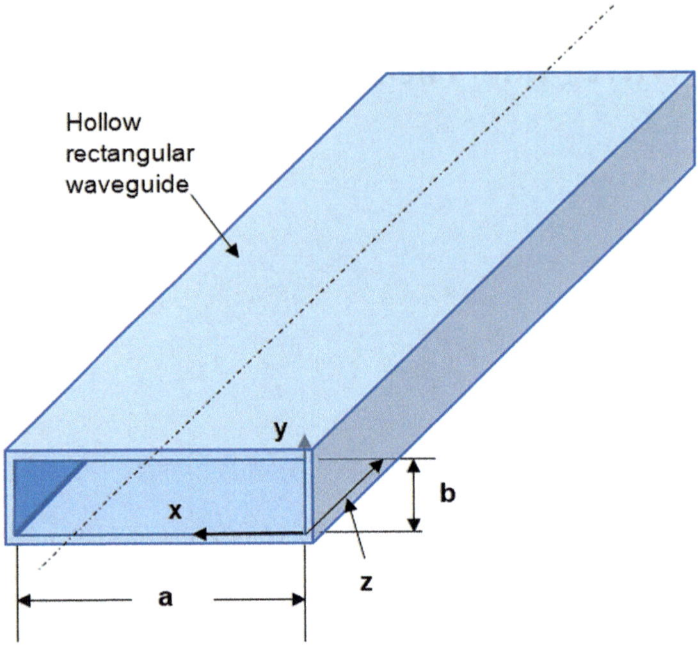

Fig. B.1 Rectangular waveguide coordinate system

are distinguished by the presence of a field component in the direction of propagation (z-direction in Fig. B.1)—a magnetic field (H_z) for TE modes and an electric field (E_z) for TM modes as indicated above. These modes in waveguide can be generated from potential functions (see Collin in Bibliography), with mathematical efficiency recommending the use of a magnetic-type Hertzian potential for TE modes and an electric type for TM modes. Employing the Collin terminology these potentials can be expressed as:

$$\mathbf{\Pi}_h = \hat{\mathbf{a}}_\mathbf{z} \Pi_h \tag{B.1}$$

and

$$\mathbf{\Pi}_e = \hat{\mathbf{a}}_\mathbf{z} \Pi_e \tag{B.2}$$

Other potential function choices are possible depending on the dictates of the problem to be solved. For example in antenna analyses the functions **A** (vector magnetic potential function) and ϕ (electric potential function) are solutions commonly used (see Chap. 2). Here we concentrate on the TE mode solutions to illustrate the typical waveguide analysis procedure. In terms of $\mathbf{\Pi}_\mathrm{h}$ the electric and magnetic fields in the waveguide are given by:

$$\mathbf{E} = -j\omega\mu_0\nabla \times \mathbf{\Pi_h} \tag{B.3}$$

$$\mathbf{H} = k_0^2\mathbf{\Pi}_h + \nabla\nabla \cdot \mathbf{\Pi}_h \tag{B.4}$$

On applying the Maxwell equations it follows that $\mathbf{\Pi_h}$ is a solution of the wave equation. That is, in phasor form:

$$\nabla^2\mathbf{\Pi}_h + k_0^2\mathbf{\Pi}_h = 0 \tag{B.5}$$

For an empty waveguide $k_0 = \omega\sqrt{\mu_0\varepsilon_0} = 2\pi/\lambda_0$ where λ_o is the free-space wavelength. Since the solution to Eq. (B.5) must represent z-directed propagating waves in the waveguide, then we can assume that:

$$\mathbf{\Pi}_h = \hat{\mathbf{a}}_\mathbf{z}\psi_h(x, y)\exp \pm\gamma z \tag{B.6}$$

where $\gamma = \alpha - j\beta$ is the complex propagation coefficient, while ψ_h is a solution of:

$$\nabla_t^2\psi_h + k_c^2\psi_h = 0 \tag{B.7}$$

∇_t operates on x and y only, while

$$k_c^2 = k_0^2 + \gamma^2 \tag{B.8}$$

In the Cartesian coordinate frame (Fig. B.1), a wave equation such as Eq. (B.7) can be solved by the method of separation of variables, which means that ψ_h can be expressed as H_z = X(x)Y(y). Consequently, on applying the boundary conditions on E and H, and hence ψ_h, at the perfectly conducting walls of the waveguide (see Fig. B.1), the solution to Eq. (B.7) is easily found to be:

$$\psi_h = C(\cos\frac{m\pi x}{a}\cos\frac{n\pi y}{b})\exp -j\beta_{mn}z \tag{B.9}$$

where $\gamma_{mn} = -j\beta mn$ for lossless waveguide.

B.2 TE Solutions for Rectangular Waveguide

With ψ_h now known it is not difficult to establish the fields within the waveguide. From Eq. (B.4) we obtain:

$$H_z = -\nabla_t^2\psi_h = k_0^2\psi_h \tag{B.10}$$

Consequently for the mnth mode we can write:

$$H_{zmn} = C \cos\frac{m\pi x}{a} \cos\frac{n\pi y}{b} \exp -j\beta_{mn}z \tag{B.11}$$

In this equation m and n are integers such that m = 0, 1, 2 … ∞, and n = 0, 1, 2 … ∞ (n = m ≠ 0). This solution also requires that:

$$k_c^2 = k_{cmn}^2 = \left(\frac{m\pi}{a}\right)^2 + \left(\frac{n\pi}{b}\right)^2 \tag{B.12}$$

Hence from Eq. (B.8) we must have

$$\gamma_{mn}^2 = \left(\frac{m\pi}{a}\right)^2 + \left(\frac{n\pi}{b}\right)^2 - k_0^2 = k_{cmn}^2 - k_0^2 \tag{B.13}$$

where $k_O = \omega/c$. In this equation k_O is the free-space propagation coefficient, k_{cmn} is the cut-off coefficient, while γ is the waveguide propagation coefficient for the mnth solution or mode. The other field components of the mode are derived from H_z using the following relations derived from the Maxwell equations:

$$\begin{aligned} H_{xmn} &= -\frac{\gamma_{mn}}{k_{cmn}^2}\frac{\partial H_{zmn}}{\partial x} \\ H_{ymn} &= -\frac{\gamma_{mn}}{k_{cmn}^2}\frac{\partial H_{zmn}}{\partial y} \\ E_{xmn} &= -\frac{j\omega\mu_0}{k_{cmn}^2}\frac{\partial H_{zmn}}{\partial y} \\ E_{ymn} &= \frac{j\omega\mu_0}{k_{cmn}^2}\frac{\partial H_{zmn}}{\partial x} \end{aligned} \tag{B.14}$$

It is important to observe, that these modal field expressions form (in mathematical terms) a *complete* set. This means that any arbitrary field distribution in the waveguide, due for example to the presence of a discontinuity such as a post or an iris, can always be represented mathematically as a combination or summation of the known modes of the empty waveguide by a process not unlike the use of Fourier series to represent arbitrary periodic waveforms with sinusoids.

B.3 Frequency Characteristics of Waveguide Modes

For fast wave solutions in uniform metallic waveguides the propagation equation has the general form:

$$\gamma_{mn}^2 = k_{cmn}^2 - k_0^2 \tag{B.15}$$

where k_{cmn} is a constant whose magnitude is governed by the size and shape of the waveguide, and $k=k_0$ is the plane wave phase constant for the interior medium. For rectangular waveguide, we have already seen that:

$$k_{cmn}^2 = \left(\frac{m\pi}{a}\right)^2 + \left(\frac{n\pi}{b}\right)^2 \tag{B.16}$$

It has other dimensionally dependant values for other waveguides, such as circular waveguide and elliptical waveguide. However, for all such waveguides the propagation equation has the same form, as expressed in Eq. (B.16).

In a lossless waveguide for which k_{cmn} is real and constant, and

$$k_0 = \omega\sqrt{\mu_0\varepsilon_0} \tag{B.17}$$

is real and linearly related to frequency, two types of modal solution exist in waveguide.

These are:

1. Those for which $k_o > k_{cmn}$ giving

$$\bar{\gamma} = j\beta_{mn} \text{ where } \beta_{mn} = \left(k_0^2 - k_{cmn}^2\right)^{1/2} \tag{B.18}$$

β_{mn} is termed the phase coefficient in rad/m for the mnth mode. The exponential multiplier in the field expression now has the form:

$$\exp(-j(\beta_{mn}z - \omega t))$$

which implies that the mode in question propagates with no attenuation in a lossless waveguide.

2. Those for which $k_o < k_{cmn}$ giving

$$\gamma_{mn} = \alpha_{mn} = \left(k_{cmn}^2 - k_0^2\right)^{1/2} \tag{B.19}$$

α_{mn} is termed the attenuation coefficient in nepers/m. The exponential multiplier in the field expressions now has the form

$$\exp(-\alpha_{mn}z)\exp(j\omega t)$$

and this implies that the mode produces a field which attenuates in z. It does not propagate. Such a mode is said to be evanescent.

The frequency at which $k_o = k_{cmn}$ is termed the cut-off frequency. It is the frequency at which, in a loss-less waveguide, a given mode changes abruptly from propagating to non-propagating. For any waveguide there is in general a separate and distinct cut-off frequency for every TE mode, and every TM mode, in the mode set. On the other hand it is not uncommon for a TE mode and a TM mode to have the same cut-off frequency. Such modes are said to be degenerate. For a TE mode and a TM mode in rectangular waveguide the cut-off frequency is evaluated as follows:

Let ω_c denote the frequency at which $k_o = k_{cmn}$. Then:

$$\begin{aligned}
\omega_c\sqrt{\mu_0\varepsilon_0} &= \sqrt{\left(\frac{m\pi}{a}\right)^2 + \left(\frac{n\pi}{b}\right)^2} \\
\text{i.e.}\quad \omega_c &= c\sqrt{\left(\frac{m\pi}{a}\right)^2 + \left(\frac{n\pi}{b}\right)^2} \\
\text{Hence } f_{cmn} &= \frac{c}{2}\sqrt{\left(\frac{m^2}{a^2} + \frac{n^2}{b^2}\right)} \\
&= \frac{c}{2ab}\sqrt{(m^2b^2 + n^2a^2)}
\end{aligned} \tag{B.20}$$

Also, since $c = f_o$, then

$$\lambda_{cmn} = \frac{2ab}{\sqrt{(m^2b^2 + n^2a^2)}} \tag{B.21}$$

B.4 TE_{10} Mode

The mode with the lowest cut-off frequency, which means that it can exist alone in a smooth empty waveguide, is the TE_{10} mode. The field expressions for this mode (in magnitude), using Eqs. (B.11) and (B.14), are:

$$\begin{aligned}
H_z &= \cos\frac{\pi x}{a} \\
H_y &= 0 \\
H_x &= j\frac{\beta_{10}a}{\pi}\sin\frac{\pi x}{a} \\
E_x &= 0 \\
E_y &= -j\frac{ka}{\pi}\eta_0\sin\frac{\pi x}{a}
\end{aligned} \tag{B.22}$$

Each component in Eq. (B.22) is of course multiplied by the wave term

$$\exp(-j(\beta_{10}z - \omega t))$$

The functional forms in Eq. (B.22) are sufficiently simple to permit the construction of field distributions in pictorial form, as shown in Fig. B.2. This is also true of the next dozen or so TE and TM modes as many texts on electromagnetic theory demonstrate [see Bibliography].

For this TE_{10} mode the cut-off frequency is given by:

$$\begin{aligned}
f_{c10} &= \frac{c}{2a} \quad and \\
\lambda_{c10} &= 2a
\end{aligned}$$

In a typical X-band waveguide (8.0–12.4 GHz) with a broad dimension of 2 cm this simple relation sets cut-off at 7.5 GHz ($k_0 a = \pi$ in Fig. B.3).

A number of waveguide properties are encapsulated within the propagation equation, which are very effectively illuminated by constructing a $k_0 a - \beta a$, or Brillouin diagram, for the waveguide. The beauty of this form of representation is that the slope of any line on the curve is representative of velocity, and Eq. (2.65) plots as a simple circle when γ (α) is real, and as a parabola, when γ (β) is imaginary. This is shown in Fig. B.3. The figure gives a quick and comprehensive picture of the most important waveguide characteristics, such as cut-off frequencies (π for TE_{10}, 2π for TE_{20} etc.), bandwidth (distance between cut-off points on $k_0 a$ axis), phase velocities (slope of line from origin to the frequency of interest on the TE_{10} mode propagation solution), group velocities (slope of the TE_{10} mode propagation solution at frequency of interest), higher order mode rates of attenuation (distance from $k_0 a$ axis to intercept on αa solution at frequency of interest), and frequency stratification of the modes. Usually only the first few TE modes are shown (TE_{10}, TE_{20}, TE_{11}, TE_{21} etc.) but in principle the complete TE_{mn} mode structure could be displayed. A similar diagram exists for TM modes. Note that the parabolic curves depicting propagating solutions lie above the $v = c$ line. Such

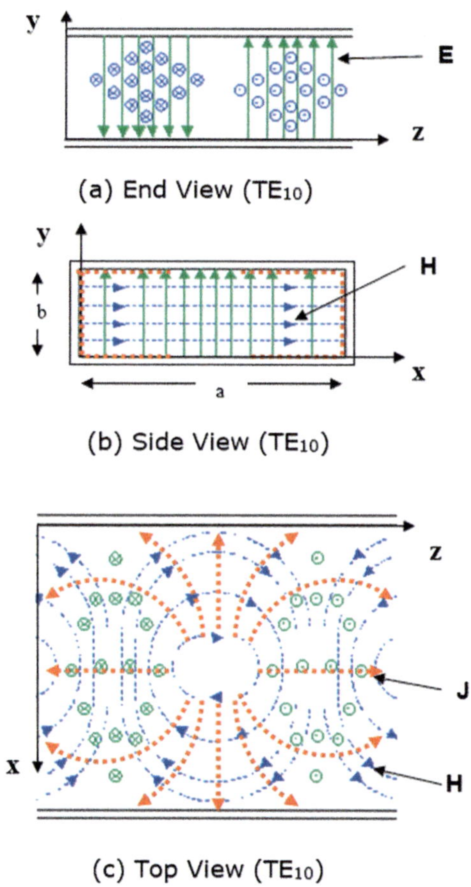

Fig. B.2 TE_{10} mode field patterns, **a** E and H-fields in the y-z plane, **b** E-field in the x-y plane, **c** E and H-fields in the x-z plane

solutions for uniform waveguide are termed 'fast' waves. Below cut-off, $k_0a = \pi$ for the TE_{10} mode, the solutions are evanescent as we have already seen and these cut-off solutions are represented by the quadrant of a circle (blue dotted curve for TE_{10} mode—it is not quite circular because of the slightly different axis scales). The attenuation coefficient α in nep/m can be deduced from the horizontal distance from the frequency of interest on the k_0a axis to the dotted curve of the mode of interest.

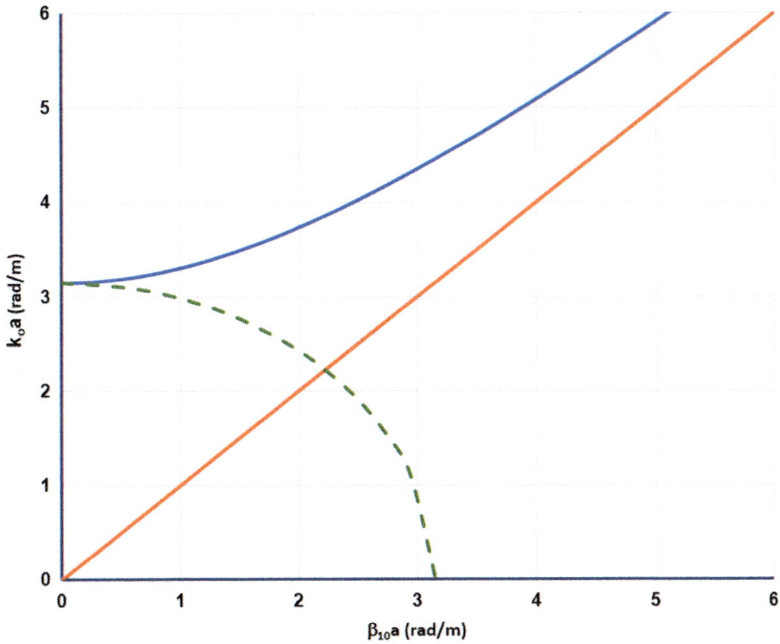

Fig. B.3 Brillouin diagram for TE_{10} mode showing frequency (k_0a) as a function of phase ($\beta_{10}a$) for rectangular waveguide (dark blue curve: a = 2 cm, b = 1 cm), and as a function of $\alpha_{10}a$ (green dotted curve). Orange line represents the speed of light line v = c

Bibliography

1. R.E. Collin, *Field Theory of Guided Waves* (McGraw-Hill Book Co., New York, 1960)
2. R. Ferrari, *An Introduction to Electromagnetic Fields* (Van Nostrand Reinhold Co. Ltd., New York, 1975)
3. C.T.A. Johnk, *Engineering Electromagnetic Fields and Waves* (Wiley, New York, 1988)
4. R.P. Feynman, R.B. Leighton, M. Sands, *Lectures in Physics, Vol. II* (Addison-Wesley Publishing Co., London, 1972)
5. P.M. Morse, H. Feshbach, *Methods of Theoretical Physics* (McGraw-Hill Book Co., New York, 1953)
6. J.D. Kraus, *Electromagnetics* (McGraw-Hill Book Co., London, 1984)

Appendix C
Waveguide Green's Function and Scattering Parameters

C.1 Green's Function for Empty Rectangular Waveguide

For the moment method solution of broadwall slot radiators in empty rectangular waveguide, the following are the relevant magnetic dyadic Green's functions for the geometry of Fig. 5.1.

(1) The exterior region usually treated as a half-space is (see Morse and Feshbach and Collin in the Bibliography):

$$\bar{G}_{mc}(\mathbf{r}/\mathbf{r_0}) = -(\bar{\mathbf{I}} + \frac{\nabla\nabla}{k_0^2})\frac{\exp -jk_0|\mathbf{r}-\mathbf{r_0}|}{2\pi|\mathbf{r}-\mathbf{r_0}|} \tag{C.1}$$

(2) The appropriate Green's function for an arbitrary slot radiator in waveguide [see Chap. 5, Eq. (5.5)] is the magnetic dyadic Green's function. However the problem can be simplified by noting that an inclined slot radiator in the broadwall of waveguide as depicted in Fig. 5.1 can generally be resolved into an x-directed (transverse) element and a z-directed (longitudinal) element. Actually in practice broadwall waveguide slot radiators are either purely transverse or wholly longitudinal. Inclined slots are hardly ever seen. This means that the magnetic dyadic Green's function for the feed waveguide can be separated into a part that is relevant to the transverse slot computation and a part that applies to the longitudinal slot.

As has already been hinted at above the magnetic dyadic Green's function for waveguide, see Collin in Bibliography, can be constructed by a summation of the complete set of modes for the rectangular box forming the waveguide. That is:

A. J. Sangster, *Compact Slot Array Antennas for Wireless Communications*, Signals and Communication Technology, https://doi.org/10.1007/978-3-030-01753-8

$$\bar{G}_{ma} = \begin{cases} \frac{1}{2j\omega\varepsilon_0}\sum\limits_{n=1}^{\infty} \mathbf{H}_n^+(\mathbf{r})\mathbf{H}_n^-(\mathbf{r}) & z > z_0 \\ \frac{1}{2j\omega\varepsilon_0}\sum\limits_{n=1}^{\infty} \mathbf{H}_n^-(\mathbf{r})\mathbf{H}_n^+(\mathbf{r}) & z < z_0 \end{cases} \tag{C.2}$$

where

$$\mathbf{H}_n^{\pm} = (\pm\mathbf{h}_n + \mathbf{h}_{zn})\exp(\mp\Gamma_{mn}z) \tag{C.3}$$

$$\mathbf{E}_n^{\pm} = (\mathbf{e}_n \pm \mathbf{e}_{zn})\exp(\mp\Gamma_{mn}z) \tag{C.4}$$

These vector fields, representing the normal mode field solutions for the waveguide, are normalised to be in accord with:

$$\iint_{S_x} [\mathbf{e}_n \times \mathbf{h}_m] \cdot \hat{\mathbf{a}}_{\mathbf{z}} dS = \delta_{nm} \tag{C.5}$$

while

$$\Gamma_{mn} = \left(\frac{m\pi}{a}\right)^2 + \left(\frac{n\pi}{b}\right)^2 - k_0^2 = k_c^2 - k_0^2 \tag{C.6}$$

Computationally friendly forms for $\mathbf{E}_n^{\pm}$ and $\mathbf{H}_{\mathbf{n}}^{\pm}$ can be constructed by employing the potential function generators as outlined in Collin. Hence for TE modes:

$$\begin{aligned} \mathbf{H}^{\pm} &= k_c^2\psi_h \exp(\mp\Gamma_{mn}z)\hat{\mathbf{a}}_{\mathbf{z}} \mp \Gamma_{mn}\exp(\mp\Gamma_{mn}z)\nabla_t\psi_h \\ \mathbf{E}^{\pm} &= j\omega\mu_0\hat{\mathbf{a}}_{\mathbf{z}} \times \nabla_t\psi_h \exp(\mp\Gamma_{mn}z) \end{aligned} \tag{C.7}$$

and for TM modes:

$$\begin{aligned} \mathbf{E}^{\pm} &= \mp k_c^2\psi_e \exp(\mp\Gamma_{mn}z)\hat{\mathbf{a}}_{\mathbf{z}} + \Gamma_{mn}\exp(\mp\Gamma_{mn}z)\nabla_t\psi_e \\ \mathbf{H}^{\pm} &= \pm j\omega\varepsilon_0\hat{\mathbf{a}}_{\mathbf{z}} \times \nabla_t\psi_e \exp(\mp\Gamma_{mn}z) \end{aligned} \tag{C.8}$$

Collin provides the following forms for the potential functions:

$$\begin{aligned} \psi_h &= \cos\left(\frac{m\pi x}{a}\right)\cos\left(\frac{n\pi y}{b}\right) \\ \psi_e &= \sin\left(\frac{m\pi x}{a}\right)\sin\left(\frac{n\pi y}{b}\right) \end{aligned} \tag{C.9}$$

By combining Eq. (C.2) through (C.9) the Green's function component relevant to transverse slot modelling is as follows

$$\bar{G}^{xx}_{ma}(\mathbf{r}/\mathbf{r}_0) = \frac{1}{2j\omega\varepsilon_0}\left\{\begin{array}{l} \sum_{m=1}^{\infty}\sum_{n=1}^{\infty}(\omega\varepsilon_0)^2\left(\frac{n\pi}{b}\right)^2 B^2_{mn}\exp(-\Gamma_{mn}|z-z_0|)\sin\frac{m\pi x}{a}\sin\frac{m\pi x_0}{a}\cos\frac{n\pi y}{b}\cos\frac{n\pi y}{b} \\ -\sum_{m=0}^{\infty}\sum_{n=0}^{\infty}\left(\frac{m\pi}{a}\right)^2\Gamma^2_{mn}A^2_{mn}\exp(-\Gamma_{mn}|z-z_0|)\sin\frac{m\pi x}{a}\sin\frac{m\pi x_0}{a}\cos\frac{n\pi y}{b}\cos\frac{n\pi y}{b} \\ n \neq m = 0 \end{array}\right\} \tag{C.10}$$

The factors A_{mn} and B_{mn} arise from the normalisation process represented by Eq. (C.10) and are given by:

$$A^2_{mn} = \frac{\varepsilon_{om}\varepsilon_{on}}{j\omega\mu_0\Gamma_{mn}k_c^2 ab} \tag{C.11}$$

$$B^2_{mn} = \frac{4}{j\omega\varepsilon_0\Gamma_{mn}k_c^2 ab} \tag{C.12}$$

where ε_{om} and ε_{on} are equal to zero if $m = 0$ or $n = 0$ otherwise they are equal to unity.

For the longitudinal slot radiator in rectangular waveguide it is shown in Collin that the sum of modes formulation produces a function for which non-convergence issues arise. However a direct solution to Eq. (5.5) in Chap. 5 together with the appropriate waveguide boundary conditions has been determined by Rahmat-Samii. The result is the following:

$$\bar{G}^{zz}_{ma}(\mathbf{r}/\mathbf{r}_0) = \sum_{m=0}^{\infty}\sum_{n=0}^{\infty}\frac{\varepsilon_{om}\varepsilon_{on}}{2ab\Gamma_{mn}}\exp(-\Gamma_{mn}|z-z_0|)\cos\frac{m\pi x}{a}\cos\frac{m\pi x_0}{a}\cos\frac{n\pi y}{b}\cos\frac{n\pi y_0}{b} \tag{C.13}$$

(3) For a waveguide with finite wall thickness (*T*) the slot is treated as a rectangular cavity of dimensions $L \times W \times T$, with a local coordinate system ξ, η, ς as indicated in Fig. 5.1. For this cavity the relevant magnetic dyadic Green's function (consistent with ξ-directed magnetic currents at open ends) can be formulated heuristically following the summation of modes approach in Collin (see Bibliography). The slot is viewed as a very short (length = T) of rectangular waveguide (*a* replaced by *L*: *b* replaced by *W*). This yields:

$$\bar{G}_{mb} = \sum_{l=0}^{\infty}\sum_{n=0}^{\infty} \frac{\varepsilon_{l0}\varepsilon_{n0}}{WL} \cos\left(\frac{l\pi(\eta+\frac{W}{2})}{W}\right) \cos\left(\frac{l\pi(\eta_0+\frac{W}{2})}{W}\right) \sin\left(\frac{n\pi(\xi+\frac{L}{2})}{L}\right) \sin\left(\frac{n\pi(\xi_0+\frac{L}{2})}{L}\right)$$
$$\times \frac{\left(1-(\frac{n\pi}{K_0 L})^2\right)}{k_g \sin(k_g T)} \begin{bmatrix} \cos(k_g\varsigma)\cos(k_g(\varsigma_0+T)) \\ \cos(k_g(\varsigma+T)\cos(k_g\varsigma_0) \end{bmatrix} \quad \varsigma \neq \varsigma_0 \tag{C.14}$$

where

$$k_g^2 = k_0^2 - (\frac{n\pi}{L})^2 - (\frac{l\pi}{W})^2 \tag{C.15}$$

C.2 Longitudinal Slot in Conventional Waveguide

For a slot in waveguide, or in any metallic surface, excitation of the slot is associated with magnetic field components aligned with its major axis, and hence current density flows normal to this axis (see Chap. 1). Consequently for the longitudinal slot in waveguide as shown in Fig. 5.1, with θ = 0°, the component of $\mathbf{H}_{10}$ which is effective in exciting the slot is the z-component H_{10z}. From Appendix B, this has the form:

$$H_{10z}^{\pm} = A_{10}\left(\frac{\pi}{a}\right)^2 \cos\frac{\pi x}{a} \exp(\mp\Gamma_{10}z) \tag{C.16}$$

with

$$A_{10} = (k_0 Z_0 \beta_{10} ab)^{-1/2}$$

for fields normalised to unity power flow. Also

$$\Gamma_{10} = j\beta_{10} = \left[\frac{\pi^2}{a^2} - k_0^2\right]^{1/2}$$

Hence substituting into Eq. (5.17) yields:

$$H_{10r}|_z = -\frac{1}{2}\sum_{s=1}^{N} \alpha_s H_{10z}^{+} \int_{x_1-w/2}^{x_1+w/2} \int_{-L/2}^{+L/2} A_{10}\left(\frac{\pi}{a}\right)^2 \cos\left(\frac{\pi x}{a}\right) \exp(-\Gamma_{10}z) \sin\left(\frac{s\pi(z+L/2)}{L}\right) dz dx \tag{C.17}$$

On combining Eqs. (C.16) and (C.17) the reflection coefficient (S_{11}) for backward scattering into the TE_{10} mode in the waveguide we obtain:

$$S_{11} = \frac{A_{10}\pi}{j\sqrt{2}a}\left[\sin\frac{\pi(x_1 + w/2)}{a} - \sin\frac{\pi(x_1 - w/2)}{a}\right]\sum_{s=1}^{N}\left[\alpha_s \frac{\frac{s\pi}{L}}{\left(\frac{s\pi}{L}\right)^2 - \beta_{10}^2}\right] \times \begin{cases} 2\cos\frac{\beta_{10}L}{2} & \text{s odd} \\ j2\sin\frac{\beta_{10}L}{2} & \text{s even} \end{cases} \tag{C.18}$$

The transmission coefficient (S_{12}) directly from the TE_{10} mode magnetic field, as above, or from the relation:

$$S_{12} = 1 - (-1)^s S_{11} \tag{C.19}$$

By conservation of energy the power radiated (p_{rad}) by the longitudinal slot can be deduced to be:

$$p_{rad} = p_{inc} - (p_t + p_r) \tag{C.20}$$

This can be expressed more conveniently as:

$$\frac{p_{rad}}{p_{inc}} = \left[1 - \left(S_{11}^2 + S_{12}^2\right)\right] \tag{C.21}$$

C.3 *T-Slot in Bifurcated Waveguide*

As far as the calculation of the waveguide scattering and power radiated is concerned the T-slot radiator differs little from the conventional longitudinal slot presented above but using moment method results presented in Chap. 5, Section 5.4 . However the derivations are slightly complicated by the fact that the T-slot scatters into both halves of the bifurcated waveguide. The internal scattering coefficients have therefore been considered to comprise two components S_{11}^+ and S_{11}^- representing backward scattering into the left hand (x-positive) and right hand (x-negative) waveguides respectively, while S_{12}^+ and S_{12}^- are the corresponding forward scattered components. With these definitions established it is not difficult to show that:

$$S_{11}^{\pm} = 2jA_{10}\sin\left(\frac{\beta_{10}w}{2}\right)\alpha_1\sigma\left(\frac{L}{2},\frac{\pi}{a}\right) \tag{C.22}$$

for s = 1 since, as we have noted in Chap. 5, the first basis function is enough to secure good results, and

$$\sigma(p, q) = \begin{cases} \frac{q \sin \beta p - \beta \sin qp}{q^2 - \beta^2} & (q \neq \beta) \\ \frac{1}{2} \frac{(\sin qp + \sin \beta p)}{q + \beta} - \mathrm{pcos}\beta\mathrm{p} & (q = \beta) \end{cases} \tag{C.23}$$

where β is the phase coefficient for the slot transmission line. In this case the transmission coefficient can be calculated from:

$$S_{12}^{\pm} = 1 - S_{11}^{\pm} \tag{C.24}$$

As in the case of the longitudinal slot power conservation can be applied to determine the power radiated by the T-slot, which generates the relation:

$$\frac{p_{rad}}{p_{inc}} = \left[1 - \left(S_{11}^{+})^2 + (S_{12}^{+}\right)^2\right] \tag{C.25}$$

This equation accommodates the double waveguide feed.

Bibliography

1. P.M. Morse, H. Feshbach, *Methods of Theoretical Physics* (McGraw-Hill Book Co., 1953)
2. R.E. Collin, *Field Theory of Guided Waves* (McGraw-Hill Book Co., 1960)
3. R.E. Collin, On the incompleteness of E and H modes in waveguides. Can. J. Phys. **51**, 1135–1140 (1973)
4. Y. Rahmatt-Samii, On the question of computation of Dyadic Green's function at the source region in waveguides and cavities. IEEE Trans. Microwave Theory Tech. **23**, 762–765 (1975)

Appendix D
Green's Function for Fin Loaded Waveguide

The Green's function for slow-waveguide periodically loaded with metallic fins can be developed heuristically from the well-established empty rectangular waveguide formulation (see Appendix C). The rectangular waveguide Green's function in Appendix C can be separated into three elements.

(1) The real part of the dominant mode term.
(2) The imaginary part of the dominant mode term.
(3) The imaginary part formed from the higher order all of which can generally be assumed to be evanescent modes.

By analogy it is not unreasonable to suggest that a Green's function for the fin-loaded rectangular waveguide will display similar components. In relation to item (3), the effect of periodic loading on the form of a potential Green's functions for a slow-wave structure has been studied elsewhere and clearly indicates that in the space $a \times b$ above the corrugated surface the evanescent modes are essentially the same as for the equivalent empty waveguide. Thus referring to Eq. (C.10) in Appendix C we can deduce that the evanescent mode contribution to the require Green's function is:

$$\bar{G}^{xx}_{\Im ma}(\mathbf{r}/\mathbf{r}_0) = \frac{1}{2j\omega\varepsilon_0}\left\{\begin{array}{l} \sum\limits_{m=1}^{\infty}\sum\limits_{n=1}^{\infty}(\omega\varepsilon_0)^2\left(\frac{n\pi}{b}\right)^2 B^2_{mn}\exp(-\Gamma_{mn}|z-z_0|)\sin\frac{m\pi x}{a}\sin\frac{m\pi x_0}{a}\cos\frac{n\pi y}{b}\cos\frac{n\pi y}{b} \\ -\sum\limits_{m=0}^{\infty}\sum\limits_{n=0}^{\infty}\left(\frac{m\pi}{a}\right)^2\Gamma^2_{mn}A^2_{mn}\exp(-\Gamma_{mn}|z-z_0|)\sin\frac{m\pi x}{a}\sin\frac{m\pi x_0}{a}\cos\frac{n\pi y}{b}\cos\frac{n\pi y}{b} \\ n\neq m=0 \\ m=1:n\neq 0 \end{array}\right\} \tag{D.1}$$

Furthermore, with reference to item (2) above, in a waveguide feeding slot radiators, we already know that the imaginary part of the dominant mode term has no influence on a scattering calculation for a centred transverse slot and can be equated to zero. This assumption applies equally to the centred transverse slot in the

A. J. Sangster, *Compact Slot Array Antennas for Wireless Communications*, Signals and Communication Technology, https://doi.org/10.1007/978-3-030-01753-8

upper broadwall of periodic waveguide. In addition to the above comparisons between conventional waveguide and periodic waveguide, the latter case has within every modal solution embedded space harmonics, in order to satisfy the periodic surface boundary condition. Fortunately, when it is operated as a feed waveguide the presence of the space harmonics can largely be ignored, since the space harmonic field components exist only in very close proximity to the fin tips and thus exhibit negligible influence on slot excitation at the radiating wall, provided b is not too small. This is unlikely, since generally b has to be large enough to permit useful power flow through the waveguide.

For the formulation of a serviceable Green's function for the periodic waveguide feed we can therefore concentrate on the contribution of the real part of the dominant HE_{11} mode namely $\bar{\bar{G}}^{xx}_{\Re ma}(\mathbf{r}/\mathbf{r}_0)$. As with the conventional waveguide formulation the xx component of the Green's function can be deduced from the x-directed magnetic potential:

$$\pi_h = \hat{\mathbf{a}}_x \psi_h \tag{D.2}$$

By analogy with the periodic waveguide field solutions presented in Chap. 8 [Eqs. (8.17)–(8.21)] it is reasonable to suggest that:

$$\pi_h = \hat{\mathbf{a}}_x \cos\frac{m\pi x}{a}\cosh\alpha_{c11}(y-b)\exp(-j\beta_{11}z) \tag{D.3}$$

For the coordinate system adopted in Fig. 8.9

Following the derivation steps detailed in Appendix C we eventually obtain:

$$\begin{aligned}\bar{\bar{G}}^{xx}_{\Re ma}(\mathbf{r}/\mathbf{r}_0) = \Bigg\{ & \hat{\mathbf{a}}_x\hat{\mathbf{a}}_x \frac{\omega\varepsilon_0\beta_{10}^2}{\beta_{11}k_0^2 ab}\left(\frac{1}{2}\frac{\sinh 2\alpha_{c11}b}{4\alpha_{c11}b}\right)^{-1}\cos\frac{\pi x}{a}\cos\frac{\pi x_o}{a} \\ & + \hat{\mathbf{a}}_z\hat{\mathbf{a}}_z \frac{\omega\varepsilon_0\beta_{11}\pi^2}{\beta_{10}k_0^2 a^3 b}\left(\frac{1}{2}\frac{\sinh 2\alpha_{c11}b}{4\alpha_{c11}b}\right)^{-1}\sin\frac{\pi x}{a}\sin\frac{\pi x_o}{a} \\ & \pm \hat{\mathbf{a}}_x\hat{\mathbf{a}}_z \frac{j\omega\varepsilon_0\pi}{k_0^2 a^2 b}\left(\frac{1}{2}\frac{\sinh 2\alpha_{c11}b}{4\alpha_{c11}b}\right)^{-1}\left(\sin\frac{\pi x}{a}\cos\frac{\pi x_o}{a} - \cos\frac{\pi x}{a}\sin\frac{\pi x_o}{a}\right)\Bigg\}\exp(-j\beta_{11}|z-z_o|)\end{aligned} \tag{D.4}$$

As noted above the third term in Eq. (D.4) does not contribute to radiation from a transverse slot in the $y = b$ wall and can be ignored. Finally

$$\bar{\bar{G}}^{xx}_{ma}(\mathbf{r}/\mathbf{r}_0) = \bar{\bar{G}}^{xx}_{\Re ma}(\mathbf{r}/\mathbf{r}_0) + \bar{\bar{G}}^{xx}_{\Im ma}(\mathbf{r}/\mathbf{r}_0) \tag{D.5}$$

Index

A. J. Sangster, *Compact Slot Array Antennas for Wireless Communications*, Signals and Communication Technology, https://doi.org/10.1007/978-3-030-01753-8

M

Printed by Printforce, the Netherlands